EXTRUDED CABLES FOR HIGH-VOLTAGE DIRECT-CURRENT TRANSMISSION

EXTRUDED CABLES FOR HIGH-VOLTAGE DIRECT-CURRENT TRANSMISSION

Advances in Research and Development

GIOVANNI MAZZANTI

MASSIMO MARZINOTTO

IEEE PRESS SERIES ON POWER ENGINEERING

Mohamed E. El-Hawary, *Series Editor*

IEEE

IEEE PRESS

WILEY

Published simultaneously in Canada.

Library of Congress Cataloging-in-Publication Data:

Mazzanti, Giovanni, 1962-
Extruded cable for high voltage direct current transmission : advances in research and development / Giovanni Mazzanti, Massimo Marzinotto.
pages cm
Includes bibliographical references.
ISBN 978-1-118-09666-6 (cloth)
1. Electric power distribution–Direct current. 2. Electric power distribution–High tension. 3. Electric cables. 4. Electric cable sheathing.
I. Marzinotto, Massimo, 1975- II. Title.
TK3111.M39 2013
333.793′2–dc23

2012042392

10 9 8 7 6 5 4 3 2 1

CONTENTS

FOREWORD

High-voltage direct-current (HVDC) transmission is not new. Edison, a pioneer of all things DC, constructed a 45-km link connecting Miesbach and Munich in 1882 using rotating DC machines at each end. A few HVDC links were constructed in the next 50 or so years, partly following the advent of mercury arc rectifiers; but they largely fell out of favor when the advantages of alternating-current (AC) transmission were realized. AC benefits from using transformers—which are straightforward to manufacture and efficient to use—to step-up and step-down voltages. The AC induction machine, while perhaps not so efficient, is rugged and low in maintenance and cheap to make. The twentieth century saw the development of electrical distribution systems connecting consumers and generators within cities and of transmission systems using national grids.

However, in the infamous words of Bob Dylan, "The times they are a-changin'." Grids are no longer national—the countries of Europe are connected, the states of the United States are connected, and the provinces of China are connected. We are moving to pan-continental grids covering typically 10 million square kilometers and distances of 4000 km or more. Yet our grids have transmission lines that rarely stretch more than a 100 km at a time. This is partly because of the capacitive and inductive limitations of AC transmission and partly because they were never designed for anything else. Our networks are getting congested: On occasions more than 90% of the power transmitted in Belgium is not being used by the Belgians. It is simply transmitted across the country connecting the neighboring countries on either side.

Our current electrical transmission networks are analogous to road networks with no motorways (or interstate highways.) We need electrical superhighways to transmit huge amounts of power long distances across continents. This is partly to connect consumers with sustainable sources of power. For example, the windy wet northern climate of Europe offers wind and hydroelectric generation, while the sunnier climes in the south offer photovoltaic and solar thermal possibilities. Generally, humans shy away from the really wet, windy, and hot climates and so long-distance transmission is often necessary to connect such regions of generation to centers of population. Nuclear power stations are often distant from cities, perhaps for reasons of safety. Similarly, the Three Gorges Dam situated near Yichang in China, which the guidebook states generates the equivalent of 10 nuclear power stations, needs to transmit vast amounts of power to Shanghai, 1000 km to the east. East–west pan-continental electrical superhighways facilitate "power lopping." The 4000-km width of a typical continent is about 10% of the circumference of the world and therefore corresponds to a 2.5-h time difference. Those in the east will be switching on their kettles for breakfast while those in the west are still asleep. Those in the west will be cooking their dinners while those in the east are doing the washing up. Transmitting power east and west can therefore reduce a region's peak power generation requirements. Electrical grids are necessary for the 21st century, and the lower electrical losses and reduced space demands afforded by HVDC transmission systems when used over longer distances are likely to afford the appropriate solution technologies.

Because our existing grids are connected together, we have also seen other problems. Cascading failures, where a failure on one network brings down a connected network leading to a domino effect bringing down successive networks, are difficult to control on massively interconnected AC network systems. Fifty-five million people were left without electricity in the northeastern and midwestern U.S. states and Ontario in August 2003 after such a cascading failure. HVDC back-to-back links between AC grids are likely to be efficacious firewalls preventing such wide-scale failures. Deregulation and privatization, smart grids, and indeed some renewable energy sources such as wind have led and will lead to faster changing demands and supplies requiring systems that can react more quickly. Without the problems of phase synchronization, this is more straightforward to achieve with DC than AC systems.

Of course, the move to HVDC brings with it some technical challenges. Since the 1980s, we have seen the successful wide-scale introduction of extruded coaxial high-voltage AC power cables with polymeric insulation (typically cross-linked polyethylene). These cables can support voltages of 10s to 100s of kilovolts and carry currents of 1000s of amps. Three-phase systems may transport all the energy from a large power station. Early problems with reliability have been overcome through the use of triple extrusion techniques for extruding "semicon" polymeric conductors contemporaneously with the insulation, and through the use of super-clean materials, the exclusion of water, and great care to avoid contaminants and voids in the insulation and protrusions from the electrodes. Due to geometrical effects, the electric field in the insulation is slightly higher nearer the center conductor, but not greatly so. Although the temperature near the center conductor can be significantly higher than that at the outer electrode, this makes little difference to the electric field in AC cables, as the field is controlled by the permittivity, which itself is not strongly dependent on temperature. The electric field is therefore defined everywhere by the applied voltage and the cable geometry; this is important since high electric fields are likely to lead to more rapid degradation and possibly to breakdown. Under DC conditions this is not necessarily the case. The electric field may be controlled by the conductivity, which is highly dependent on temperature, and may be enhanced by electrical charge ("space charge") that can accumulate in the insulation. An in-depth understanding of these phenomena is therefore required to manufacture and operate HVDC systems using extruded power cables.

With the resurgence of interest in using HVDC transmission systems, this book is therefore timely and will, I am sure, be of great interest to manufacturers and operators of such systems. Although not all the problems are solved, especially with the quest to operate at higher voltage and powers, the book will help to guide the way forward. It will also provide an excellent introduction for researchers in this area and those developing new cable materials and systems.

I was therefore very pleased to be asked by Giovanni Mazzanti at the 2012 IEEE CEIDP to write this foreword and I would like to congratulate the authors on a job well done!

John Fothergill

City University London

ACKNOWLEDGMENTS

First of all we sincerely thank some persons belonging to the IEEE DEIS (Dielectrics and Electrical Insulation Society) who—as reference members for both their role and their outstanding scientific contribution to this society—have encouraged our idea of writing this book since the beginning and have fostered its publication by Wiley-IEEE Press, namely Simon Rowland, Reuben Hackam, and Len Dissado. Moreover, we thank John Fothergill very much for its excellent foreword to this book.

We also thank all the peer referees that have reviewed our sample paragraphs before the acceptance of the manuscript for publication by Wiley-IEEE Press, as well as after the completion of the manuscript. They have spent a lot of time and attention in reading carefully this lengthy stuff and have provided precious suggestions for improving the quality of the book. Moreover, we are grateful to the Wiley-IEEE Press staff for the patience and care devoted to the editing and printing of our manuscript.

We greatly acknowledge the following people (in alphabetical order), who have provided valuable contributions to this book in terms of photos and design schemes from the main manufacturers of HVDC extruded cable systems in the world:

Shinya Asai (J-Power System Corporation)

Fabio Bocchi (Prysmian Powerlink)

Fabien Charles (General Cable - Silec)

Marc Jeroense (ABB High Voltage Cables)

Bengt Jonnerfeld (former ABB High Voltage Cables)

Cornelia Kohn (ABB High Voltage Cables)

Mohammed Mammeri (General Cable - Silec)

Marco Marelli (Prysmian Powerlink)

Josep Martinez (General Cable - Silec)

Pierre Mirebeau (Nexans)

Giorgio Mussi (Hitachi Cable Europe)

Akitoshi Watanabe (Viscas)

We also acknowledge Stefania De Felice for the valuable help with the graphics for the cover of the book.

Finally, we acknowledge our families for the patience and support they gave us during the lengthy writing of this book.

LIST OF SYMBOLS AND ACRONYMS

List of Symbols

A:	fraction of degraded moieties in the DMM model
$A_{eq}(E)$:	value of A at equilibrium between forward and backward reactions in the DMM model
A^*:	critical fraction of degraded moieties in the DMM model
a:	temperature coefficient of the electrical conductivity of an insulation
a_0, a_1:	best-fitting coefficients that express a linear dependence of a on electric field
b_0, b_1:	best-fitting coefficients that express a linear dependence of b on temperature
$\mathbf{a}_n$:	outwardly directed unit vector normal to the surface
$B = \Delta W/k_B$:	thermal endurance constant typical of the insulating material
B_q:	proportionality constant between stored electromechanical energy per moiety and its contribution to the degradation free energy barrier in the DMM model
b:	electric field coefficient of the electrical conductivity of an insulation
b_{ET}:	parameter that accounts for the electrothermal synergism
b_{MT}:	parameter that accounts for the thermomechanical synergism
b_{EM}:	parameter that accounts for electromechanical synergism
b_q:	exponent that rules the power law field dependence of the charge centers in the DMM model
C:	capacitance
C':	constant depending on material properties in the DMM model
C_0:	capacitance of a noncompressed sample
C_2:	capacitance seen by the current amplifier in the double-capacitor configuration of the TSM
C_E:	parameter of the Eyring model that depends on material, electric field, and temperature
C_q:	constant that rules the power law field dependence of the charge centers in the DMM model

c: electric field coefficient of the electrical conductivity of an insulation (alternative to b or γ)
D_{dis}: thickness of the slab where the whole hot-electron cooling process occurs
D_e: outer diameter of the cable
d: thickness of the dielectric
d: distance between the plus cable and the minus cable of a bipolar cable system
d_M: number of directions in which a defect can move in a cubic crystal structure
d: linear dimension of a void along the field direction
d_C: critical size of the damaged zone in the polymer
$\mathbf{E}$: electric field (vector)
E: electric field (module of $\mathbf{E}$)
E': conventional electrical stress
E_0: reference electric stress or design electrical stress of a power cable
E_{AC}: (capacitive) electric field in AC regime
E_c: electric field at the conductor
E_{DC}: electric field in DC regime
E_i: maximum impulsive electric field
E_{inj}: field level above which charge injection takes place from a point cathode in an insulator
E_l: Laplacian field
E_{MAX}: maximum field in a cable
E_{MIN}: minimum field in a cable
E_m: mean electric field (stress)
$E_{m,\mathrm{AC}}$: design mean electric field in the AC cable
$E_{m,\mathrm{AC},1.70}$: mean electric field in the AC cable at a constant rms voltage $1.70U_0$
$E_{m,\mathrm{DC}}$: design mean electric field in the DC cable
$E_{m,\mathrm{DC},1.45}$: mean electric field in the DC cable at a constant voltage $1.45U_0$
E_{mc}: (Poissonian) threshold field for the onset of the FLSC model
E_t: total electric field
$E_{t,s}$: total electric field when the impulse has the same polarity of the DC voltage
$E_{t,o}$: total electric field when the impulse has the opposite polarity of the DC voltage
E_V: electric field inside a void
$E_{w,1}$: electric field that the cable in a laboratory shall withstand
$E_{w,2}$: electric field that the cable installed in a field shall withstand
$e_p(t)$: externally applied pulsed electric field
F: cumulative failure probability

F_e:	force acting on electrons in the presence of an electric field
F_{eff}:	fraction of avalanche hot electrons effective in chemical damage through DEA
F_{hot}:	hot fraction of electrons impinging the polymer surface
GA:	electron affinity of a polymer (matrix)
GAd:	electron affinity of the defect
G_a:	activation energy governing the temperature dependence of the conductivity of an insulation
G_b:	energy for bond breaking
G_C:	energy level of the bottom of the conduction band (mobility edge)
G_F:	energy of the Fermi level
G_f:	energy for bond forming
G_{hop}:	activation energy associated with the hopping transport of electrons through the insulation volume
G_t:	energy level of traps
G_V:	energy level of the top of the valence band
$G(\varepsilon_r)$:	a function of relative permittivity ε_r containing the dependence of ε_r with pressure
g_C:	geometric capacitive field factor
g_R:	geometric resistive field factor
H:	burial depth of a cable
h:	hazard function
h_c:	overvoltage reduction factor due to the cable line
h_{EM}:	life exponent in the exponential life model (EM)
h_o:	height to ground of the overhead conductor
$h_P = 6.62 \times 10^{-34}$ J/s:	Planck constant
I:	current
I_0:	TS current measured on a cable due to the space charge accumulated after conditioning
I_C:	cable ampacity
I_{DC}:	DC cable ampacity
I_f:	TS current measured on a cable before conditioning
$i(t)$:	time-varying current
$\mathbf{J}$:	current density (vector)
J:	current density (module of $\mathbf{J}$)
J_e:	current density of electrons
J_C:	conduction current density
K_1:	deterioration coefficient
K_2:	temperature coefficient
K_3:	safety factor for uncertainties
K_{TDC}:	temperature rise coefficient
K_{PDC}:	polarity reversal coefficient
K_ρ:	calibration factor in the PEA method
k_1:	parameter of the Endicott model
k_2:	parameter of the Endicott model

$k_B = 1.38 \times 10^{-23}$ J/K: Boltzmann constant
k_E: constant in the impulse test IPM, depending on insulating material and specimen geometry
k_s: safety coefficient
L: time-to-failure (or lifetime, or life)
l: cable length
l_1: length of the cable tested in laboratory (or "small" cable)
l_2: length of the cable installed in field (or "large" cable)
M: mechanical load
M: number of model parameters
M': conventional mechanical stress
M_0: mechanical load at design conditions
m: mechanical endurance coefficient (MEC)
m_e: true or *free electron* mass
m_{eff}: "effective" electron mass
m_0: value of MEC m at reference temperature T_0 and reference electric field E_0
N: number of applied stresses
N_b: number of breakable bonds
N_{CH}: number of CH bonds in the slab of thickness D_{dis}
N_c: number of charge carrier species
N_{eff}: effective density of states for the conduction band
N_{el}: number of electrons produced across a void from one single injected electron
N_l: enlargement factor (over length)
N_m: number of "moieties" influenced on average by each microcavity in the DMM model
N_t: number density of traps
n: life exponent, or voltage endurance coefficient (VEC)
$n(t)$: number of electrons accumulated at the PE–void interface at a given time t
$n(w;E_V,t)$: energy distribution of electrons in a void (at time t and field E_V)
$n(w;E_V)$: equilibrium energy distribution function of electrons in a void
n_o: intrinsic charge density in the dielectric
n_1: injection component of the charge density in a dielectric
n_c: number density of conduction band electrons
n_{ch}: volume density of charge carriers, i.e., number of charge carriers per unit volume
n_e: number density of electrons
n_i: volume density of the ith ($i = 1, \ldots, N_c$) charge carrier, or life exponent in the impulse test IPM
$n_{\mathrm{inj}}(t)$: number of injected electrons at time t
n_t: number density of *occupied* trap states
$\mathbf{P}$: polarization (vector)
P: polarization (module of $\mathbf{P}$)

P_0: design failure probability of a power cable
P_{DC}: DC transmission power
$P_{b,1}$: breakdown probability at the electric field value $E_{w,1}$
$P_{b,2}$: breakdown probability at the electric field value $E_{w,2}$
P_{trans}: probability that the nearest-neighbor transition takes place in time dt
p: pressure
$p(z,t)$: pressure wave function of space z and time t
$p_1, p_2, \ldots, p_M$: life model parameters
Q_1: charge induced on an electrode (located at $x=0$)
Q_2: charge induced on an electrode (located at $x=d$)
Q_i: charge situated in the bulk of an insulator at an abscissa x_i
Q_ρ: charge per unit length of a cable
$q(t)$: time-varying charge
q_C: amount of charge of spherical charge centers in the DMM model
$q_e = 1.6 \times 10^{-19}$ C: electric charge of an electron
q_i: charge of the ith ($i = 1, \ldots, N_c$) charge carrier
R: reliability function
R_{DC}: conductor electrical resistance in DC regime
R_{dis}: damage growth rate in PE
R_{el}: rate of electrons colliding with the polymer surface after being injected into a void
r: radial coordinate in the insulation of a power cable
r_1: inner radius of the insulation of a power cable
r_2: outer radius of the insulation of a power cable
r_{FLSC}: radius of the charge injecting tip in the field-limited space charge model
r_h: radius of the overhead conductor
r_i: inner radius of the insulation of a power cable
r_{i1}: inner insulation layer radius of a small cable
r_{i2}: inner insulation layer radius of a large cable.
r_m: position of the maximum barrier height due to applied field in the Poole–Frenkel mechanism
r_o: outer radius of the insulation of a power cable
r_{o1}: outer insulation layer radius of a small cable
r_{o2}: outer insulation layer radius of a large cable
r_0: radius of spherical charge centers in the DMM model
S: surface or cross section
$S(f)$: system response function
$S_1, S_2, \ldots, S_N$: values of the N stresses applied to the insulation
T: absolute (thermodynamic) temperature
T_0: reference temperature or design temperature of a power cable
T_1: thermal resistance of the insulation thickness
T_4: thermal resistance of the surrounding medium
T': conventional thermal stress

T_{av}: average temperature of a sample containing a temperature gradient
T_H: reference temperature higher than the operating temperature
T_i: absolute temperature at the inner radius of cable insulation
T_i: thermal resistances of the different cable layers and of the laying environment
T_o: absolute temperature at the outer radius of cable insulation
t: time, or time to failure
t_0: design life of a power cable
t_2: time to failure at voltage V_2
t_1: time to failure at voltage V_1
t_{DC}: time duration of DC test voltage
t_{dis}: time to disruption of the slab of thickness D_{dis}
t_F: 100 Fth percentile of time to failure
t_{inj}: time to the injection of the first electron
U: voltage
U_0: rated voltage, i.e., design voltage of the cable
U_c: effective amplitude of a voltage impulse superimposed to the rated DC voltage
U_{DC}: (preexisting) DC service voltage or DC test voltage
$U_{DC,cal}$: DC calibration voltage in the PEA method
U_h: maximum overvoltage that impinges to the cable entrance
U_i: amplitude of voltage impulse (to ground)
U_p: (resulting) impulse peak voltage
U_T: voltage threshold below which no electrical aging takes place
$U_{w,o}$: maximum absolute peak value of a withstand impulse voltage with opposite polarity to superimposed DC voltage
$U_{w,s}$: maximum absolute peak value of a withstand impulse voltage with the same polarity as superimposed DC voltage
u_S: speed of sound
V_0: voltage applied to a sample between plane parallel electrodes
V_2: 63.2th percentile of the breakdown voltage for a full-size cable
V_{cal}: oscilloscope signal during calibration in the PEA technique
V_{DC}: DC voltage
$V_{DC,1}$: DC voltage applied to the small cable
$V_{DC,2}$: DC voltage applied to the large cable
$V_P(t)$: electrical pulse of short duration applied in the PWP method
V_{TFL}: trap-filled limit voltage
V_{tr}: transition voltage above which SCLC current dominates over the ohmic component
$v(t)$: voltage signal or instantaneous voltage
v_e: drift velocity of electrons
$v_s(t)$: detected space charge signal
W_C: Joule losses in the conductor

W_{C1}: conductor losses in the small cable
W_{C2}: conductor losses in the large cable
W_{Cmax}: maximum conductor losses (when $\delta = \delta_{max}$)
w: parameter of the Endicott model
w_e: electron energy
X: compressibility of a material
x: linear space coordinate (abscissa)
Z_1: surge impedance of the overhead line
Z_2: surge impedance of the cable
Z_b: acoustic impedance of electrode
Z_s: acoustic impedance of sample
z: linear space coordinate in the propagation of a pressure wave
z_f: location of the wavefront of a pressure wave

α: difference between the coefficient of thermal expansion and temperature coefficient of dielectric constant
$\alpha(E_V)$: 1st Townsend coefficient
α_E: electrostriction coefficient
$\alpha_{E,1}$: scale parameter of the dielectric strength Weibull distribution for the small cable 1
$\alpha_{E,2}$: scale parameter of the dielectric strength Weibull distribution for the large cable 2
α_n: number of impulses that correspond to a 63.2% breakdown probability
α_p: temperature coefficient of permanent polarization
α_t: scale parameter of the Weibull distribution of failure times
α_x: coefficient of thermal expansion
α_ε: temperature coefficient of dielectric constant
$\beta(E_V)$: field-dependent multiplication rate of electrons in a void
β_E: shape parameter of the dielectric strength distributions for the small and large cable
β_t: shape parameter of the Weibull distribution of failure times
γ: electric field coefficient of the electrical conductivity of an insulation (alternative to b and c)
γ_r: voltage refraction factor
Δ: free energy difference between reactant and product state
ΔG: Gibbs free energy of activation
ΔG_t: energy gap between the bottom of the conduction band and the energy level of traps (trap depth)
ΔH: activation enthalpy per moiety
ΔS: activation entropy per moiety
ΔT: temperature drop or "gradient" across a layer
$\Delta T(x,t)$: temperature change vs. space x and time t
Δt_P: width of the HV pulse in the PEA method
Δt_S: transit time of acoustic waves across in the PEA method
$\Delta V(t)$: voltage change across a sample due to nonuniform thermal expansion vs. time t

ΔV_m: barrier height reduction due to applied field in Schottky injection and Poole–Frenkel mechanism
ΔW: activation energy of the main thermal degradation reaction
δ: field inversion coefficient
δ_{max}: maximum value of field inversion coefficient (at full load)
δ_S: elemental strain in the DMM model
ε: (di)electric permittivity or dielectric constant
ε_0: permittivity of free space
ε_r: relative permittivity of a dielectric
θ: temperature in Celsius degrees
θ: ratio of free charge carriers to those trapped
θ_a: ambient (soil) temperature in Celsius degrees
θ_c: conductor temperature in Celsius degrees
θ_s: screen temperature in Celsius degrees
λ: thickness of the shell on the cavity wall where the N_m moieties lie in the DMM model
$\lambda_{T,d}$: thermal conductivity of the insulation (W $m^{-1}K^{-1}$)
μ: mobility of charge carriers[1]
μ_e: mobility of electrons
μ_i: mobility of the ith ($i = 1, \ldots, N_c$) charge carrier
μ_∞: high-field limiting value of μ in the FLSC model
ν: vibrational frequency around a defect site in a plane orthogonal to the direction of movement
ν_0: attempt to escape frequency
ρ: electric space-charge density (or distribution)
ρ_0: electrical resistivity of the insulation at reference temperature T_0 and electric stress E_0
ρ_E: electrical resistivity
ρ_P: equivalent charge density of permanent polarization
ρ_T: total space-charge density
$\rho_{T,S}$: thermal resistivity of the soil
$\rho_{T,d}$: thermal resistivity of the dielectric
σ: electrical conductivity
$\sigma(x)$: "internal" conductivity at coordinate x within insulation sample thickness
$\sigma(0)$: surface charges on the electrode at $x = 0$
$\sigma(d)$: surface charges on the electrode at $x = d$
σ_0: insulation conductivity at reference temperature T_0 and electric stress E_0
σ_∞: reference value of ionic conductivity at very high temperature T ($\rightarrow\infty$)
σ_{cal}: surface charge in the PEA calibration procedure
τ: time constant in the evolution of transient DC electric field

[1] Electrons, unless otherwise specified.

ϕ:	coulombic potential barrier (or activation energy) for Schottky injection
ϕ_m:	work function of a metal

List of Acronyms

AC:	alternate current
ALT:	accelerated life tests
B2B:	back-to-back (transmission)
CBM:	conduction band minimum
CDF:	cross-cut dust figure
CSC:	current source converters
DC:	direct current
DEA:	dissociative electron attachment
DMM:	Dissado, Mazzanti, and Montanari
EELS:	electron energy loss spectroscopy
EHV:	extra-high voltage
EHVDC:	extra-high-voltage direct current
EM:	exponential model
EPDM:	ethylene–propylene diene-monomer
EPR:	ethylene–propylene rubber
ETT:	electric triggered thyristor
EVA:	ethylene vinyl acetate
FEM:	finite-element method
FEP:	fluorinated ethylene–propylene
FGM:	field grading material
FLSC:	field-limited space charge
GF:	gas filled
HDPE:	high-density polyethylene
HIC:	halving interval in Celsius degrees
HPOF:	high pressure oil filled
HV:	high voltage
HVAC:	high-voltage alternate current
HVDC:	high-voltage direct current
IGBT:	insulated gate bipolar transistors
IPM:	inverse power model
IRC:	integrated return conductor
LCC:	line-commutated current source converters
LDPE:	low-density polyethylene
LDT:	long-distance transmission
LLDPE:	linear low-density polyethylene
LIMM:	laser intensity modulation method
LIPL:	lightning impulse protective level
LIPP:	laser-induced pressure pulse
LM:	Levenberg–Marquardt
MEC:	mechanical endurance coefficient

MIND: mass impregnated nondraining
MI-PPL: polypropylene paper laminate
MWS: Maxwell–Wagner–Sillars
MTTF: mean time to failure
MVAC: medium-voltage alternate current
OF: oil filled
OIT: oxidation induction time
PD: partial discharge(s)
PE: polyethylene
PEA: pulsed electroacoustic
PET: polyethylene terephthalate
PI: polyimide
PP: polypropylene
PPL: polypropylene paper laminate
PPLP: polypropylene paper laminate
PQ: prequalification
PSD: photostimulated discharge
PST: phase-shifting transformers
PVC: polyvinylchloride
PVDF: poly-vinyli-dene fluoride
PWP: pressure wave propagation (or pressure wave pulse)
R&D: research and development
SAPEI: SArdegna–PEnisola Italiana (Sardinia–Italian Peninsula)
SCFF: self-contained fluid filled
SCLC: space-charge limited conduction
SCOF: self-contained oil filled
SCR: silicon-controlled rectifiers
SEM: scanning electro-microscope
STATCOM: static synchronous compensator
SVC: static var compensator
TB: technical brochure
TEM: transmission electro-microscope
TI: temperature index
TL: thermoluminescence
TR-XLPE: (water) tree-retardant cross-linked polyethylene
TS: thermal step
TSC: thermally stimulated currents
TSDC: thermally stimulated discharge currents
TSPC: thermally stimulated polarization currents
TSM: thermal step method
VBM: valence band maximum
VEC: voltage endurance coefficient
VSC: voltage source converter
WG: working group
XLPE: cross-linked polyethylene
XPS: X-ray photoelectron spectroscopy

INTRODUCTION

High-voltage direct-current (HVDC) extruded cable systems are a technology whose leading edge is currently fast moving forward and is experiencing a sharply growing market cycle in a strategic area—that is, HVDC transmission—which is likely to have profound effects on electricity transmission strategy throughout the world over the next few decades.

As far as we know, there is no book in the international technical-scientific literature at present that focuses on HVDC extruded cable systems. For this reason, we wrote this book to fill a need in view of the large expansion envisaged in the HVDC transmission, with the goal of describing the state of the art of research and development activities concerning HVDC extruded cable systems. If we have achieved this goal somehow, this is thanks to the vital support of both milestone papers coming from the scientific literature and precious material kindly provided by the main manufacturers of HVDC cable systems worldwide. Therefore, we are deeply indebted to—and we sincerely acknowledge—all those who have either written those milestone papers, or made the material from manufacturers available, thereby making the completion of this book possible. For this reason, we were extra careful in the compilation of the reference lists at the end of each chapter.

However, as a matter of fact the technical and scientific literature on HVDC extruded cable systems is really broad; thus, it is very hard to cite all the possible sources of information about this topic. In addition, to keep the treatment within reasonable length limits, we have selected among the many potential items of this book those that are more innovative and peculiar of HVDC extruded cable systems, thereby only mentioning some important aspects of HV cable technology. Hence, we sincerely apologize if some contributions, although outstanding, have not been included in this book for these reasons. Moreover, we encourage those who feel we have omitted something important to let us know: such suggestions will be very useful to us in the future.

We have done our best to let the huge amount of reference material speak directly to the reader. Therefore, we have often provided citations and figures directly from such reference material in order to avoid expressing our personal interpretations or ideas as much as possible and let in turn the reader form his or her own opinions.

For all these reasons, it can be argued that the book is intended first of all for electrical power engineers from electrical utilities and transmission system operators, as well as from the cable manufacturing industry. More broadly, it is devoted to people from both the industrial world and the academic community working in the area of high-voltage DC transmission and distribution, from practicing engineers to researchers. Again, the book may also be useful as background reading material for graduate students in the power area, as well as in the senior year of an undergraduate degree with a power option.

Giovanni Mazzanti
Massimo Marzinotto

CHAPTER 1

INTRODUCTION

High-voltage direct-current (HVDC) power transmission (see Fig. 1.1) is becoming more and more competitive compared to high-voltage alternate-current (HVAC) power transmission (see Fig. 1.2), especially for bulk power transmission over long distances, as pointed out, for example, in [1–5] and in Chapter 2. This is due to many technical and commercial advantages shared by both HVDC transmission systems employing traditional current source converters (CSC) and HVDC transmission systems using the more recently developed voltage source converters (VSC; see Fig. 1.3).

As to the typology of transmission lines (both HVAC and HVDC), as is well known they can be either overhead lines or cable lines. The overhead lines are much cheaper; but the cable lines are attractive for crossing wide metropolitan areas or long distances in the open sea. In particular HVDC cable lines often constitute the only technical solution feasible for long-distance sea crossing [1]. Compared to overhead lines, cable lines offer no visual impact and a reduced environmental impact, at least as perceived by the general public.[1]

In particular, cable lines can be an interesting technical option when railway or motorway tunnels are already available. In that case, the integration of a cable line into an existing transportation corridor requires less laying expenses than a traditional underground cable line, thereby greatly cutting cable line costs. Of course, this integration is not an easy task and requires great care from engineering and technical viewpoints, since multidisciplinary issues have to be tackled, such as installation and geometric compatibility, electromagnetic compatibility, external effects of a possible pole-to-ground fault, reliability of the synergistic cable–structures approach, fire behavior, and so on [6].

As far as the insulation used in HVDC transmission cables is concerned, at present most HVDC installations in service all over the world use paper–oil insulated

[1] The general public does perceive that cable lines have no impact on the environment, contrary to overhead lines, whose visual impact is evident. In fact, strictly speaking, the reduced environmental impact of underground transmission cable lines with respect to overhead lines is questionable since land cable lines mostly require excavation activities in the soil to a depth from ~1 to ~2 m all along their path, and a constant maintenance activity of the vegetation over the cable service life in order to avoid the growth of trees along the right of way, which could interfere with cables. In the case of overhead lines, only the height of trees within the right of way has to be tailored to avoid contact with power conductors. Also underwater cables require laying activities that might have some impact on the environment.

Extruded Cables for High-Voltage Direct-Current Transmission: Advances in Research and Development, First Edition. Giovanni Mazzanti and Massimo Marzinotto.

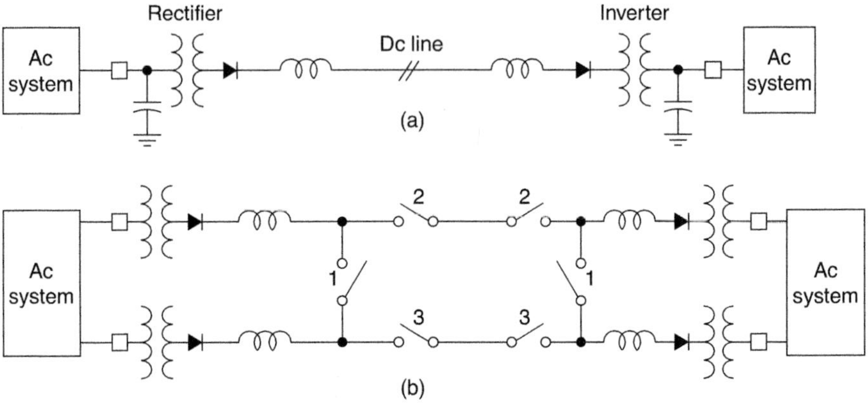

Figure 1.1 Single-line diagrams of HVDC transmission systems: (*a*) monopolar system and (*b*) bipolar system (after [1]).

cables. In particular, the following typologies of paper–oil insulation have been used successfully in underground (marine or land) HVDC power transmission systems since the mid-1950s [3]:

1. Mass impregnated nondraining (MIND)
2. Self-contained oil filled (SCOF)
3. High-pressure oil filled (HPOF)
4. Gas filled (GF)

Self-contained oil filled cables (also known as Emanueli cables) have been employed in Europe, while HPOF and GF cables have found wider application in the

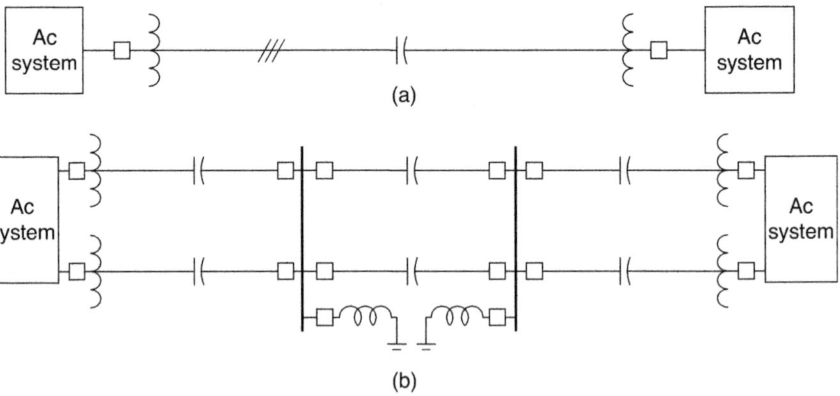

Figure 1.2 Single-line diagrams of HVAC transmission systems: (*a*) single-circuit three-phase AC line and (*b*) double-circuit three-phase AC line. The figure shows series capacitors and shunt reactors required for compensating, respectively, the series inductive and shunt capacitive reactance of long lines (after [1]).

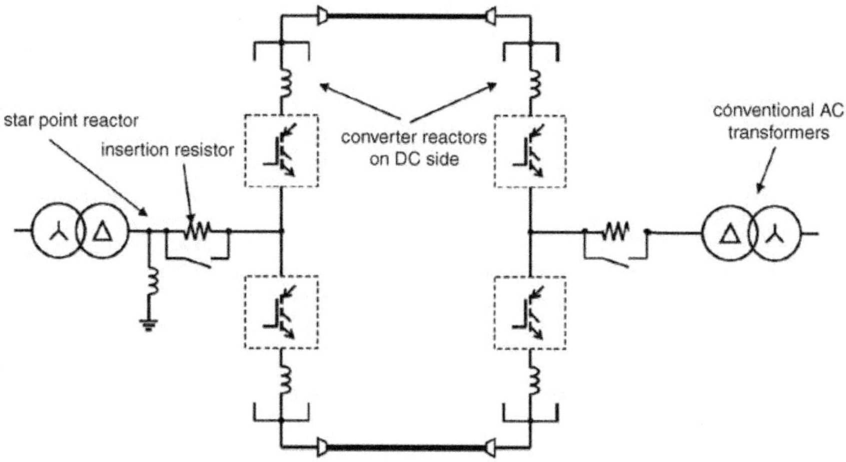

Figure 1.3 Single-line diagram of HVDC VSC transmission system (after [2]).

United States. Until recently most HVDC cables used in high-capacity power transmission utilized MIND insulation [7], since these cable systems have shown very high service reliability, notwithstanding the limitation of cable system service temperature. A better understanding of their vulnerability to load cycling [8] has paved the way for even higher transmission capacity.

The SCOF, HPOF, and GF cables have operational limitations that affect both cable system service temperatures and installation lengths; in addition, environmental issues associated with the use of oil-filled cables [3] should not be neglected. Moreover, paper–oil insulated cables have a rather complex and expensive manufacturing process [7]. For these reasons, some manufacturers are developing HVDC cable systems very similar to the traditional MIND cables that utilize polypropylene paper laminate (PPL) insulation and impregnating compounds that have nondraining characteristics. These cable systems are designed to operate at conductor temperatures of 80–90°C. However, they are not yet a fully established technology [3].

It is now widely believed that extruded HVDC cables offer significant advantages over the traditional paper insulated cable types, for example [3, 9]:

1. A higher conductor temperature can be used, giving a more compact cable for the same power rating (in fact, conductor temperature is limited by the maximum allowable temperature of the insulation). A temperature limit in the region of 55°C is applied to a traditional MIND cable to reduce the risk of contraction voids forming when the system load is reduced. Moreover, higher temperatures tend to reduce oil viscosity, involving a risk of oil migration that might cause a lower oil concentration in the upper insulation layers, with obvious consequences. The typical operating temperature of a polymeric cross-linked polyethylene (XLPE) insulated AC cable system is 90°C, and current extruded insulation for HVDC cables in some cases is ready to operate at such high temperatures (see Chapter 3 and Chapter 5).

2. Lighter moisture barriers can be used, giving a lighter cable.
3. Jointing of extruded cables is simpler.
4. With respect to SCOF, HPOF, and GF oil–paper insulated cables, two more advantages can be found [3]:
   a. Increased system lengths. With no requirement to limit oil/gas pressure transients, extruded dielectric HVDC cable systems could be installed in extremely long lengths.
   b. Reduced maintenance. There are no oil/gas system feed tanks, gauges, or alarm systems to maintain.

Therefore, due to such advantages of extruded insulation versus paper–oil insulation, there is an incentive for use of polymeric insulation for HVDC power cables with higher transmission capacity. Since the early 1970s many attempts have been made to introduce extruded insulation, but until the end of the last millennium they were unsuccessful. The first prototypes of HVDC extruded cables with polymeric insulation realized with traditional thermoplastic PE [10] or XLPE [3, 9] compounds failed. The cable systems were found to fail prematurely at operating voltages and were particularly weak when tested under polarity reversal conditions. It became evident that significant developments were required to achieve the functional electrical properties required for reliable HVDC transmission systems.

Since then, a huge amount of research and development (R&D) studies were performed for developing HVDC extruded cables. These studies are widely attested in the international technical-scientific literature (see, e.g., [3, 7, 9–28] among a huge reference list).

As a result of steady improvements of material technology and extrusion methods, the situation has now changed [9]. Furthermore, new experimental methods to study the field distribution under DC stress combined with a better theoretical understanding have made the development of HVDC extruded cables possible. In addition space-charge profile measurement techniques were set up [15–20] and new PE-based materials capable of withstanding HVDC were produced [9, 21].

The first visible result of these achievements was the commissioning in 1999 of the world's first HVDC transmission system using extruded polymeric insulated cables in the Swedish island of Gotland [22]. This installation uses voltage source converters (VSC) with insulated gate bipolar transistors (IGBT), which are advanced semiconductor devices (see Chapter 2). In this way, one of the major problems with polymeric insulated HVDC cables—namely, polarity reversal withstand [3, 9, 10, 23]—is avoided. Polarity reversal withstand is extremely important for the resistance of the insulation to DC breakdown, as it has been shown that the breakdown resistance is reduced by approximately 10% on polarity reversal [9]. The insulation is an XLPE compound for HVDC.

Consequently, until 2011 HVDC extruded cable systems have found their exclusive role with VSC technology: from the world's longest underground transmission system (180 km)—namely, the Australian Murraylink project [24]—to undersea cable projects (see Chapter 7) [25–27]. The marine applications should be emphasized as this is a growing area for application also in offshore wind platforms [26, 29]. Until 2011, the top-rated HVDC extruded cable system as to the voltage was the Trans Bay Cable (see Fig. 1.4 and Section 7.2.7), a ±200 kV DC, 400 MW, VSC bipolar undersea cable across the San Francisco Bay in California [13, 26, 27]. In 2012, the voltage rating of the Trans Bay Cable was equated by the East-West Interconnector, a ±200 kV DC, 500 MW, VSC bipolar land and undersea extruded cable system that connects the British and the Irish grids. Available HVDC extruded cable systems with VSC technology have reached a voltage of 300 kV and more, and a power of around 1000 MW (see, e.g., [25, 28]), but at present, the top rated HVDC extruded cable system worldwide uses CSC converter stations: it is the Hokkaido-Honshu link in Japan, a ±250 kV DC, 600 MW, CSC bipolar land and undersea intertie (see Section 7.2.8).

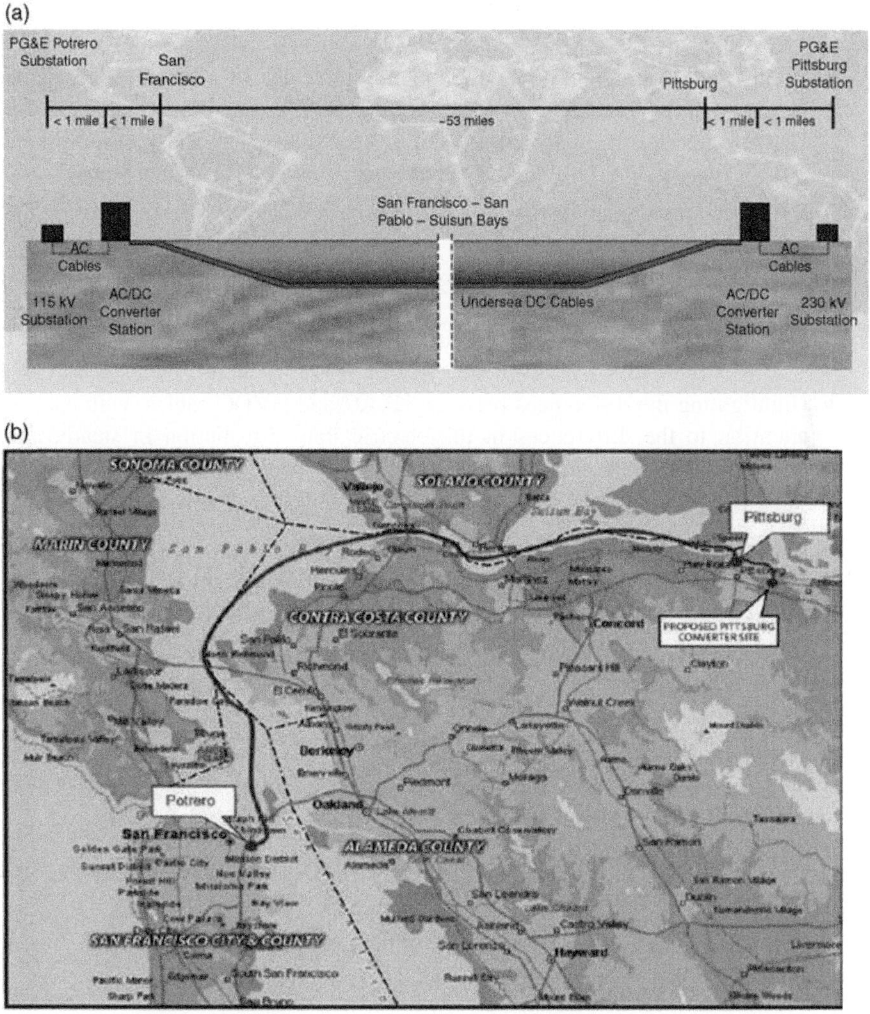

Figure 1.4 Sketch of the main characteristics of the Trans Bay Cable system (USA): (*a*) cross section of the link and (*b*) map of the cable route (courtesy of Prysmian).

With the dawn of the new millennium a new era has started for HVDC extruded cable technology, a technology dormant since the 1970s. This is the result of enduring R&D efforts in insulation materials and systems carried out by scientists and engineers all over the world, so as to put HVDC extruded cable systems in the limelight. An encouraging future is round the corner: higher voltage levels, larger power transfer, and available polarity reversal.

The book is organized as follows. In Chapter 2, the fundamentals of HVDC cable transmission are discussed:

- Describing briefly the historical evolution of HVDC power transmission and outlining the advantages/disadvantages with respect to traditional HVAC transmission
- Comparing HVDC transmission systems with HVAC transmission systems from the economic viewpoint
- Analyzing the configurations and operating modes of HVDC transmission systems with special focus on converters (of both the line-commutated current source type and the self-commutated voltage source type) and the ultimate goal of singling out the effect of converter typology on HVDC cable insulation
- Reviewing the main types of HVDC transmission cables with emphasis on the differences between underground and undersea cables, on one hand, and between oil–paper insulated and extruded cables on the other.

In Chapter 3, the main principles of HVDC extruded cable design are illustrated:

- Highlighting the differences between HVAC and HVDC cables, with special attention to the differences in the electric field distribution in steady and transient conditions
- Treating the influence of temperature of the laying environment on the steady field of an HVDC extruded cable
- Describing the effects of impulses superimposed on DC voltage and outlining a novel statistical approach to impulse voltage test levels for HVDC extruded cables
- Introducing the problems involved by the modification of the stress distribution due to space-charge effects
- Listing the PE-based dielectrics for HVDC extruded cables with special focus on the morphology of PE and its influence on the electrical properties of cable insulation.

In Chapter 4, the key issue of space-charge storage in extruded HVDC insulation and the relevant manifold effects are tackled:

- Charge injection and transport in insulating polymers
- Low-field conduction mechanisms (both ohmic and ionic conduction)
- High-field conduction mechanisms (charge injection from electrodes and bulk-controlled high-field conduction mechanisms)

Then, the processes that result in space-charge accumulation are illustrated:

- Charge generation (via electronic charge injection, field-assisted thermal ionization of impurities, spatially inhomogeneous electric polarization, and steady DC current coupled with a spatially varying ratio of permittivity to conductivity)
- Charge trapping (via physical defects and chemical defects)

Thereafter, the space-charge measurement methods for HVDC extruded insulation are treated extensively, by performing a comparative review of the various methods introduced over the years, classified into two categories:

- Thermal methods, that is, thermal pulse method, laser intensity modulation method (LIMM), and thermal step method (TSM)
- Pressure pulse methods, that is, pressure wave propagation (PWP) method, laser-induced pressure pulse (LIPP) method,[2] and pulsed electroacoustic (PEA) method

Subsequently, the up-to-date development of the best-suited techniques for measuring space charges in extruded insulation for HVDC cables are reported, namely, the TSM, the PEA, and the PWP techniques, with a final synthetic comparison between the best space-charge measurement methods for power cables: PEA versus TSM.

In Chapter 5, the improvements to the design of extruded HVDC cable systems brought about by the most recent research and development activities are discussed. The R&D trends in improving extruded polymeric insulation for HVDC cable systems are analyzed and various approaches used historically are mentioned. Special attention is dedicated to the most common and up-to-date approaches that aim at the suppression of space charge in the insulation by:

- Modifying the characteristics of the electrode/insulation interfaces
- Modifying the bulk insulation characteristics via additives or fillers

Then, further requirements for the improvement of HVDC extruded cable design are considered and the improved design of practical HVDC extruded cables, accessories (joints and terminations), and overall systems is illustrated in some detail, resorting to both literature data and material provided by courtesy of the main manufacturers.

In Chapter 6, the matter of HVDC extruded cable insulation life modeling is tackled:

- Recalling the fundamentals of life modeling and reliability estimates of power cables, which consist in the traditional approach to insulation life modeling and in the probabilistic framework of HVDC extruded insulation

[2] As argued in Section 4.4.2, in practice the LIPP is the far more common implementation of the PWP method at present.

- Focusing on the theory and applications of life models conceived for extruded HVDC cables, that are either based on space-charge effects [i.e., the field limited space-charge model and the space-charge Dissado–Mazzanti–Montanari (DMM) model] or on the interaction between space charges and partial discharges (i.e., the life model based on damage growth from microvoids)
- Introducing the long-debated question of whether space charge is a cause or an effect of aging, by means of life models that do not regard space charge as a direct cause of aging.

In Chapter 7, the main HVDC extruded cable system realizations in the world are illustrated by presenting the main operational extruded systems worldwide.

REFERENCES

1. E.W. Kimbark. *Direct Current Transmission*, Vol. 1. Wiley, New York, 1971.
2. K. Friedrich. "Modern HVDC PLUS Application of VSC in Modular Multilevel Converter Topology." 2010 IEEE International Symposium on Industrial Electronics (ISIE 2010), Bari, Italy, July 4–7, 2010.
3. EPRI Report No. 1008720. DC Cable Systems with Extruded Dielectrics. Palo Alto, California, December 2004.
4. F. Rüter, J.L. Parpal, and S.G. Swingler. "A Review of HVDC Extruded Cable Systems." Presented at the International Council on Large Electric Systems (CIGRÉ), 2000 Session, Panel 2, Paper 2-03, Paris, France, 2000.
5. M.P. Bahrman and B.K. Johnson. "The ABCs of HVDC Transmission Technologies." *IEEE Power Energy Mag.*, Vol. 5, No. 2, pp. 32–44, March–April 2007.
6. R. Benato, R. Caldon, L. Fellin, and E. Di Bartolomeo. "Linee elettriche in cavo c.a. e c.c. in infrastrutture ferroviarie/stradali esistenti e future: un utilizzo affidabile delle sinergie." Proceedings of the 2009 AEIT Conference, Catania (Italy), September 27–29, 2009 (in Italian).
7. E. Ildstad, J. Sletbak, B.R. Nyberg, and J.E. Larsen. "Factors Affecting the Choice of Insulation System for Extruded HVDC Power Cables." Paper D1-203, CIGRE Session 2004, Paris (France), 2004.
8. G. Evenset. "Cavitation as a Precursor to Breakdown of Mass-Impregnated HVDC Cables." Ph.D. Thesis, NTNU, 1999.
9. T.L. Hanley, R.P. Burford, R.J. Fleming, and K.W. Barber. "A General Review of Polymeric Insulation for Use in HVDC Cables." *IEEE Electr. Insul. Mag.*, Vol. 19, No. 1, pp. 14–24, January–February 2003.
10. M. Salah Khalil. "International Research and Development Trends and Problems of HVDC Cables with Polymeric Insulation." *IEEE Electr. Insul. Mag.*, Vol. 13, No. 6, pp. 35–47, November–December 1997.
11. R.N. Hampton. "Some of the Considerations for Materials Operating under High-Voltage, Direct-Current Stresses." *IEEE Electr. Insul. Mag.*, Vol. 24, No. 1, pp. 5–13, January–February 2008.
12. T. Sörqvist, T. Christen, M. Jeroense, V. Mondiet, and R. Papazyan. "HVDC-Light® Cable Systems—Highlighting the Accessories." 21st Nordic Insulation Symposium (NORDIS 2009), Gothenburg (Sweden), June 15–17, 2009.

13. T. Westerweller, K. Friedrich, U. Armonies, A. Orini, D. Parquet, and S. Wehn."Trans Bay Cable—World's First HVDC System Using Multilevel Voltage-Sourced Converter." CIGRE B4 session paper B4-101, Paris (France), 2010.

14. Y. Sakai, H. Niinobe, and T. Hirasawa. "Development of Pre-molded Accessories for HVDC Extruded Cable System." Paper A.2.3, 8th International Conference on Insulated Power Cables (Jicable'11), Versailles—France, June 19–23, 2011.

15. D. Fabiani, G.C. Montanari, C. Laurent, G. Teyssedre, P.H.F. Morshuis, R. Bodega, L.A. Dissado, A. Campus, and U.H. Nilsson. "Polymeric HVDC Cable Design and Space Charge Accumulation. Part 1: Insulation/Semicon Interface." *IEEE Electr. Insul. Mag.*, Vol. 23, No. 6, pp. 11–19, November–December 2007.

16. S. Delpino, D. Fabiani, G.C. Montanari, C. Laurent, G. Teyssedre, P.H.F. Morshuis, R. Bodega, and L.A. Dissado. "Polymeric HVDC Cable Design and Space Charge Accumulation. Part 2: Insulation Interfaces." *IEEE Electr. Insul. Mag.*, Vol. 24, No. 1, pp. 14–24, January–February 2008.

17. D. Fabiani, G.C. Montanari, C. Laurent, G. Teyssedre, P.H.F. Morshuis, R. Bodega, and L.A. Dissado. "HVDC Cable Design and Space Charge Accumulation. Part 3: Effect of Temperature Gradient." *IEEE Electr. Insul. Mag.*, Vol. 24, No. 2, pp. 5–14, March–April 2008.

18. J. Castellon, P. Notingher, S. Agnel, A. Toureille, F. Brame, P. Mirebeau, and J. Matallana. "Electric Field and Space Charge Measurements in Thick Power Cable Isulation." *IEEE Electr. Insul. Mag.* Vol. 25, No. 3, pp. 30–42, May–June 2009.

19. T. Takada. "Acoustic and Optical Methods for Measuring Electric Charge Distributions in Dielectrics." *IEEE Trans. Dielectr. Electr. Insul*, Vol. 6, No. 5, pp. 519–547, October 1999.

20. S. Hole. "Recent Developments in the Pressure Wave Propagation Method." *IEEE Electr. Insul. Mag.*, Vol. 25, No. 3, pp. 7–20, May–June 2009.

21. V. Englund, P.O. Hagstrand, U.H. Nilsson, A. Farkas, C.O. Olsson, and G. Dominguez. "Development of High Performance Polymeric Materials for HVDC Cables." Paper A.2.5, 8th International Conference on Insulated Power Cables (Jicable'11), Versailles (France), June 19–23, 2011.

22. M. Byggeth, K. Johannesson, L. Palmqvist, J. Jonsson, C. Törnkvist, C. Liljegren, and U. Axelsson. "The Development of an Extruded HVDC Cable System and Its First Application in the Gotland HVDC Light Project." Paper B.7.5. pp. 538–542, International Conference on Insulated Power Cables (Jicable'99), Versailles—France, 1999.

23. J. Densley. "New Material Developments in HVDC." Panel 2, Paper 2-00, CIGRE Session 2000, Paris, France, 2000.

24. T.B. Worzyk, M. Jeroense, T. Sörqvist, and M. Farr. "The Murraylink Project—The First Commercial 150 kV Extruded HVDC Cable System." Paper A.7.4., pp. 217–222, International Conference on Insulated Power Cables (Jicable'03), Versailles—France, 2003.

25. B. Jacobson, Y. Jiang-Häfner, P. Rey, G. Asplund, M. Jeroense, A. Gustafsson, and M. Bergkvist. "HVDC with Voltage Source Converters and Extruded Cables for up to ±300 kV and 1000 MW." Paper B4-105, CIGRE Session 2006, Paris, France, 2006.

26. N. Kuljaca, M. Marelli, A. Micheletti, and E. Zaccone. "Cavi sottomarini per impianti di generazione offshore—stato dell'arte e sviluppi futuri." Atti del Convegno Nazionale AEIT 2011, Milano, June 27–29, 2011(in Italian).

27. M. Bacchini R. Grampa, M. Marelli, T. Westerweller, and N. Kelley. “200 kV DC Extruded Cables Crossing the San Francisco Bay.” CIGRE Session paper B1–105, Paris 2010.

28. M. Jeroense, M. Bergkvist, A. Gustafsson, H. Rapp, T. Sörqvist, and J. Svahn. “Increased Voltage for the HVDC Light® Product Range—A Complete Solution.” Paper B.4.7, pp. 425–430, International Conference on Insulated Power Cables (Jicable’07), Versailles—France, 2007.

29. T. Worzyk. *Submarine Power Cables*. Spinger, Berlin, 2009.

CHAPTER 2

FUNDAMENTALS OF HVDC CABLE TRANSMISSION

This chapter recalls the fundamentals of HVDC cable transmission. The main contents of the chapter are as follows:

- A brief description of the historical evolution of HVDC power transmission (Section 2.1)
- The basic principles for an economic comparison between HVDC transmission systems and traditional HVAC transmission systems (Section 2.2)
- An analysis of the configurations and operating modes of HVDC transmission systems (Section 2.3)
- An analysis of the converters commonly used in HVDC transmission systems, that is, the line-commutated current source converters and the self-commutated voltage source converters, with the ultimate goal of singling out the effect of converter typology on HVDC cable insulation (Section 2.4)
- An overview of the main types of HVDC cables, by emphasizing the differences between underground and undersea transmission cables on one hand and between oil–paper insulated and extruded cables on the other (Section 2.5)

2.1 HISTORICAL EVOLUTION OF HVDC POWER TRANSMISSION

The transmission and distribution of electrical energy started with direct current, thanks to Thomas Alva Edison. In 1876 he established his research laboratory at Menlo Park, New Jersey. A commercial dynamo (DC generator) appeared in 1879, an incandescent lamp the same year, and in 1882 the Pearl Street Station began providing electrical energy to New York City's financial district (via $\approx$ 30 km of underground cables!). The energy source was a cluster of six "jumbo dynamos" of 100 kW, each capable of lighting 1200 light bulbs. The electric era had begun, using direct current [1–3]. In the same year 1882, a 50-km-long 2-kV DC transmission line was built between Miesbach and Munich in Germany [4]. At that time, conversion

Extruded Cables for High-Voltage Direct-Current Transmission: Advances in Research and Development, First Edition. Giovanni Mazzanti and Massimo Marzinotto.

between reasonable consumer voltages and higher DC transmission voltages could only be realized by means of rotating DC machines.

However, the dominance of DC was brief: George Westinghouse joined Nikola Tesla and others to exploit the performance advantages of transformers and induction motors. Alternating-current generation, transmission, and utilization started to be—and are still—dominant [2, 3]. Indeed, in AC systems voltage conversion is achieved simply—with low losses and little maintenance—via AC transformers that allow high power and insulation levels within one single unit. Furthermore, three-phase synchronous generators are superior to DC generators in every respect. For these reasons, AC technology was introduced at a very early stage in the development of electrical power systems, and it was soon accepted as the only feasible technology for generation, transmission, and distribution of electrical energy [3, 4].

On the other hand—as outlined in the milestone study [5], a valuable reference for this chapter—HVAC transmission has disadvantages that make DC transmission attractive, namely [4, 5]:

1. Inductive and capacitive elements of overhead lines and cables limit the transmission capacity and the transmission distance of AC transmission links.
2. Such limitation is of particular significance for cables. Depending on the required transmission capacity, the system frequency and voltage, as well as the evaluation of losses, the achievable transmission distance for an HVAC cable will be in the range of 40–100 km with adapted reactive compensations (typically it is ~70 km at 220 kV, ~40 km at 380 kV) [1, 6]. It will mainly be limited by the charging current (see Fig. 2.1; which indicates the cable typologies that can be conveniently used in relation to the length, power, and voltage of the link and the type of transmission system [7]). Moreover, it is shown in [8] that more stringent constraints may be imposed in some cases by the no-load regime, especially when the HVAC cable line is connected to nodes

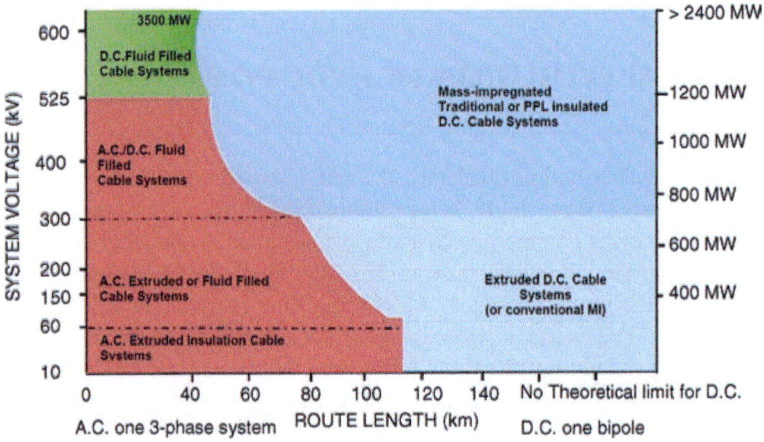

Figure 2.1 Sketch of the cable typologies that can be conveniently used in relation to the length, power, and voltage of the link and the type of transmission system (courtesy of Prysmian).

with a very low short-circuit power: In this case, a maximum length of the cable line exists in order that the maximum phase-to-ground voltage level of the system ($U_m/\sqrt{3}$[1]) is not exceeded at the receiving end of the line, especially if reactive power is not compensated.

3. Direct connection between two AC systems with different frequencies is not possible.
4. Weak connections with few lines between two independent AC systems can, under particular conditions, give rise to system instability in one or both systems or to undesirable power flow scenarios.

Conversely, the advantages of bulk power transmission systems using HVDC cables can be summarized as follows (see, e.g., [5, 9]):

1. The lower energy losses with the same current in the conductor. Indeed, the skin effect and the proximity effect in the conductor are absent (so that the section of the conductor is fully exploited), as well as the additional losses in the conducting layers outside the conductor (metallic screen or sheath, armoring, etc.) [5]. Dielectric losses are also absent in practice if leakage currents can be neglected, as is mostly the case. The average higher electric field stress of HVDC cables leads also to a higher utilization of the cable [10] (see Fig. 2.2).
2. The lower line costs with the same transmitted power (see Section 2.2) [5].
3. HVDC cable lengths are not limited by charging currents and no reactive compensation (for the cable itself) is required at the end stations and/or at intermediate points as in the case of AC transmission systems.

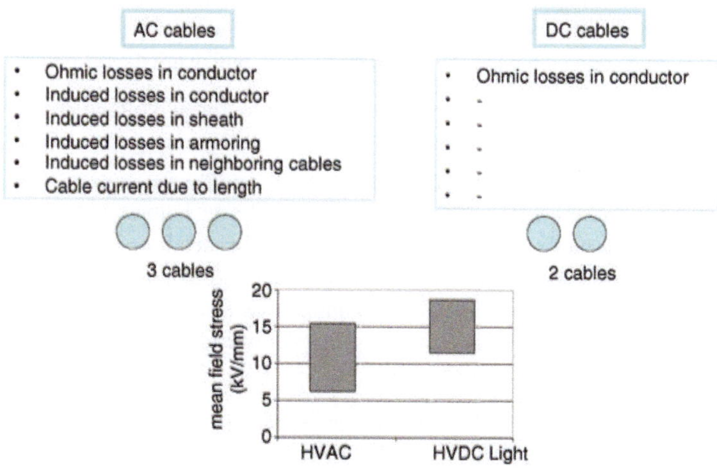

Figure 2.2 AC and DC cable comparison (after [10], courtesy of ABB).

[1] For example, when considering a cable connection to a node with short-circuit current equal to 10 kA, the maximum length of a typical 380-kV double-circuit XLPE insulated cable line—in order not to exceed the maximum phase-to-ground voltage level of the system $U_m/\sqrt{3} = 420/\sqrt{3} = 245$ kV at the receiving end of the line—turns out to be equal to 15 km [8].

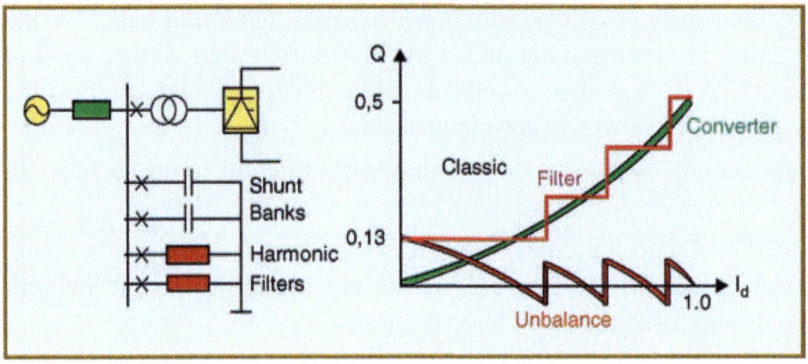

Figure 2.3 Reactive power compensation for conventional HVDC converter stations (after [5]).

4. HVDC cables are more economical when compared to the corresponding HVAC cables for the same power rating due to conductor and insulation savings.
5. DC transmission does not have an impact on rising the short-circuit power at the terminal station.
6. The power flow in a DC scheme can be easily controlled at high speed. Thus a DC link supports the stability of the interconnected AC system.
7. A unique practical feasibility of HVDC cables is their use as undersea interties in bulk power transmission.

Therefore, over the past century research and development efforts were continuously devoted to the technological improvement of DC transmission as a complement and supplement to AC transmission. Such efforts had the main goal of reducing progressively the main disadvantages of HVDC transmission, which are the following:

1. The high cost and losses of converter stations, and as far as voltage source converter (VSC) technology is concerned, their limited operating experience.
2. The non-negligible reactive power demand from AC nodes for conventional HVDC converter stations (i.e., current source converters; see below) that can be roughly estimated as half of the transmitted active power (see Fig. 2.3) [5] and the injection of current and voltage harmonics, especially (but not exclusively[2]) on the AC side of converter stations. This requires—as a sum of the technical solutions commonly chosen for the various types of converters—the installation

[2] In an HVDC system, operation of rectifier and inverter stations produce also harmonics in the DC current and DC voltage of the HVDC transmission line, which may cause interference on the telephone and communication lines running parallel to the DC line. To reduce the interference from HVDC converters, utilities specify the allowable interference level, while the supplier of HVDC systems installs filters to meet the requirements. To limit the harmonic currents flowing into power lines, passive filters are commonly used both on the AC side and the DC side of an HVDC station. During recent years, active filters using high-frequency converters have been studied as complements or alternatives to the passive filters [11]. As a last but not least remark about harmonics in HVDC systems, it must be pointed out that DC harmonics may have deleterious effects on HVDC cables, due to the possible overheating of conductors that involves a higher-than-expected operating temperature of the insulation.

Figure 2.4 Conventional HVDC converter station with AC filters in the foreground and valve hall in the background (after [5]).

of capacitor banks on the AC side and of power harmonics filters on the AC and DC side of converter stations (see Figs. 2.3–2.5) [11].

Hence, it is clear that an essential step for the development of HVDC transmission and its integration into the AC transmission systems is the development of high-power AC/DC and DC/AC conversion systems that have acceptable costs, performances, and reliability.

In the history of HVDC transmission, the first AC/DC and DC/AC conversion systems developed were the so-called *line-commutated current source converters* (commonly denoted with the acronym LCC) belonging to the family of current source converters (CSC), examined in detail in Section 2.4 and schematically illustrated in Figure 2.5 [5].

The invention of mercury arc rectifiers in the 1930's made the design of LCCs possible. In 1941, the first contract for a commercial HVDC system was signed in

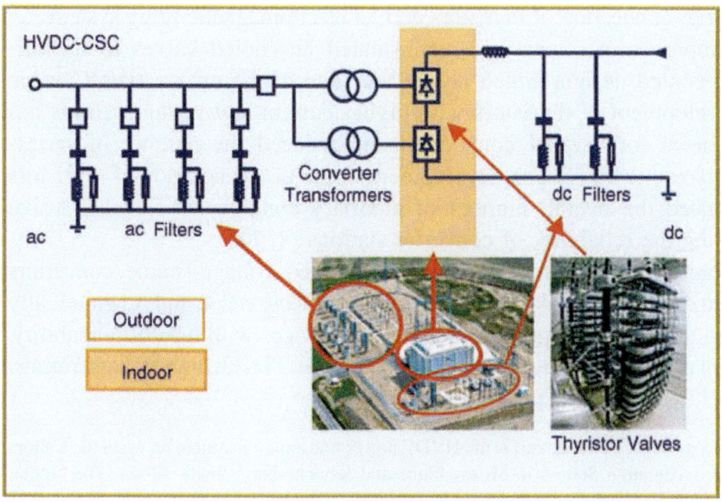

Figure 2.5 Conventional HVDC with current-source converters (after [5]).

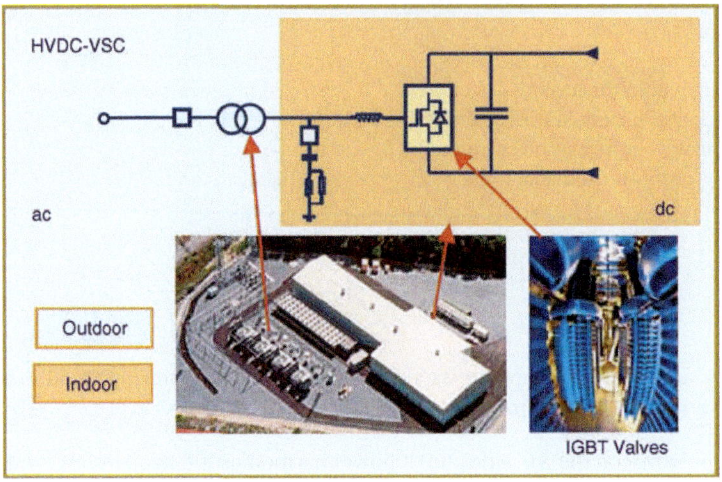

Figure 2.6 HVDC with voltage source converters (after [5]).

Germany: a 115 km long underground cable supplying 60 MW power to the city of Berlin. The system featured ±200 kV voltage rating and 150 A cable ampacity and was ready to be energized in 1945, but it was never put into operation [4] due to the events related to World War II. Since then, several large HVDC systems have been realized with mercury arc valves.

The replacement of mercury arc valves by valves with controlled diodes—Silicon-controlled rectifiers (SCR) or thyristors—was the next major development. The first thyristor valves were put into operation in the late 1970s. The outdoor valves for Cahora Bassa HVDC power transmission system[3] were designed with oil-immersed thyristors with parallel–series connection of thyristors and an electromagnetic firing system.

Further improvements went via air-insulated air-cooled valves to the air-insulated water-cooled design, which is still the state of the art for HVDC valve design. The development of thyristors with higher current and voltage ratings has eliminated the need for parallel connection and reduced the number of series-connected thyristors per valve. The development of the last-generation of thyristors has further reduced the overall number of auxiliary components for each valve thereby increasing the reliability of converter stations [4, 12].

Much more recently than CSC, self-commutated voltage-source converters (VSC) have been developed, also examined in detail in Section 2.4, and schematically illustrated in Fig. 2.6. VSCs require semiconductor devices with turn-off capability. The development of insulated gate bipolar transistors (IGBT) with high-voltage ratings

[3] Cahora Bassa (also spelled Cabora-Bassa) is the HVDC power transmission system between the Cahora Bassa Hydroelectric Generation Station in Mozambique and Johannesburg, South Africa. The bipolar HVDC power line can transmit 1920 MW at ±533 kV and 1800 A. Thyristor valves are used, which unlike other HVDC schemes are mounted outdoors and not in a hall. The system was commissioned in three stages starting in March 1977 with four bridges, and in full operation of eight bridges in June 1979.

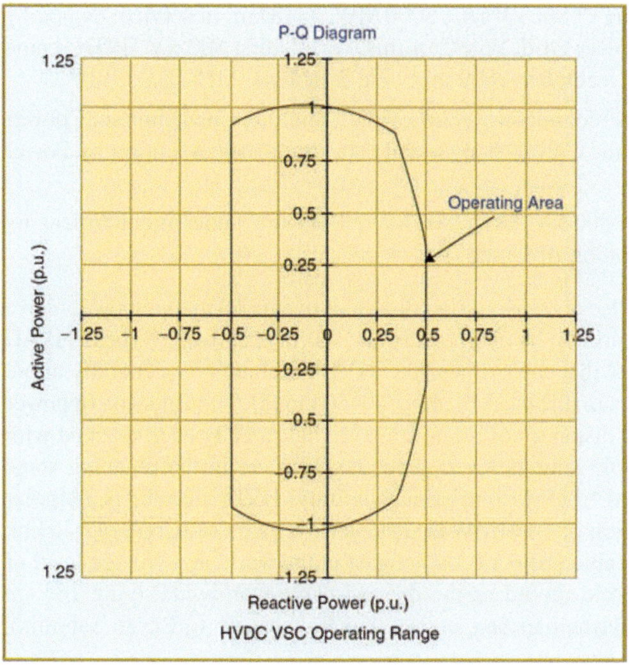

Figure 2.7 Operating range for voltage source converter HVDC transmission (after [5]).

has accelerated the development of voltage-sourced converters for HVDC applications, first in the lower power range [4] and later on in the hundreds of megawatts range [13]. The main characteristics of VSCs are four-quadrant operation capability (an example is reported in Fig. 2.7), but higher losses than conventional CSCs [4, 5].

The first commercial applications of HVDC transmission were HVDC cable lines realized in order to overcome the technical limitations related to the reactive power demand of HVAC cables for considerable distances, especially in the case of submarine transmission, when intermediate power factor correction units cannot be realized. Subsequently, HVDC transmission has been applied even for long-distance overhead lines, offering cost-effective solutions compared to HVAC technology as the transmission distance increases (see Section 2.2) [12].

As mentioned earlier, the technology for HVDC power transmission over long distances was introduced in the second half of the last century. From the first installations, which allowed the transmission of less than 100 MW of power, transmission capacity has increased steadily toward ever higher values. To date, the state of the art in many networks features transmission performances above 3 GW over large distances with only one bipolar overhead DC line at the ultra-high-voltage DC (UHVDC) level, that is, above 500 kV [12]. The advantages of UHVDC transmission have driven the following projects that are operating, under execution, or under procurement [12]:

1. Xiangjiaba-Shanghai, ±800 kV DC, 6400 MW, 1953 km, in service, owned by State Grid Corporation of China, currently the world's largest UHV DC transmission system, which started bipolar operation in June 2010.

2. Yunnan-Guangdong, ±800 kV DC, 5000 MW, 1418 km, in service, owned by China Southern Power Grid. Pole 1 of this world's first 800-kV HVDC came into operation in December 2009 and pole 2 in June 2010.
3. NER/ER-Agra Interconnector, ±800 kV DC, multiterminal, nominal power 6000 MW, 1728 km, with 33% overload capacity, under execution by Power Grid of India.
4. Jinping-Sunnan, ±800-kV DC, 7200 MW, 1935 km, under procurement by State Grid Corporation of China.

At present, generally for transmission distances over 600 km (and power above 1000 MW), power transmission by overhead HVDC is cheaper than overhead HVAC (see Section 2.2). As to the cable lines, the HVDC solution is generally more convenient than HVAC from distances of about 40–50 km [5]. Transmission of power up to 600–800 MW over distances of around 300 km has already been reached with undersea cables, and cable lengths up to about 1000 km are in the planning stage [13]. In particular the SAPEI (SArdinia PEninsula Italy) HVDC intertie is a bipolar system for the transmission of 1000 MW at ±500 kV over a cable length of 440 km. For the time being, this cable project is the longest in the world at a voltage level of 500 kV and holds the world record as the deepest cable worldwide, lying 1650 m below sea level in the Tyrrhenian Sea, connecting Fiumesanto to Borgo Sabotino.

2.2 ECONOMIC COMPARISON BETWEEN HVAC AND HVDC TRANSMISSION SYSTEMS

For a given transmission task, feasibility studies are carried out before the final decision on implementation of an HVAC or HVDC system can be taken. Figure 2.8 shows typical cost comparison curves between AC and DC transmission considering [1, 4]:

1. AC versus DC station terminal costs
2. AC versus DC line costs
3. AC versus DC capitalized value of losses

As can be seen from Figure 2.8, whenever long-distance transmission is discussed, the concept of "break-even distance" between HVAC and HVDC transmission systems arises: the break-even distance is the value of transmission line length such that the savings in HVDC line and loss costs offset the higher converter station costs. Indeed, the DC curve is not as steep as the AC curve because of the considerably lower per-unit-length (pul) values of HVDC line costs and HVDC capitalized loss costs. These considerably lower line costs come from the fact that—consider for the sake of illustration a HVDC bipolar line—the bipolar line uses only two insulated sets of conductors rather than three. Moreover, conductors are not affected by AC losses. For overhead HVDC lines, this results in narrower rights-of-way, smaller towers, and lower line losses than for overhead HVAC lines of comparable capacity: A rough approximation of the savings in line construction

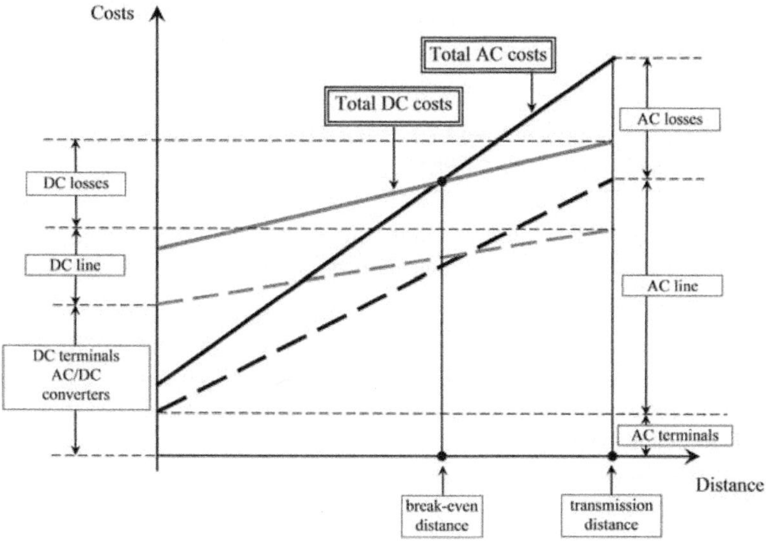

Figure 2.8 Qualitative plot comparing the costs of HVAC (black lines) and HVDC (gray lines) transmission systems. Dashed and solid lines represent, respectively, the costs without and with actualized energy losses.

is 30%.[4] For long AC lines the cost of intermediate reactive power compensation has to be taken into account, too [5].

For overhead lines, the break-even distance is in the range of 500–800 km [4] (with a trend to decrease toward the 500 to 600-km range [5, 13]) and depends first of all, of course, on the rated voltage and on the power to be transmitted, as well as on a number of other factors, for example, country-specific cost elements, interest rates for project financing, loss evaluation, cost of right-of-way, and the like. Although break-even distance is influenced by the costs of right-of-way and line construction, the concept itself should be regarded with care and is somehow misleading, since in many cases more AC lines are needed to deliver the same power over the same distance due to system stability limitations. As to the cable lines, the HVDC solution is generally more convenient than HVAC from distances of about 40–50 km, as pointed out above [5, 13, 14].

It must be emphasized that, different from AC lines, in a HVDC link the voltage level is not imposed by the existing grid, but it has to be chosen by the project designer within a preselected voltage level range through an optimization procedure performed on a technical-economic basis: The investment cost of the transmission line typically increases with its voltage, while the losses decrease with it; the optimization procedure aims at comparing and minimizing the total cost [15]. Anyway, it is important to outline

[4] Similar savings in line construction hold also for HVDC versus HVAC cable lines. However, these savings could be lower for extruded insulation HVDC cable lines with rated voltage ≥300 kV [14], which are close to the current technological limit as regards this type of cable. Therefore, at 300 kV (or higher voltage levels) extruded insulation HVDC cable lines may require innovative design and manufacturing solutions for the cable and accessories that can affect the pul cost of the cable system as a whole (see Chapters 3 and 5).

that general rules for the estimation of the break-even distance cannot be found since its estimation depends on many factors that are also not technological.

2.3 CONFIGURATIONS AND OPERATING MODES OF HVDC TRANSMISSION SYSTEMS

During its development, HVDC transmission has provided solutions to various problems, giving rise to the following applications, illustrated schematically in Figure 2.9:

1. *HVDC Back-to-Back (B2B) Transmission* The expression "back-to-back" indicates that the rectifier and the inverter are located in the same station and interconnected without a transmission line. Back-to-back converters are mainly used for power transmission between adjacent AC grids that cannot be synchronized. They can also be used within a meshed grid in order to achieve a defined power flow [4]. Back-to-back converters can work to increase the stability in the transit of power in long AC lines, too (see Fig. 2.9*a*) [12].
2. *HVDC Long-Distance Transmission (LDT)* HVDC LDT systems (see Fig. 2.9*b*) often provide a more economical alternative to AC transmission for long-distance bulk power delivery from huge resources, located in remote and scarcely populated areas, to large load centers at a great distance from them.[5]
3. *Hybrid HVDC Transmission* HVDC transmission integrated in the complex interconnected HVAC grid[6] (see Fig. 2.9*c*) [16, 17].

The reasons for these hybrid solutions are basically the lower transmission costs, as well as the ability to bypass heavily loaded AC systems and introduce elements of discontinuity in the power grid. These elements are crucial in limiting the "domino effect" in system blackouts [12, 18], so that asynchronous HVDC links act as an effective "firewall" against the propagation of cascading outages in one network from passing to another network [5]. This is one of the main benefits of using HVDC transmission, which—both in the B2B and in the LDT configuration—is

[5] A few examples are the above-cited Cahora Bassa power plant; the Itaipu hydroelectric power plant on a tributary of the Parana River in South America, having 12,600 MW rated power, from which three transmission lines radiate: the first is a 220-kV AC line to Paraguay, the second is a 760-kV AC line to Brazil, the third is a 600-kV DC line to Brazil as well; the Three Gorges hydroelectric power plant on the Yang-Tze River in China.

[6] An example of a hybrid HVDC transmission is the Pacific Intertie. The Pacific Intertie is a ±500-kV DC bipolar overhead line linking the Pacific Northwest to the Los Angeles area. The line capacity is 3100 MW to serve two to three million Los Angeles households and is 48.7% of the Los Angeles Department of Water and Power (LADWP) electrical system's peak capacity. The Intertie originates near the Columbia River at the Celilo Converter Station on Bonneville Power Administration's grid outside The Dalles (Oregon) and is connected exclusively to the Sylmar (California) Converter Station north of Los Angeles, which is owned by five utility companies and managed by LADWP. The Intertie is capable of transmitting power in either direction, but power on the Intertie flows mostly from north to south. The Pacific Intertie can be referred to as a hybrid HVDC transmission due to the fact that it works in parallel with three 500-kV AC lines, which carry up to 4800 MW. Two AC lines run from the John Day Dam (Oregon) to Lugo (California), near Los Angeles, the other AC line runs from Celilo Station (Oregon) to Tracy-Telsa (California), near San Francisco (for further details see [16]).

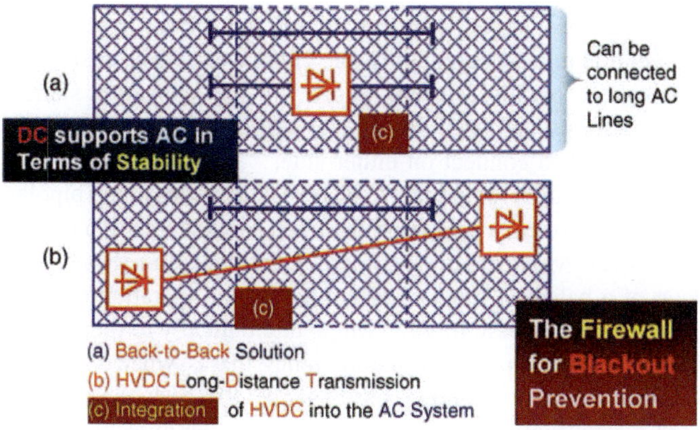

Figure 2.9 Main applications of HVDC transmission (after [12]).

able to control the power flow in a very quick way, supporting in this case the grid, thereby preventing automatically blackouts in case of cascading outages; this is not possible with AC synchronous links [12]. Figure 2.10 shows the various common system configurations and operating modes used for CSC HVDC transmission [5].

Monopolar systems are the simplest and least expensive systems for moderate power transfers since only two converters (one per each terminal) and one high-voltage insulated cable or line conductor are required. For current return such systems can be used with electrodes (both land or sea electrodes) or with dedicated medium-voltage insulated conductors (both with overhead lines or with cable lines).

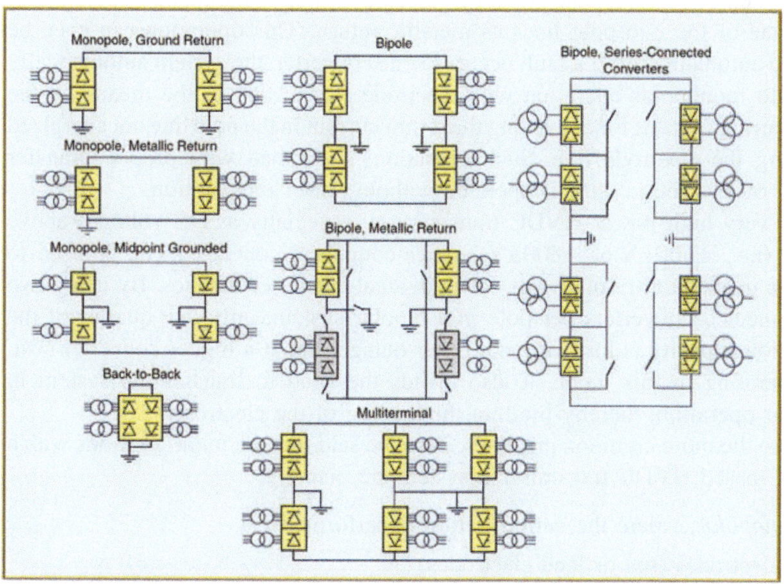

Figure 2.10 HVDC configurations and operating modes—CSC technology (after [5]).

The latter is usually referred to as "metallic return," while the former is called an electrode return. When metallic return is used, only one of the two converter stations is grounded so as to avoid a parallel current return via the converter station grounding. Back-to-back stations are used for interconnection of asynchronous networks and AC lines are used to connect on either side.

The most common configuration for modern CSC HVDC transmission is bipolar with a single 12-pulse converter for each pole terminal. Each converter station is then made of two monopolar systems with a common point of interconnection: the neutral. This gives the advantage of a balanced bipolar system in normal operation without electrode or metallic return and the possibility to switch automatically to monopolar operation with electrode or metallic return during contingency, that is, when one of the two pole lines or one of the four converter bridges have a fault. Obviously the nominal power is halved when the system switches from bipolar to monopolar operation. Such flexibility belongs to bipolar systems that are capable of maintaining the system in operation during single-fault contingency within the DC system. Bipolar systems are also able to guarantee, in some cases, the transmission for a double-fault contingency within the DC system, but in such a case with a temporary interruption of the power flow: It is the case of a system running in monopolar operation for a fault in one of the two pole lines and facing a second outage in one of the two converters in operation. In such a case, if forecasted, once the DC system is off, the connections at each station terminal can be rearranged and then the system resumes the monopolar operation with the healthy pole line and the other two converters that were put out of service when the first fault occurred to the pole line.

Furthermore, bipolar systems equipped with electrodes can even use metallic return instead of electrode return when a fault occurs at one of the four converters: Indeed in such a case, since the system is forced to switch to monopolar operation, it can use one of the two pole lines as metallic return. This operation can even be performed automatically: If a fault occurs in one converter, the system automatically switches to monopolar operation with electrode return, and in the meantime the system rearranges itself for conveying the return current in the pole line not energized by opening the electrode line. Such operation, performed with proper "transfer switches" on the neutral bus, is operated without power interruption.

For very high power HVDC transmission, especially at DC voltages above ±500 kV (i.e., ±600 kV or ±800 kV), series-connected converters can be used to reduce the unavailability of energy for individual converter outages. By using two series-connected converters per pole in a bipolar system, only one quarter of the transmission capacity is lost for a converter outage, but at a higher converter cost. When operating in this mode, it also avoids the need to transfer the system in monopolar operation, thereby limiting the loading of the electrodes [5].

As to the more common practices, it can be said that the main solutions with a CSC LCC-based HVDC transmission system are, namely:

- *Monopolar*, where the current return is performed via
  - Electrodes (sea or land electrodes)
  - Overhead/cable line (with a medium-voltage insulation level) usually referred to as metallic return

- *Bipolar*
  - With a third path for the current return to be used in emergency conditions via electrodes or metallic return
  - Without a third path for the current return (referred to as "stiff bipolar system")

On the other hand, VSC-based HVDC transmission is usually arranged according to the following three schemes:

- *Symmetric Monopolar (or "stiff bipolar" system)* One converter per station with two poles output operated at $+U_0$ and $-U_0$. For cable lines, in this case a fault in one of the two cables or in one of the two stations entails the shutdown of the whole system.
- *Asymmetric Monopolar* One converter per station with one pole and electrode or metallic return. For cable lines, in this case a fault in one of the two cables or in one of the two stations entails the shutdown of the whole system.
- *Bipolar* Two converters per station series connected with two poles output and electrode or metallic return between the midpoints of the two series-connected converters. For cable lines, in case of a fault in one of the two cables or in one of the two stations, the scheme switches to an asymmetric monopole at half power. This configuration is the same as that adopted when using bipolar CSC converters. Although to our knowledge, it has never been realized to date, it will surely be the future of VSC converter systems since it can guarantee a higher reliability level.

2.4 CSC AND VSC CONVERTERS

2.4.1 Operation of a Line-Commutated Current Source Converter (LCC–CSC)

Conventional HVDC transmission uses LCC–CSC (simply referred to as LCC in this section) with thyristor valves. Such converters need a steady and synchronous voltage source to operate. The basic building block presently employed in HVDC conversion is the three-phase, full-wave bridge referred to as a 6-pulse or Graetz bridge. The term 6-pulse is due to six switching operations or commutations[7] per period, resulting in a characteristic harmonic ripple of six times the fundamental frequency in the DC output voltage. Each 6-pulse bridge consists of six controlled switching elements or thyristor valves. In turn, each valve consists of a suitable number of series-connected thyristors to achieve the required DC voltage rating [5].

The DC terminals of two 6-pulse bridges with AC voltage source phase-shifted by 30° can be connected in series to increase the DC voltage and eliminate some of the characteristic AC current and DC voltage harmonics. This is referred to as 12-pulse operation and represents the state-of-the-art technology for LCC HVDC

[7] A commutation is the current transfer from one phase to another in a synchronized firing sequence of the thyristor valves.

systems. In 12-pulse operation, the characteristic AC current and DC voltage harmonics have frequencies of $12n \pm 1$ and $12n$, respectively ($n = 1,2, \ldots$). The 30° phase-shifting is achieved by feeding one bridge through a transformer with a star-connected secondary and the other bridge through a transformer with a delta-connected secondary (see Fig. 2.11). HVDC transmission schemes utilize 12-pulse converters to reduce the harmonic filtering requirements needed for 6-pulse operation, for example, fifth and seventh harmonics filtered on the AC side and sixth harmonic on the DC side. This is because, although these harmonic currents still flow through the valves and the transformer windings, they are 180° out of phase and cancel out on the primary side of the converter transformer [5]. Figure 2.12 shows the thyristor valve arrangement for a 12-pulse converter with three quadruple valves, one for each phase. Each thyristor valve is built up with series-connected thyristor modules. The thyristor technology is both direct light-triggered thyristor (LTT) [4, 12] and the electric-triggered thyristor (ETT), the latter especially for very high power thyristors.

Line-commutated converters require a relatively strong synchronous voltage source in order to commutate correctly. The three-phase symmetrical short-circuit capacity available from the network at the converter connection point should be at least three times the converter rating for converter operation without other countermeasures. If such a ratio is between three and two, some countermeasure will be taken, for example, third harmonic filters. If such a ratio is below two, even countermeasures cannot guarantee a correct converter operation [1].

LCCs can only operate with the AC current lagging the voltage; thus the conversion process demands reactive power. Reactive power is supplied by the AC filters—which look capacitive at the fundamental frequency—or shunt capacitor banks. Any unbalance in reactive power between these local sources and the converter must be accommodated by the AC system and maintained

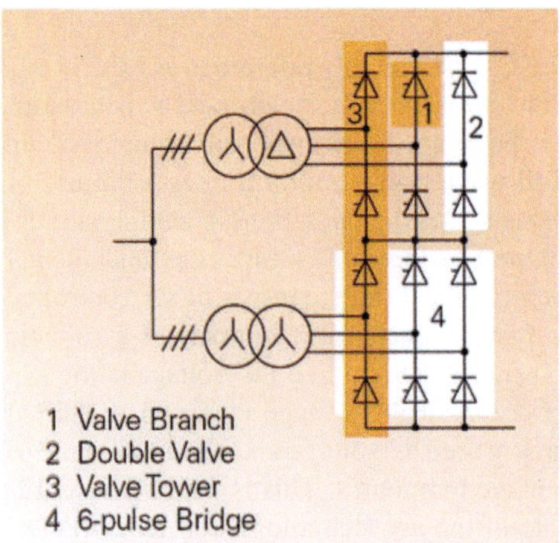

Figure 2.11 Arrangement of the transformers and of the valve branches in a 12 pulse bridge (after [5]).

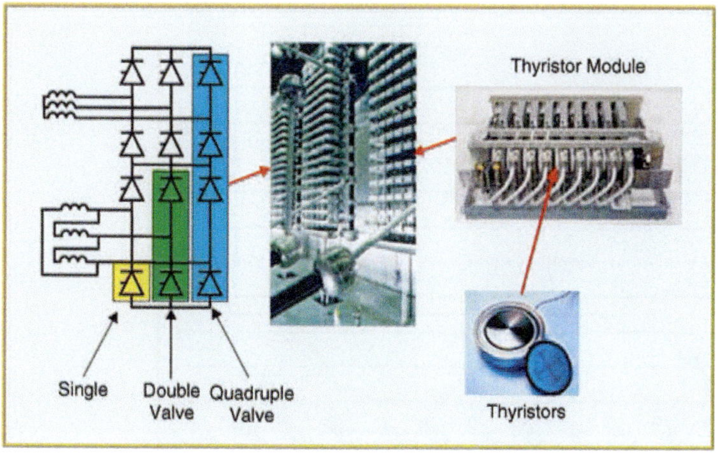

Figure 2.12 Thyristor valve arrangement for a 12-pulse converter with three quadruple valves, one for each phase (after [5]).

within a given range to match the AC voltage variation constraints. The weaker the AC system or the further the converter is away from generation, the tighter the reactive power exchange must be to keep the AC voltage within the required range. Reactive power demand, reactive power compensation, and reactive power exchange with the AC network as a function of DC load current have already been illustrated qualitatively for a conventional HVDC converter station in Figure 2.3 [5].

2.4.2 Operation of a Self-Commutated Voltage Source Converter (VSC)

HVDC transmission using VSCs can be of three types:

1. Full pulse-width modulation (PWM) [5, 14];
2. Partial PWM and multilevel step wave;
3. Full multilevel step wave [4, 12].

VSC-based systems are self-commutated with insulated-gate bipolar transistor (IGBT) valves. Since the first VSC HVDC was introduced in the late 1990s, the progression to higher voltage and power ratings for IGBT valve VSC converters has roughly paralleled that for thyristor valve CSC converters in the 1970s [5]. Figure 2.13 illustrates solid-state converter development for the two different types of converter technologies using thyristor valves and IGBT valves.

The overall system performances can take advantage of HVDC transmission with VSCs. VSC technology can rapidly control both active and reactive power independently of one another. Reactive power can also be regulated at each terminal independent of the DC transmission voltage level, thereby achieving total flexibility to place converters anywhere in the AC network, since no constraint is associated

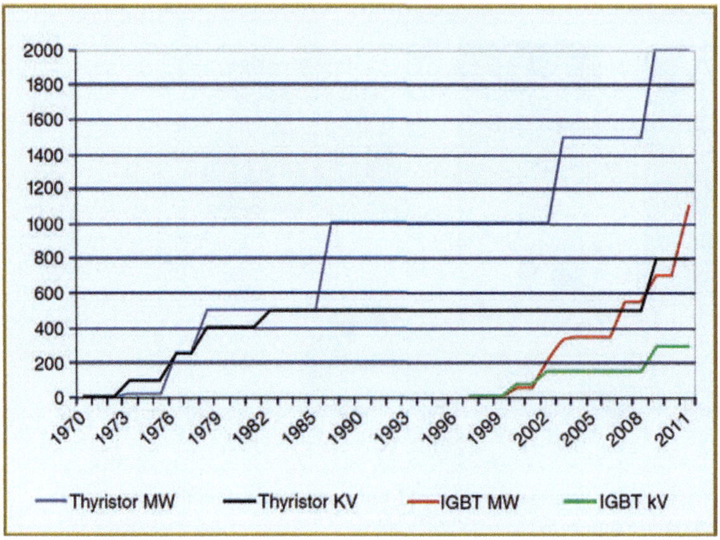

Figure 2.13 Solid-state converter development (after [5]).

with the minimum network short-circuit capacity. Moreover, each VSC station can work as a static synchronous compensator (STATCOM) when the DC link is out of service. Self-commutated VSCs even permit black start, namely the converter can synthesize a balanced set of three-phase voltages as a virtual synchronous generator. The dynamic support of the AC voltage at each converter terminal improves the voltage stability and can increase the transfer capability of the sending- and receiving-end AC systems, thereby enhancing the transfer capability of the DC link itself [5].

Figure 2.14 shows the IGBT converter valve arrangement for a VSC station, while the active and reactive power operating range for a converter station with VSC

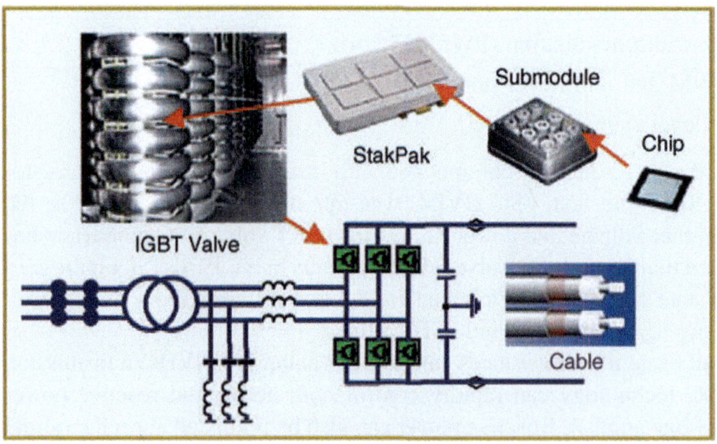

Figure 2.14 HVDC VSC with IGBT valve converter arrangement (after [5]).

was shown in Figure 2.7. Unlike conventional (CSC) HVDC transmission, the converters do not demand reactive power for their own operation and can actually control their reactive power to regulate the AC system voltage just as a generator.

2.4.3 CSC versus VSC: How Do They Affect Cable Insulation?

HVDC systems are sometimes called to support a grid portion interconnected at one end, for example, for a load rejection. In such a case, a fast reversal of power flow direction becomes necessary. In CSC systems such power flow direction reversal is accomplished via a change in the pole voltage polarity since the current flow direction cannot be reversed. On the contrary, in VSC systems the same pole voltage polarity is always maintained, whereas the current flow direction is reversed in order to reverse the power flow from one converter station to the other.

Such a basic difference has an impact as far as the insulation of the cable system is considered. Indeed, one of the main limitations that has hampered the development of extruded cables since the 1970s was the ability of the insulation to withstand polarity reversal. Today the polarity reversal still represents a challenge for extruded insulation cables for HVDC application. For this reason, to date HVDC cable lines belonging to HVDC systems with CSC have MIND insulation (with the exception of the Hokkaido-Honshu link, see Section 7.2.8), while HVDC cables lines belonging to HVDC systems with VSC can have both MIND and extruded insulation.

However, it must be pointed out that some manufacturers have developed or are developing new HVDC extruded cable systems through long-term load cycle tests that include polarity reversals (see Section 5.6.5 and Section 7.2.8).

2.5 CABLES FOR HVDC TRANSMISSION

2.5.1 Underground and Undersea Cable Transmission

Contrary to AC cables, there is no physical restriction limiting the distance or power level for HVDC underground or undersea cables. Underground cables can be used on shared rights-of-way with other utilities without reliability concerns about the use of common corridors, provided that the needed engineering and technical aspects are resolved [5, 19, 20].

Using HVDC transmission for underground or undersea cable systems involves considerable savings in the costs of installed cables and losses. Depending on the power level to be transmitted, these savings can offset the higher converter station costs at distances of 40–50 km or more [5, 13] (see Sections 2.1 and 2.2). Furthermore, there is a drop-off in cable capacity with AC transmission over distance due to its reactive component of charging current, since cables have higher capacitances and lower inductances than AC overhead lines. This can be compensated by intermediate shunt reactors for underground cables at increased expense, but this solution is unpractical—or even unfeasible—for undersea cables (see Chapter 1) [5].

For a given cable conductor cross section, the line losses with HVDC cables can be about half those of AC cables. As highlighted above, this is due to AC cables requiring more conductors (three phases), carrying the reactive component of current, skin, and proximity effect, and induced currents in the cable sheath and armor [5].

With a cable system, the need to balance unequal loadings or the risk of postcontingency overloads often necessitates use of series-connected reactors or phase-shifting transformers (PST). These potential problems do not exist with a controlled HVDC cable system [5].

HVDC extruded cables with prefabricated joints used with VSC-based transmission are lighter, more flexible, and easier to joint than the mass-impregnated oil–paper cables used for conventional HVDC transmission, thereby making them prone to land cable applications where transport limitations and extra jointing costs can raise installation costs. The lower-cost cable installations made possible by the HVDC extruded cables and prefabricated joints makes long-distance underground transmission economically feasible for use in areas with rights-of-way constraints or subject to authorization problems and delays with overhead lines [5].

As pointed out above, transmission systems crossing the sea are an important application for HVDC. In undersea applications HVDC is the preferred technology to overcome distances longer than some tens of kilometers and transmission capacities from several hundreds to more than a thousand megawatts (for bipolar systems). For the undersea transmission part, a special cable is required, suitable not only for DC current and voltage but also for the mechanical stresses associated with the laying procedure, the hydrostatic pressure of water, and the possible damage by tides and occasional impacts with anchors, fishing gears (e.g., trawlers), and so on [4, 19].

The cable arrangements are an important area of development for undersea HVDC cable systems. For monopolar transmission systems, the return path, as mentioned above, is either the sea or a second cable (see metallic return in Fig. 2.10). The first solution requires reliable and environmentally friendly electrode systems, whereas the second one has severe impact on the costs of the overall transmission scheme [1, 19].

Therefore, cables are sometimes developed with an integrated return conductor [4]. In such a case, the cable core is the traditional design cable (essentially with mass-impregnated insulation or with extruded insulation; see Section 2.5.2). As to the return conductor (often XLPE insulated), it can be wound outside the lead sheath as an a "second screen," thereby also forming part of the armor, together with the flat steel wire layer on the outside of the return conductor insulation. A cable of this type with MIND insulation of the cable core was installed for the Northern Ireland–Scotland (Moyle) HVDC interconnection with a rated voltage of 250 kV and rated power 250 MW [4], while ±250-kV HVDC coaxial cables with extruded insulation of the cable core were also developed recently (see Section 5.6.5) [21].

Alternatively, the XLPE insulated metallic return cable can be simultaneously laid (and buried, if needed) on the sea bottom in close bundle together with the

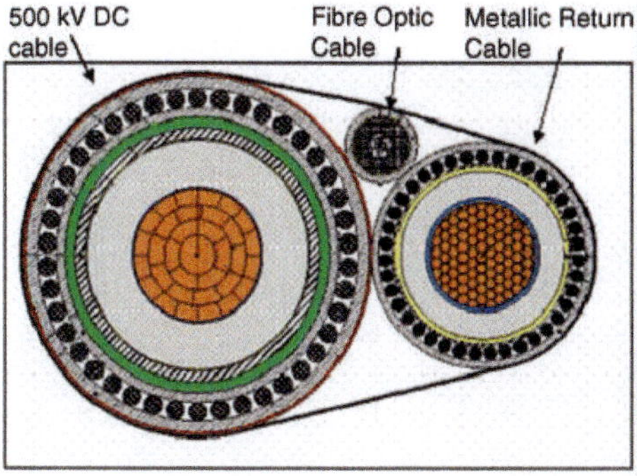

CONSTRUCTIONAL DETAILS OF 500 kV HVDC CABLE			
1 - Copper Conductor: Diameter of circular conductor:		mm^2 mm	2100 51.9
2 - Conductor Screen			
3 - Insulation: mass impregnated, paper Thickness:		mm	19.5
4 - Insulation Screen			
5 - Lead Alloy Sheath:			
6 - PE Sheath			
7 - Reinforcement: Galvanised steel tapes			
8 - Bedding: Polyester Tapes			
9 - Armour: Galvanised steel wires			
10 - Serving: Polypropylene strings			
Overall diameter of the cable		mm	131
Weight of cable	In air	kg/m	53.5
	In water		44

Figure 2.15 Bundle of three cables—a 500-kV HVDC cable, a medium-voltage metallic return cable, and a fiber-optic cable—for the HVDC Neptune Regional Transmission System (courtesy of Prysmian).

HVDC cable and a single fiber-optic cable (see Fig. 2.15) [22]. This is the solution chosen for the Neptune Regional Transmission System, a 105-km-long HVDC interconnection between Sayreville, New Jersey, and Nassau County on Long Island, New York, via undersea and underground cables with a monopolar HVDC—metallic return scheme—rated terminal voltage is 500 kV DC, bidirectional power transfer capability is 660 MW, and converter stations house LCC converters with direct LTT valves. When operation was started at the end of June 2007, the link was the first installed using paper-insulated, mass-impregnated cables at this high-voltage level (500 kV) and size (2100 mm^2) [22].

The main details of the HVDC cable used in the Neptune Intertie are listed in Figure 2.15. The medium-voltage metallic return cable consists of a 2000-mm^2 copper conductor, DC cross-linked polyethylene (DC-XLPE) insulation, copper screen, polyethylene sheath, and galvanized steel armoring; overall cable diameter is ~100 mm and weight is, respectively, 30.4 kg/m in air and 23.2 kg/m in water. For the land portion on Long Island, the metallic return cable was split into two cables in parallel, laid on the two sides of the HVDC cable. This configuration was chosen in order to minimize the magnetic field due to the direct current flowing in the cables to a value less than 20 μT. The fiber-optic cable is a standard single-wire armored undersea cable; it is equipped with 24 low-attenuation, single-mode fibers [22].

2.5.2 Different HVDC Cable Types

As can be argued from the above considerations, as well as from Chapter 1, different types of HVDC cables are available, depending mainly on cable insulation. From Chapter 3 on, focus is on HVDC cables with extruded insulation, and cable structure will be analyzed in more depth. Here, a short review including other types of cables (and the relevant insulation) is provided. On the whole, the following typologies of

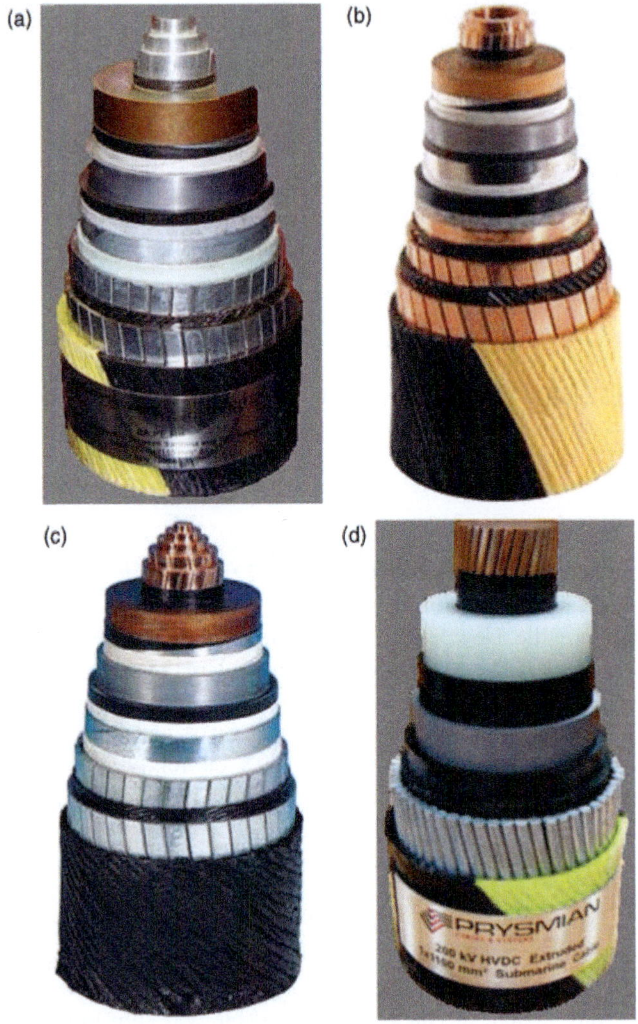

Figure 2.16 Main typologies of HVDC cables used in undersea HVDC power transmission systems: (*a*) mass-impregnated nondraining (MIND) cables, (*b*) self-contained oil-filled (SCOF) cables, (*c*) polypropylene paper laminate (PPL, MI-PPL, PPLP) or lapped thin-film insulated cables, (*d*) polymer-insulated or extruded-insulation cables (courtesy of Prysmian).

HVDC cables have been mainly used in undersea (see Fig. 2.16) or land HVDC power transmission systems [4, 19, 20, 23, 24]:

1. Mass impregnated nondraining (MIND) cables (see Fig. 2.16a);
2. Oil-filled (OF) cables, which are divided in turn into self-contained oil-filled (SCOF) cables (see Fig. 2.16b) and high pressure oil-filled (HPOF) cables;
3. Polypropylene paper laminate (PPL, MI-PPL, PPLP) or lapped thin-film insulated cables (see Fig. 2.16c);
4. Polymer-insulated or extruded insulation cables (see Fig. 2.16d).

2.5.2.1 Mass-Impregnated Nondraining (MIND) Cables [4, 19, 20, 23–25]

At present, this cable type is used in most of the HVDC applications. It consists of different layers, as shown in Figure 2.17.

1. Compressed or compact round conductors of copper or aluminum shaped wires;
2. Conductor semicon of carbon-loaded paper;
3. Insulation, that is, pure cellulose paper impregnated with oil and resin;
4. Insulation semicon of carbon-loaded paper;
5. Metallic sheath;
6. Thermoplastic sheath;
7. Reinforcement of steel tapes (only for undersea cables);
8. Bedding (only for undersea cables);
9. Armor of steel flat wires (only for undersea cables).

The copper or aluminum conductor has a compressed or a compact round shape. The conductor is covered by oil and resin-impregnated papers; the resin serves for providing the oil with nondraining characteristics. The conductor semicon and

Figure 2.17 Mass-impregnated nondraining (MIND) HVDC cable for the deep water section of SAPEI undersea link (courtesy of Prysmian).

insulation semicon layers are of carbon-loaded papers, whereas the outer layer consists of copper or aluminum tapes and/or woven fabrics. The fully impregnated cable is then lead sheathed to keep the outside environment away from the insulation. The next layer is the anticorrosion protection, which consists of extruded high-density polyethylene (HDPE).

Galvanized steel tapes are applied around the HDPE layer to prevent the cable from permanent deformation during cable loading. Over the steel tapes a polypropylene string is applied followed by galvanized steel wire armor, for protecting the cable from additional stresses associated with the undersea environment (see Section 2.5.1).

Typically, this technology is available for voltages up to 500 kV and a transmission capacity of up to 800 MW in one cable with installation depths of up to 1640 m below sea level at present [24] and nearly unlimited transmission lengths. The transmission capacity of mass-impregnated cables is limited by the conductor temperature, that typically should not exceed 55°C.

2.5.2.2 *Oil-Filled (OF) Cables [4, 19, 20, 23–25] (see Fig. 2.18)* In comparison to mass-impregnated cables, the conductor is insulated by paper impregnated with a low-viscosity oil and incorporates a longitudinal duct to permit oil flow along the cable. The other layers are much the same as listed above.

Oil-filled cables are suitable for both AC and DC voltages with DC voltages up to 600 kV DC. Due to the required oil flow along the cable, the transmission line lengths are, however, limited to <100 km [4, 23].

Oil-filled cables are in turn divided into [19, 20, 23]:

1. Self-contained oil filled (SCOF) cables often referred to also as self-contained fluid-filled (SCFF) or low-pressure oil-filled (LPOF) cables, which use low-pressure oil (see Fig. 2.18);
2. High-pressure oil-filled (HPOF) cables, which use high pressure oil.

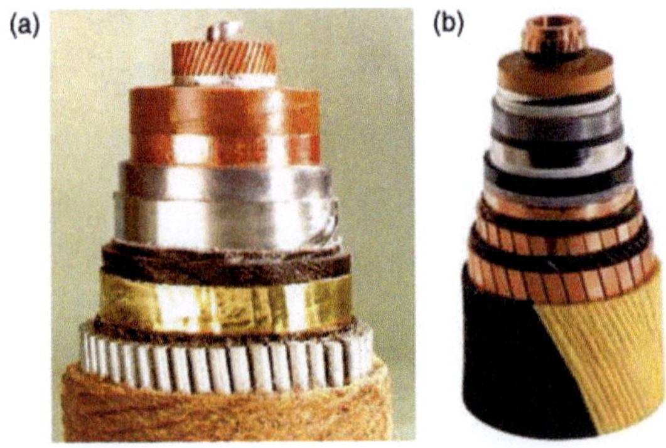

Figure 2.18 Two types of self-contained oil-filled (SCOF) HVDC cable: (a) after [25] and (b) courtesy of Prysmian.

Different from SCOF cables, whose conductor is realized with a hollow shape for the oil flow, the HPOF cables are laid inside a pipe filled with oil at high pressure.

2.5.2.3 *Polypropylene Paper Laminate (known under a few different acronyms, i.e., PPL, MI-PPL, PPLP) [19, 20, 23, 24] or Lapped Thin-Film Insulated Cable [4]* Some manufacturers have developed HVDC cable systems that utilize PPL insulation and impregnating compounds that have nondraining characteristics for HVDC with a voltage rating of 600 kV. These cable systems are designed to operate at conductor temperatures of 80–90°C [23]. It is claimed that this type of cable can sustain up to 60% higher electrical stresses in operation, making it suitable for very long and deep undersea cables [4]. However, they are not yet a fully established technology in DC systems [23].

2.5.2.4 *Polymer-Insulated or Extruded-Insulation Cable [4, 19, 20, 23–25] (see Fig. 2.19)* Until the mid-1990s paper–oil insulated cables dominated the HVDC cable market all over the world, as shown in Table 2.1. Since then, to overcome the disadvantages of these cable types, cable suppliers made a renewed R&D effort to replace paper–oil insulation with extruded insulation (see Chapter 1). The conductor is a stranded copper or aluminum conductor insulated by an extruded insulation layer. The extruded insulation is polymeric and mostly based on polyethylene compounds, among which the preferred ones are low-density polyethilene (LDPE) and particularly cross-linked polyethilene (XLPE) [4, 23–25] of a special kind for best performance under HVDC condition (see [9, 21, 24, 25–27] and Chapter 5): This special kind of XLPE is commonly referred to as DC-XLPE. The polymeric insulation is suitable for a higher conductor temperature than MIND.

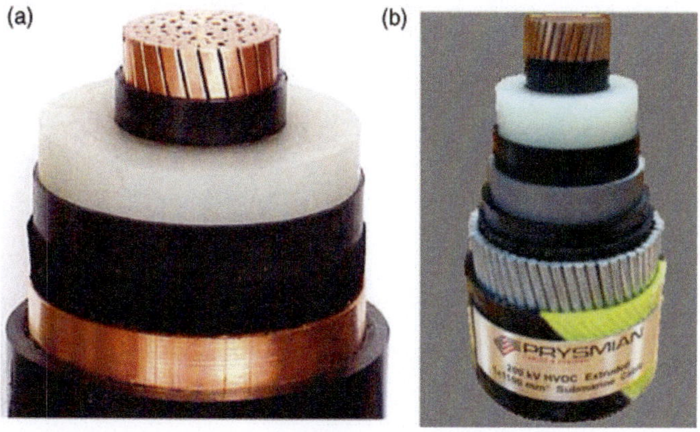

Figure 2.19 Two types of XLPE-insulated HVDC cable: (*a*) land cable after [25] and (*b*) undersea cable courtesy of Prysmian.

Table 2.1 Some Significant Information about HVDC Cables Worldwide until mid-1990s

	Name		Commissioning Year	Voltage U_0 (kV)	Power (1 cable) (MW)	No. of Cables	Conductor Type[a]	Conductor Size (mm^2)	Insulation Type[b]	Insulation Thickness (mm)	Length (km)
1	Gotland 1[c]		1954	100	20	[e]	Cu s	90	MI	7.0	100
			1970	150	30						
2	Cross Channel[c]	Sea	1961	100	80	2	Cu r	340/390	MI	9.0	51
		Land	1961	100	80	2	Cu r	605		9.0	
3	SA.CO.I.		1965	200	100	[d]	Cu o	420	MI	11.8	119
			1992	200	100						
4	Cook Strait 1		1965	250	300	2 + 1	Cu h	520	GF	14	39
5	Konti Skan:										
	Sweden Laesoe[c]	Sea	1965/74	285	300	1	Cu r	625	MI	15	64
		Land		285	300	1	Cu r	800	MI	15	7
	Laesoe[c]	Sea	1965	285	250	1	Cu r	2 × 310	OF(2)	12.4	23
		Land	1965	285	250	1	Cu r	2 × 310	OF(2)	12.4	1
6	Vancouver 1		1969	300	156	2 + 1	Cu o	400	MI	18.5	4 + 27
7	Kingsnorth[c]		1971	266	320	2[f]	Cu h	800	OF	10.3	84
8	Mallorca/Menorca[c]		1972	200	100	4	Al h	500	OF	8.4	44
9	Skagerrak	Sea	1976	262.5	250	2	Cu r	800	MI	16	125
		Land	1976	262.5	250	2	Cu r	1000	MI	14	
10	Vancouver 2		1976	300	185	2	Cu h	400	OF	11.7	35
11	Hokkaido Honshu	Sea	1980/93	250	300	2[f]	Cu h	600	OF	14.5	42/43
		Land	1980/93	250	300	2[f]	Cu h	900	OF	13	1
12	Gotland 2/3		1983/87	150	160	2[f]	Cu r	800	MI	8	100

13	IF A 2000)	Sea	1986	270/280	250	8	Cu r	900	MI	12.3	50
	Cross Channel[c]	Land	1986	270/280	250	8	Cu h	800	OF	10.75	18 + 8
14	Konti Skan 2/3		1988/91	285	300	2	Cu r	1200	MI	15	64
15	Fenno Skan		1989	400	500	1	Cu r	1200	MI	17.5	200
16	Cook Strait 2		1991	350	500	2 + 1	Cu r	1400	MI	17.5	40
17	St. Lawrence		1993	500	625	6	Cu h	1400	OF	26.8	5.1
18	Hawali		tested	300	250	2 + 1	Al h	1600	OF	10.9	67 + 154
19	Skagerrak 3		1993	350	500	1	Cu r	1400	MI	18.0	125
20	Sheju	Sea	1993	180	150	2	Cu r	800	MI	9.5	95
		Land	1993	180	150	2	Cu r	800	MI	9.5	5
21	Baltic		1994	450	600	1	Cu r	1600	MI	19.0	250
22	Kontek	Sea	1995	400	600	1	Cu r	2 × 800	OF(2)	16.5	55
		Land	1995	400	600	1	Cu r	2 × 800	OF(2)	16.5	120

[a]Legend for conductor type h = hollow, o = oval, r = round stranded, s = round solid.
[b]Legend for impregnation types: MI = mass impregnated, OF = oil filled (or fluid filled), GF = gas filled.
[c]Out of service.
[d]Two cores flat cable.
[e]Tested for DC and AC and operated at 138 kV.
[f]Plus neutral.

Source: After [9].

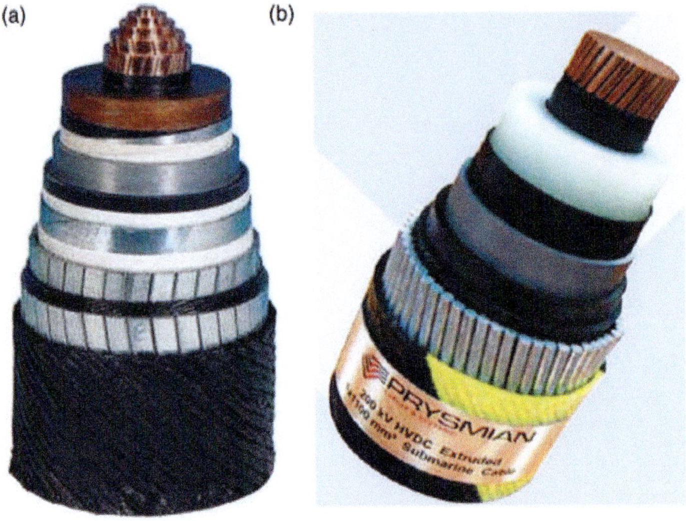

Figure 2.20 (a) Typical undersea HVDC cable with lapped insulation (courtesy of Prysmian) and (b) Typical undersea HVDC cable with extruded insulation (courtesy of Prysmian).

2.5.3 HVDC Cable Insulation

As is obvious from the types of HVDC cables described in Section 2.5.2, as well as from several observations made in this chapter, HVDC cables can be divided into two main categories depending on the type of insulation,[8] namely:

1. Oil–paper insulation cables, also referred to as lapped insulation cables, or lapped paper cables, or simply lapped cables, that is, made of several insulating layers [24, 25, 28]), see Figure 2.20a;
2. Extruded insulation cables; see Figure 2.20b.

The former cables are those with more established experience and their recent developments include the possibility of increasing the power transmitted through the use of a paper–prolipropilene laminate in place of paper, thereby giving rise to the PPL cables [23–25]. Extruded insulation cables are less well established, not only as regards the design and construction but also in terms of experience in the installation, operation, and maintenance, but the research and innovation efforts are enabling their production for use with increasing voltage and power ratings up to ~300 kV and ~1000 MW at present (see, e.g., [14, 29]). Oil–paper insulation and extruded insulation are analyzed in greater detail in the next two Sections.

[8] For the sake of completeness, in Chapter 1 also gas-filled (GF) cables were mentioned. Here they are not considered since their utilization in DC systems is relevant to one single application, that is, the Cook Strait 1 Intertie (1965): see Table 2.1.

2.5.3.1 *Oil–Paper (or Lapped) Insulation* The oil–paper insulation is achieved by wrapping strips of pure cellulose paper onto the conductor, applied in helical layers to reach the total design thickness of the insulation. Then the insulation is impregnated with mixtures or mineral base fluids to impart and improve the dielectric properties [20, 24].

The calculation of the voltage gradient applied to the dielectric of an HVDC cable is very complex since in the absence of thermal gradient it is "almost capacitive", while in the presence of thermal gradient it depends on the conductivity of the insulation, which in turn is a function of the local temperature and also (though less markedly) of the local electric field [9, 23, 25] (see Chapter 3). Therefore, when the insulation is subjected to a temperature gradient due to current flow and Joule heating in the conductor, the electric field distribution inside the insulation changes significantly compared to the initial isothermal cable. This is the so-called voltage inversion of electric field profiles in HVDC cables, which is discussed in detail in Section 3.1.3.2. In the case of oil–paper insulated cables, however, as pointed out in [24] the distribution of the electric field inside the insulation is very regular and smooth in all longitudinal sections of the cable, due to the very nature of insulation layers. This prevents the formation of space-charge storage areas or centers and makes the oil–paper insulation inherently suitable for use with the HVDC LCC technology.

Still relating to the lapped cables, for very high voltage links (say, >300 kV) and high power; an alternative solution is represented by the already mentioned PPL insulation; see Section 2.5.2. In the PPL insulation, the traditional paper insulation is replaced with a laminate consisting of alternating layers of paper and polypropylene, thereby improving the characteristics of the dielectric and—given the lower quantity of impregnating fluid and the lower mobility of the same—increasing the operating temperature of the cable from 55°C (the maximum conductor temperature for continuous operation in the traditional MIND cables) to 80–85°C, with the relevant benefits on the power transmission or on the reduction of the equivalent conductor cross section [24].

2.5.3.2 *Extruded Insulation* Polymers investigated and tested so far for the construction of extruded insulation HVDC cables can be grouped into two categories: the pure materials and materials with proper additives. The latter can be further divided into additives to modify the thermal characteristics of materials and additive materials to reduce the accumulation of space charge (see Chapters 4 and 5).

In the case of pure polymeric materials, the dielectrics considered so far for the construction of HVDC cables are mainly LDPE, DC-XLPE, and HDPE, although the latter is much prone to the accumulation of space charges than the other two. These materials have been tested in a wide range of values of temperature and applied electric field, thereby showing a strong dependence of breakdown voltage on temperature (see Section 5.1.2). Moreover, the phenomena of field inversion and space-charge accumulation are the cause of a significant distortion of the electric field with respect to the capacitive field distribution typical of AC cables (see Chapters 3 and 4).

For these reasons, in the production of HVDC cables with polymeric insulation cable manufacturers seem to have abandoned the pure polymeric insulation in favor of insulation with properly selected additives [9, 26], as extensively discussed in the next chapters, particularly in Chapter 5.

Extrusion is a technique to deposit a uniform and compact layer of polymer around the conductor, in between the two layers of semiconductive screens. According to the latest and best-performing technique, extruded insulation of high-voltage cables (both for AC and DC applications) is obtained through one single extrusion process of the entire insulation thickness plus inner and outer semicons, followed immediately by a cross-linking phase of the insulation to give it the appropriate thermomechanical properties. This procedure is done through a so-called triple extrusion line: the bare conductor enters the triple extrusion head, where insulation and shields are applied in sequence. The output is the insulated conductor (cable core) that enters the vulcanization (or cross-linking) pipe, maintained at high pressures and temperatures, where thermochemical treatments (cross-linking) take place. The cable passes along the lengthy vulcanization pipeline, where it stays for a sufficient time in order to have the finished cable through a rapid and continuous production.

In order to have a perfect vulcanization of the cable while preventing the cable core from sliding on the vulcanization pipe walls, straight horizontal vulcanization pipes are sometimes avoided in favor of either pipes curved according to a catenary profile or straight vertical pipes.

The achievement of an extruded insulation that is completely nonporous and free of defects (which might store space charge and start severe degradation mechanisms, e.g., partial discharges) is not trivial and requires great care in the technological application of the extruded dielectric, as well as strict production controls.

Although the insulating layer thus formed may appear perfectly compact and homogeneous, nevertheless there may be areas within the extruded dielectric where additives or by-products resulting from cross-linking are present in different amounts, thereby involving an irregular distribution of the dielectric stress in the insulation and often causing the formation and growth of storage centers of space charge that remains trapped within the dielectric, as extensively discussed in Chapter 3 and even more in Chapter 4 [9, 23–25]. It can be argued that the presence of these space-charge storage centers gives rise to a so-called Poissonian field that adds up to the preexisting electrical field [30–32] and may result in unacceptable stress on the dielectric, especially when combined with a change of polarity (polarity reversal).

As highlighted in Section 2.4, the advent of VSC technology has enabled to reverse the power flow without reversing voltage polarity, thereby facilitating the use of extruded insulation cables, which, however, are also being qualified by a few manufacturers for use in conjunction with CSC systems through the adoption of special dielectrics [24–27, 29].

REFERENCES

1. E.W. Kimbark. *Direct Current Transmission*, Vol. 1, Wiley, New York, 1971.
2. E. Uhlman. *Power Transmission by Direct Current*. Springer, Berlin/Hiedelberg, 1975.
3. W. Long and S. Nilsson. "HVDC Transmission: Yesterday and Today." *IEEE Power Energy Mag.*, Vol. 5, No. 2, pp. 22–31, March–April 2007.

4. *High Voltage Direct Current Transmission: Proven Technology for Power Exchange.* Siemens AG—Power Transmission and Distribution: High Voltage Division, Erlangen, Germany, 2003.
5. M.P. Bahrman and B.K. Johnson. "The ABCs of HVDC Transmission Technologies." *IEEE Power Energy Mag.*, Vol. 5, No. 2, pp. 32–44, March–April 2007.
6. R.L. Hauth, P.J. Tatro, B.D. Railing, B.K. Johnson, J.R. Stewart, J.L. Fink, *HVDC Power Transmission Assessment*, Oak Ridge National Laboratory (ORNL), managed by Lockheed Martin Energy Systems for DOE, Report No. ORNL/Sub/95-SR893/1, March 1997.
7. A. Orini, M. Marelli, and E. Zaccone."I collegamenti in cavo HVDC: stato dell'arte e prospettive future." AEIT Workshop—HVDC Transmission: Technology, State of the Art and Practical Experiences, Rome (Italy), November 27, 2008 (in Italian).
8. R. Benato, M. Del Brenna, C. Di Mario, A. Lorenzoni, and E. Zaccone. "Analisi economica comparata tra linee di trasmissione in cavo XLPE e aeree." *AEIT J.*, Vol. 92, No. 9, pp. 20–29, September 2005 (in Italian).
9. M. Salah Khalil. "International Research and Development Trends and Problems of HVDC Cables with Polymeric Insulation." *IEEE Electr. Insul. Mag.*, Vol. 13, No. 6, pp. 35–47, November–December 1997.
10. M. Jeroense. "HVDC, the Next Generation of Transmission. Highlights with Focus on Extruded Cable Systems." Proc. 2008 International Symposium on Electrical Insulating Materials (ISEIM 2008), pp. 10–15, Yokkaichi, Japan, September 7–11, 2008.
11. Y. Jun, L. Zehong, and S. Yan. "Studies on Design of DC Filters of Three Gorges to Changzhou ±500 kV HVDC Link." Proceedings of 1998 International Conference on Power System Technology (POWERCON '98), Vol. 1, pp. 460–465, August 18–21, 1998.
12. T.J. Hammons, V.F. Lescale, K. Uecker, M. Haeusler, D. Retzmann, K. Staschus, and S. Lepy. "State of the Art in Ultrahigh-Voltage Transmission." *Proc. IEEE*, Vol. 100, No. 2, pp. 360–390, 2012.
13. R. Bazzocchi and G. Cordioli. "Recenti sviluppi della tecnica HVDC." Proceedings of the 2009 AEIT Conference, Catania (Italy), September 27–29, 2009 (in Italian).
14. B. Jacobson, Y. Jiang-Häfner, P. Rey, G. Asplund, M. Jeroense, A. Gustafsson, and M. Bergkvist. "HVDC with Voltage Source Converters and Extruded Cables for up to ±300 kV and 1000 MW." Paper B4-105, CIGRE Session 2006, Paris (France), 2006.
15. A. Iliceto, M. Marzinotto, S. Malgarotti, and I. Valadè."Voltage Level Selection for HVDC Interconnections: Methodology Description and a Practical Application." Workshop—HVDC Transmission: Technology, State of the Art and Practical Experiences, Rome (Italy), November 27, 2008.
16. W. Litzenberger and P. Lips. "Pacific HVDC Intertie." *IEEE Power Energy Mag.*, Vol. 5, No. 2, pp. 45–51, 2007.
17. D. Povh, D. Retzmann, and J. Kreusel. "Integrated AC/DC Transmission Systems—Benefits of Power Electronics for Security and Sustainability of Power Supply." 16th Power Systems Computation Conference (PSCC 2008), Glasgow, July 14–17 2008. Survey Paper 2, Part 1 and Part 2.
18. D. Fabiani, G. Mazzanti, and C. Biserni. "L'Italia e i blackout: la lezione del 28 settembre 2003." *Power Tech.*, Vol. 6, pp. 18–28, November 2005 (in Italian).
19. T. Worzyk. *Submarine Power Cables.* Spinger, Berlin, 2009.
20. R. Bartnikas, Ed. *Power and Communication Cables*, IEEE Press Series on Power Engineering, New York, 2000.

21. S. Hirano, J. Haraguchi, T. Mizuno, and Y. Nagoya. "Development of DC +/− 250 kV Coaxially-Integrated Return Conductor Extruded Cable." Paper B.7.3, 2007 International Conference on Insulated Power Cables (Jicable'07), Versailles—France, 2007.

22. E. Stern, J. Nash, C. Schoeniger, C. Bartzsch, G. Acquaotta, M. Bacchini, and A. Orini." The Neptune Regional Transmission System 500 kV HVDC Project." Paper B4-118, CIGRE Session 2008, Paris (France), 2008.

23. EPRI Report No. 1008720. "DC Cable Systems with Extruded Dielectrics." Palo Alto, California, December 2004.

24. M. Marelli, A. Orini and E. Zaccone. "Cavi sottomarini in corrente continua: soluzioni per il Mediterraneo." Proceedings of the 2009 AEIT Conference, Catania (Italy), September 27–29, 2009 (in Italian).

25. R.N. Hampton. "Some of the Considerations for Materials Operating under High-Voltage, Direct-Current Stresses." *IEEE Electr. Insul. Maga.*, Vol. 24, No. 1, pp. 5–13, January–February 2008.

26. T.L. Hanley, R.P. Burford, R.J. Fleming, and K.W. Barber. "A General Review of Polymeric Insulation for Use in HVDC Cables." *IEEE Electr. Insul. Maga.*, Vol. 19, No. 1, pp. 14–24, January–February 2003.

27. J.O. Boström, A. Campus, U. Nilsson, P. Carstensen, A. Gustafsson, A. Farkas, and K. Johannesson. "Cross-Linked Polyethylene Materials for DC Power Cables." 5th International Conference on Insulated Power Cables (JiCable'99), Paper C2.12, June 20–24, 1999.

28. M. Naidu and V. Kamaraju. *High Voltage Engineering*, 3rd ed. McGraw-Hill, New York, 2004.

29. H. Niinobe, N. Ishii, T. Nakajima A. Watanabe, H. Tanaka, and H. Kon."HVDC and Power Electronic Systems for Overhead Line and Insulated Cable Applications." CIGRE 2012 San Francisco Colloquium, Paper B.1–9.

30. L.A. Dissado, and J.C. Fothergill. *Electrical Degradation and Breakdown in Polymers*, Peregrinus, London, 1992.

31. J.C. Fothergill, L.A. Dissado, eds., *Space Charge in Solid Dielectrics*. Dielectrics Society, Leicester, UK, 1998.

32. G. Mazzanti, G.C. Montanari, and L. Dissado. "Electrical Aging and Life Models: The Role of Space Charge." *IEEE Trans. on Dielectr. Electr. Insul.*, Vol. 12, No. 5, pp. 876–890, October 2005.

CHAPTER 3

MAIN PRINCIPLES OF HVDC EXTRUDED CABLE DESIGN

This chapter illustrates in detail the basics of HVDC extruded cable design. The main contents of the chapter are as follows:

- A detailed analysis of the differences between HVAC and HVDC cables, starting from the diversities in the structure and formation, but with special attention to the dissimilarities in the electric field distribution in steady conditions (Section 3.1)
- A detailed analysis of the DC electric field distribution in the typical transient conditions encountered during the service of an HVDC cable, including switching-on/raising/switching-off the voltage and the load, as well as reversing the voltage polarity (Section 3.2)
- An investigation about the influence of the laying environment temperature on the steady field of an HVDC extruded cable (Section 3.3)
- An investigation about the effects of impulses superimposed to the DC voltage (Section 3.4), which gives rise to a novel statistical approach to the selection of impulse voltage test levels for HVDC extruded cables (Section 3.5)
- A preliminary and introductory overview of the problems involved by the modification of the stress distribution due to space-charge effects (Section 3.6; note that these problems are extensively treated in Chapter 4)
- A review of the polyethylene-based dielectrics that have been considered for designing and manufacturing HVDC extruded cables (Section 3.7)
- A short review of the morphology of polyethylene from the viewpoint of its influence on the performances of HVDC cable insulation (Section 3.8)

Extruded Cables for High-Voltage Direct-Current Transmission: Advances in Research and Development, First Edition. Giovanni Mazzanti and Massimo Marzinotto.

3.1 DIFFERENCES BETWEEN HVAC AND HVDC EXTRUDED CABLES

3.1.1 Differences in the Structure

As to their structure, the differences between extruded cables designed for a HVAC system and those designed for a HVDC system are negligible and can hardly be detected visually. Indeed, the same structural components are required to enable power to be transmitted using HVDC and HVAC extruded cables, with most components performing similar tasks [1] (see Fig. 3.1*a*, relevant to an HVAC extruded cable and Fig. 3.1*b*, relevant to an HVDC extruded cable).

The main exception is represented by the extruded insulation of the cable: Indeed, the electrical design of HVDC cable system insulation is fundamentally different and more complex than for an HVAC cable system, as discussed in detail later on in this chapter [1–5].

As to the elements that make up a single-core polymer-insulated HV cable, they are very similar to those already listed for a mass-impregnated cable for land application in Chapter 2 [4, 5], namely those illustrated in Figure 3.2.

3.1.2 Typical Formations of HVDC Extruded Cables

In this chapter a description of the typical formations of both land and marine HVDC extruded cables is given.

3.1.2.1 Conductor Conductors can be either made of copper or aluminum. Copper is usually the preferred solution since it guarantees, for a given ampacity, cables with small diameter and reduced Joule losses. On the other hand, when cable weight is an important constraint, for example, when a land cable has to be installed

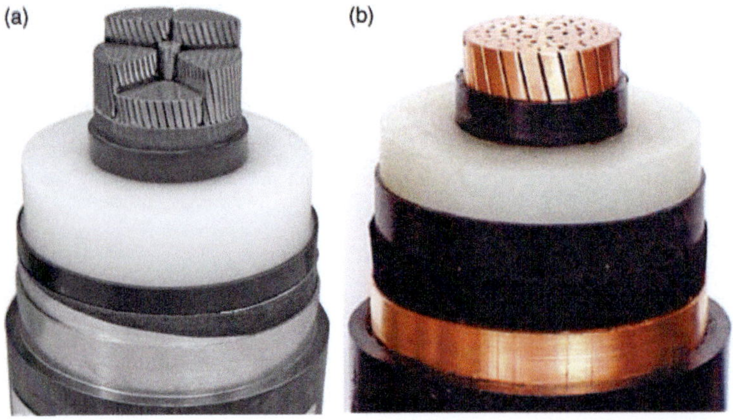

Figure 3.1 (*a*) XLPE-insulated land HVAC cable, with Milliken conductor (courtesy of Prysmian) and (*b*) XLPE-insulated land HVDC cable with profiled conductor (after [3]).

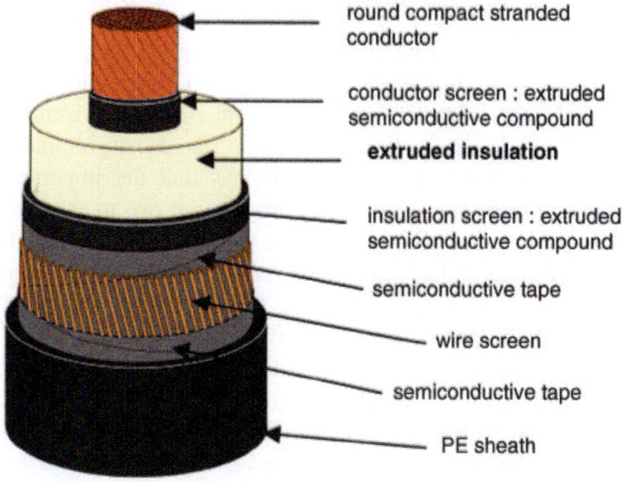

Figure 3.2 High-voltage extruded-insulation cable structure for land applications (after [6]).

over long viaducts or an undersea cable has to be installed in very deep waters, the aluminum conductor may become a necessary step. For example, although a MIND cable, the SAPEI intertie, an undersea link, reaches an installation depth of 1650 m. Each pole is 436 km long and has a 420-km-long marine section made of four cable lengths, joined to each other: the two innermost lengths, laid on a deeper sea bottom, have aluminum conductors (water depths over 400 m), while the two outermost, laid on a shallower sea bottom, have copper conductors (water depths up to 400 m). The decision to use aluminum for the two innermost sections was taken as an effort to manage a lighter cable during the installation in the deep-water section of the link. The conductor is a metallic rope,[1] and it can be either stranded from round wires (see Fig. 3.2) or profiled with layer-cake-shaped wires (see Fig. 3.1*b*).

If stranded from round wires, it consists of a series of thin round wires twisted together in a compact arrangement. Stranded copper conductors yield more flexible cables than stranded aluminum. If profiled with layer-cake-shaped wires, the conductor forms a rope particularly compact and difficult to penetrate longitudinally by the water. In this way, some manufacturers realize "solid-type" conductors (usually made of aluminum), which solve the problem of longitudinal water penetration through the interstices of the ordinary stranded conductors. Such benefit is achieved with the disadvantage of giving the cable a higher stiffness, thus a more troublesome installation.

3.1.2.2 Inner Semiconductive Layer Above the conductor, a very thin layer of semiconductive compound is applied, namely the *conductor semiconductive screen* (also referred to as conductor shield, or conductor semicon, or inner semicon

[1] Solid conductors consisting of one single massive wire are quite stiff and are usually employed only for cross sections up to ~400 mm^2.

in the following), whose function is to contribute to the radial electric field and to adhere as closely as possible to the insulation, thereby eliminating or incorporating within itself any gap at the conductor–insulation interface. Indeed, as the conductor is stranded, its surface is only approximately cylindrical and is characterized by surface roughness. In the absence of the conductor semicon, the most prominent asperities would become electric field intensification points, and the imperfect contact between conductor and insulation would very likely give rise to voids at the conductor–insulator interface, in a position where the radial field is very high (maximum in AC regime, as well as in DC regime for zero or moderate load; see Section 3.1.3.2); therefore, partial discharges stemming from such asperities and diffusing into the voids at the conductor–insulation interface would be very likely. Furthermore, apart geometrical asperities, conductor and insulation are chemically and physically different, and without the semicon layer their facing surfaces would not adhere properly to each other, thereby enhancing further the formation of air-filled microgaps, dangerous for the reason mentioned above. On the contrary, the inner semicon between conductor and insulation acts as an interface whose electrical characteristics are closer to the conductor than to the insulation, while its chemical characteristics are closer to the insulation than to the conductor; hence the irregular surface of the conductor is replaced with a perfectly round and smooth conductive surface that adheres perfectly to the insulation.

Semicons are usually made of the same base compound used for the insulation, filled with carbon black particles that guarantee a volume resistivity in the range of 100–1000 Ω·m (the volume electrical resistivity of such compound is strongly dependent on temperature, decreasing as temperature increases). It must be emphasized that, according to the up-to-date technology, the inner semiconducting layer, the insulation layer (see Section 3.1.2.3), and the outer semiconducting layer (see Section 3.1.2.4) are all extruded in one single operation by means of a triple extrusion head.

3.1.2.3 *Insulation Layer* The insulation layer, the heart of a high-voltage cable, is designed to withstand the rated voltage of the cable, U_0. The insulation is extruded directly onto the conductor screen. *Extra-clean dry curing of extruded insulation*, according to the latest technology, is recommended for achieving the best performing insulation. Further details about the extruded insulation are skipped here because cable insulation will be discussed throughout the whole book.

3.1.2.4 *Outer Semiconductive Layer* Above the insulation, a second semiconductive layer of very small thickness is applied, which also has the function (as the conductor semicon) to contribute to the radial field and to create a void-free smooth interface between the insulation and the adjacent metallic screen. As in the case of the inner semiconductive layer, this interface is needed since the metal does not adhere perfectly to the extruded polymeric dielectric, due to the different expansion coefficients of these two materials. Moreover, in HVDC cables subjected to high load the maximum electrical gradient might occur at the insulator–outer semicon interface. Therefore this interface might operate in the presence of particularly large values of electric field. As for the inner semicons, outer semicons are usually made of the same base compound used for the insulation filled with carbon black particles that guarantee a volume resistivity in the range of 100–1000 Ω·m. In any case the compounds employed for the outer semicon can be other than those for the inner semicon, thus with different values of the electrical resistivity.

3.1.2.5 Metallic Screen The electric (or electrostatic) metallic screen is applied onto the insulation semicon. Besides further contributing to the radial field in the insulation layer, it serves also to strengthen mechanically the cable and to provide a reference for the DC voltage, and is always grounded. In this way, the electric field is entirely confined within the insulation (the only layer with the specific task of withstanding the voltage). In addition to the above-mentioned task of limiting the electric field in the insulation, the electric screen accomplishes the safety action of providing a low-resistance path to the short-circuit current, avoiding in this manner short-circuit current in the soil as much as possible. Moreover, if the electric screen is made of a continuous metallic sheath, it also acts as a radial moisture barrier, thereby avoiding the occurrence of water treeing.[2]

Material and structure of the metallic screen vary widely, depending on the applications and manufacturers, as shown in Figure 3.3 for XLPE insulated cables.

[2] Although the exact definition of water treeing is a debated matter, it can be said that "water trees are degradation structures in a polymer that: 1. are permanent, 2. have grown due to at least humidity and an electric field, 3. have a lower electrical strength than the original polymer when wet, but . . . are not a short circuit or local breakdown path, and 4. are substantially more hydrophilic than the original polymer" (R. Ross. "Inception and Propagation Mechanisms of Water Treeing." *IEEE Trans. Dielectr. Electr. Insul.*, Vol. 5, No. 5, pp. 660–680 October 1998). Water trees resemble a bush (bow-tie trees) or a fan (vented trees). A wide variety of mechanisms has been proposed for explaining the growth of water trees, for example, "1. capillary action, 2. osmosis, 3. Coulomb forces, 4. dielectrophoresis, 5. thermal degradation, 6. partial discharges, 7. chemical degradation" (E.F. Steennis and F.H. Kreuger. "Water Treeing in Polyethylene Cables." *IEEE Trans. Electr. Insul.*, Vol. 25, No. 5, pp. 989–1028, October 1990). Many studies have described many aspects of water treeing in polymeric insulating materials (for a broad and sound review, see the two previously quoted studies). This topic is not treated in the present book since, as explained here—HVDC extruded cables feature commonly one or more radial moisture barriers that avoid the occurrence of water treeing in the insulation; hence this phenomenon is not peculiar to such cables. On the other hand the so called "wet design" is under consideration also for HVDC cables in the case of dynamic cables for use with floating off-shore platforms in deep water environment (T. Andersen, F. Mauseth, E. Ildstadt, "Effect of water content on the conductivity of XLPE insulation", Proc. 2012 IEEE CEIDP, pp. 649–653, Montreal (Canada), 13th–17th October 2012), even if there are strong indications that polymeric HVDC cables should be made water tight in any case to prevent water tree growth (F. Mauseth, M. Amundsen, A. Lind, "Water tree growth of wet XLPE insulation stressed with DC and high frequency AC", Proc. 2012 IEEE CEIDP, pp. 692–695, Montreal (Canada), 13th–17th October 2012). Nevertheless, an investigation that provides interesting indications about water treeing in DC polymeric cables is reported in Y. Ohki [Aiming at a More Rigorous Understanding in Electrical Insulating Materials Research." *IEEE Trans. Dielectr. Electr. Insul.*, Vol. 15, No. 5, pp. 1201–1214, October 2008 (Whitehead Memorial Lecture)]. This article analyzes the effects of superposition of a high-frequency voltage to various lower-frequency voltages, including DC voltage, on the growth mechanism of water trees in low-density polyethylene (LDPE). The results are explained resorting to the space-charge behaviour of the polymer in relation to the number of zero crossings, highlighting that the water tree growth seems to be proportional to this number.

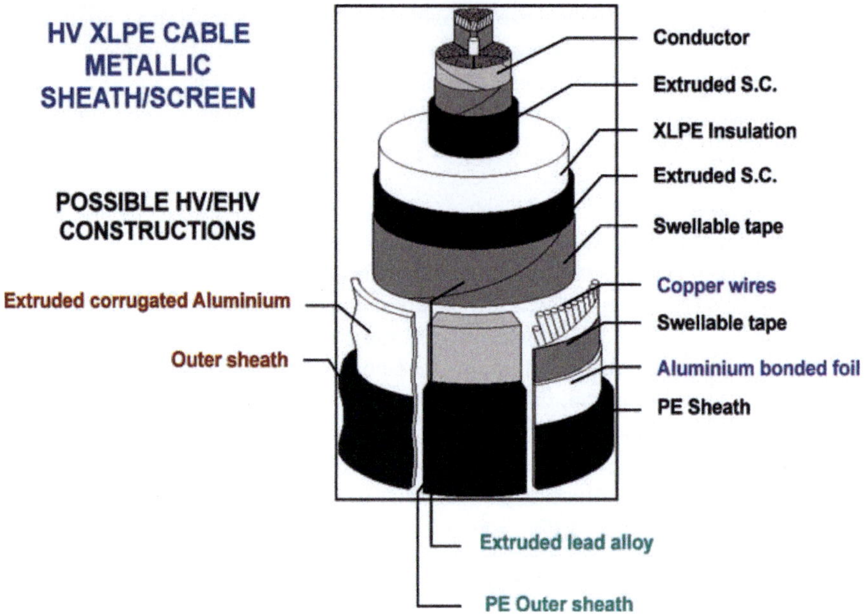

Figure 3.3 Some types of metallic screen built in XLPE-insulated cables (courtesy of Prysmian).

Due to the permeability of XLPE compounds to moisture,[3] the two main solutions for the metallic screens are:

1. Wires (usually made of copper) beneath a thin continuous metallic sheath (usually made of aluminum), the former with the main action to carry the short-circuit current, the latter to guarantee the needed barrier against water and humidity
2. A continuous metallic sheath (made of aluminum, copper or lead)

If a continuous metallic sheath of aluminum is used, it can be either an extruded layer with corrugated shape, in order to achieve good cable flexibility for transportation and laying, or a metallic foil bent in a cylindrical shape. In the latter case, the aluminum foil can be either welded with a minimum of edges overlapping, or alternatively welded "top-to-top," that is, without overlapping the edges. Glued edges with a minimum of overlapping are sometimes used instead of welding. Extruded lead sheaths are commonly used for undersea applications.

3.1.2.6 Water Sealing (or Blocking) Systems Hygroscopic powders or tapes are sometimes used as water sealing systems. Powders are dispersed in stranded

[3] Note that polymeric materials are often completely water-tight, for example, the so-called tree retardant XLPE, or TR-XLPE, but still allow for diffusion of moisture [7].

conductors. Tapes are wound both on stranded conductors and on the metallic screen. In case of cable fault with damage of the protective thermoplastic oversheath, or due to an external action, or puncturing due to an overvoltage propagating on the metallic screen—ground mode—the hygroscopic tapes prevent longitudinal water propagation over the metallic screen. Furthermore, in case of severe cable damage that might expose the conductor to external pollutants - soil, moisture, water, etc. for land cables, seawater for submarine cables - the hygroscopic powder can prevent the longitudinal propagation of water through the interstices of the stranded conductor, avoiding that moisture interacts with the inner semiconductive layer and, further on, with the insulation itself under electric field. Hygroscopic tapes wound on the conductor in this case do not prevent the moisture propagation along the conductor interstices, but the interaction of water with the inner semiconductive layer and consequently with the insulation.

3.1.2.7 Protective Thermoplastic Sheath Outside the metallic electric screen, a protective thermoplastic sheath is used to protect the metallic screen from moisture and in general from any external agents that can give rise to corrosion through galvanic and/or electrolytic actions, as well as to help reduce possible damages that may arise when the cable is laid. Protective thermoplastic sheath materials are PE-based compounds and sometimes polyvinylchloride (PVC)-based compounds. For land cables the protective thermoplastic sheath is usually the outermost layer; if this is the case, it is called a "jacket" or simply a "thermoplastic sheath." Furthermore, for land applications a thin semiconductive layer is always applied onto the outer thermoplastic sheath; this allows the thermoplastic sheath test to be performed in order to check its integrity both immediately after installation and during routine maintenance operations.

3.1.2.8 Armoring In undersea cables, a metallic armor—made up of one or more (usually two) layers of steel wires or tapes—is applied outside the protective thermoplastic sheath in order to attain the required mechanical strength of the cable during laying as well as its service life. In particular, the cable is designed to not only withstand being unloaded from the vessel but also during the possible recovery in case of problems. Armoring works as a protective shell against possible external actions when the cable is laid on the seabed due to anchoring, trawling, erosion due to tides and waves, abrasion because of moving materials on the seabed, or free-hanging cable sections that may vibrate under certain conditions. Underwater slides and turbidity currents usually triggered by seismic events, differential sea bottom movement at seismic fault zones, fish (e.g., sharks) or mammal bites, ocean dumping of dredged material or garbage, other installations including pipes and telecommunication cables, the influence of other existing cables, and the like are typical external actions that may cause faults in undersea cables [8]. Bedding made of synthetic yarns (typically polypropylene yarns; see Fig. 3.4 [9]) is always applied outside and inside an armor layer, for example, between adjacent armor layers, between an armor layer and underneath the protective thermoplastic sheath layer, and outside the outermost armor layer (see Section 3.1.2.9).

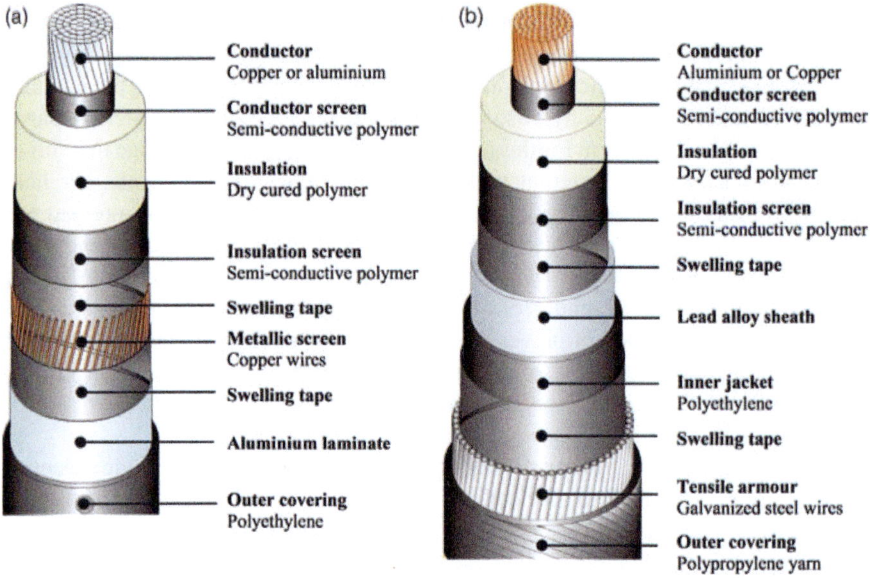

Figure 3.4 Structure of HVDC XLPE cables for land (*a*) and sea (*b*) usage in conjunction with VSC converters (courtesy of ABB).

3.1.2.9 Jacketing (or External Sheath) The external sheath is made of a thermoplastic jacket for land cables (see Section 3.1.2.7), while for undersea cables it is made of a bedding of synthetic yarns (see Section 3.1.2.8). A thermoplastic jacket is not applicable to undersea cables because of its insufficient grip, while a bedding of synthetic yarns improves the grip during cable pay-off from the vessel.

3.1.3 Differences in the Electric Field Distribution

3.1.3.1 Electric Field Distribution Within the Insulation of an HVAC Cable The electric field distribution within the insulation of an HVAC cable depends on the capacitive field developed in the cable geometry. Such fields can be easily calculated by assuming that the electrical permittivity, ε, of the insulation is constant with temperature—thus it remains constant across the dielectric as the cable is loaded and heated to its operating temperature [1]. This assumption is acceptable in practice since ε does change slightly with temperature, but the overall effect on the AC electric field distribution is negligible. Indeed, the permittivity changes by $<1\%$ over the working temperature range [3, 10].

Under the above hypothesis, it can be easily demonstrated that the electrical field strength, or stress, at any radius r within the dielectric is given by the following relationship [1, 3] (see also Fig. 3.5):

$$E_{AC}(r) = \frac{U_0}{r \ln (r_o/r_i)} \tag{3.1}$$

where

$E_{AC}(r)$ = AC electric field strength at radius r (kV/mm or MV/m)

U_0 = design voltage of the cable, that is, voltage across the dielectric (kV); in the AC regime, it is the phase-to-ground voltage

r = radius at position of interest (mm)

r_i = cable insulation inner radius (mm)

r_o = cable insulation outer radius (mm)

In some designs of paper insulated cable systems, the dielectric is made of papers with graded relative permittivities, and Eq. (3.1) should be modified accordingly [1]. It can be argued from Eq. (3.1) and Figure 3.5 that the maximum field E_{MAX} in an HVAC cable occurs always at the conductor–insulation interface, whereas the minimum field E_{MIN} occurs always at the insulation–metallic screen interface.

3.1.3.2 Steady Electric Field Distribution Within the Insulation of an HVDC Cable It is well known that the electric field under DC stress is temperature and time dependent. This is due to the fact that the resistivity of the dielectric in turn

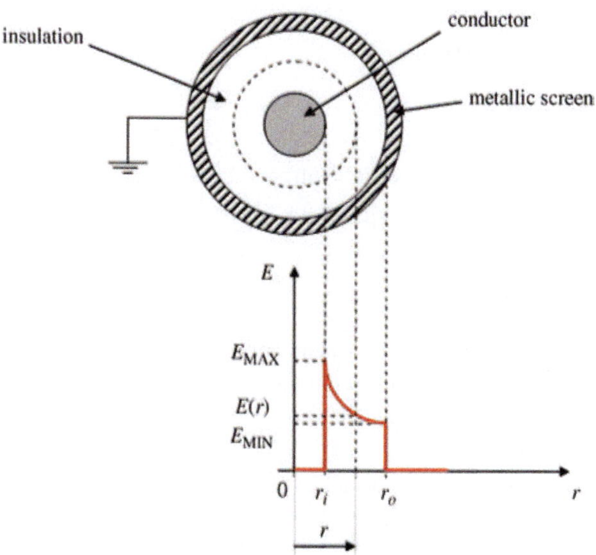

Figure 3.5 Electric field distribution within the insulation of a HVAC cable (also referred to as "Laplace," or "capacitive," or "permittivity-dependent" field distribution).

depends on temperature and electric field [1, 3, 11–13]. The electric field distribution in an HVDC cable, *under steady-state DC conditions* [11, 12], is determined by the local electrical resistance and thus by the electrical resistivity/conductivity of the insulation [1, 3, 10, 14, 15]. It can be shown that if the DC insulation resistivity were constant with temperature and electric field, then the field distribution in an HVDC cable would be identical to that of an HVAC cable of the same geometry, that is, "Laplacian" or "capacitive." This is more or less the case when an HVDC cable system is energized first at zero load, and no temperature gradient takes place across the cable [1]; thus the field distribution of a DC cable at zero load could be referred to as "quasi-Laplacian" or "quasi-capacitive."[4]

Time becomes apparent as the cable is also loaded and Joule heating of the conductor starts. Then, the DC electric field starts changing from the quasi-capacitive distribution occurring when only voltage is applied (see above) to a purely resistive distribution and becomes stable only after—strictly speaking—an infinite amount of time. Mathematically, this is described by the following equations [12, 13]:

$$\text{Electrostatic electric field} \qquad \mathbf{E} = -\nabla V \tag{3.2}$$

$$\text{Ohm's law for current density} \qquad \mathbf{J} = \sigma \mathbf{E} \tag{3.3}$$

$$\text{Gauss law for space charge} \qquad \nabla \cdot (\varepsilon_0 \varepsilon_r \mathbf{E}) = \rho \tag{3.4}$$

$$\text{Current continuity equation} \qquad \nabla \cdot \mathbf{J} = -\frac{\partial \rho}{\partial t} \tag{3.5}$$

where

$\mathbf{E}$ = electric field (kV/mm or MV/m)
V = voltage (kV)
$\mathbf{J}$ = current density (A/m^2)
σ = electric conductivity of the dielectric (S/m)
ε_0 = dielectric permittivity of the free space, 8.87×10^{-12} F/m
ε_r = relative permittivity of the dielectric
ρ = electric charge density (C/m^3)

For a deeper insight, which helps also in understanding the features of the HVDC field introduced later on, let us develop relationships (3.2)–(3.5) for long-term steady-state conditions, thereby achieving that conduction and space-charge

[4] Actually a leakage current through the insulation exists also at zero load due to the conduction current (Joule and polarization currents) and to the relevant Joule heating of the insulation. However, such current is usually very small at the operating voltages of the cable if the cable is not loaded, so that the relevant temperature gradient can be neglected at zero load [3,12]. Nevertheless, the field distribution of a DC cable at zero load is not exactly coincident with that of an HVAC cable of the same geometry because of the dependence of insulation resistivity on electric field. This issue will be made clearer later on in the present section.

formation in HVDC cables are strictly related, as shown in [16, 17]. When a DC voltage is applied to a specimen, the insulation is initially free of space charges ($\rho = 0$) and Eq. (3.4) becomes

$$\nabla \cdot (\varepsilon_0 \varepsilon_r \mathbf{E}) = 0 \tag{3.6}$$

Thus, the initial field distribution is determined by the variation of the permittivity inside the insulation [see Eq. (3.1)]. When a DC voltage has been applied for a long time, no further space-charge accumulation takes place, current density reaches its steady-state value and—on the basis of Eq. (3.3)—Eq. (3.5) can be rewritten as follows:

$$\nabla \cdot \mathbf{J} = \nabla \cdot (\sigma \mathbf{E}) = 0 \tag{3.7}$$

This means that the field distribution is determined by the variation of the conductivity inside the insulation. Consequently, if the permittivity and the conductivity vary, Eq. (3.6) and (3.7) are incompatible, hence a space charge must be established at steady state in order that both Eqs. (3.4) and (3.7) are satisfied [16]. As a consequence, Eqs. (3.7), (3.3) and (3.4) yield that at steady state it holds [12, 18]:

$$\nabla \cdot (\varepsilon_0 \varepsilon_r \mathbf{E}) = \nabla \cdot \left(\varepsilon_0 \varepsilon_r \frac{\mathbf{J}}{\sigma} \right) = \mathbf{J} \cdot \nabla \left(\frac{\varepsilon_0 \varepsilon_r}{\sigma} \right) + \frac{\varepsilon_0 \varepsilon_r}{\sigma} \nabla \cdot \mathbf{J} = \mathbf{J} \cdot \nabla \left(\frac{\varepsilon_0 \varepsilon_r}{\sigma} \right) = \rho \tag{3.8}$$

Hence

$$\mathbf{J} \cdot \nabla \left(\frac{\varepsilon_0 \varepsilon_r}{\sigma} \right) = \sigma \mathbf{E} \cdot \nabla \left(\frac{\varepsilon}{\sigma} \right) = \rho \tag{3.9}$$

As proved by relationship (3.9), the electromagnetic theory predicts that a dielectric sample where a steady DC current of density **J** is flowing, and where the ratio permittivity/conductivity varies with position, will acquire a space-charge density ρ [18]. Since usually in extruded polymers for HVDC cables the permittivity can be deemed as constant over a broad temperature interval, it can be argued that a spatial variation in the ratio permittivity/conductivity—and the associated space-charge distribution—has usually to be addressed to spatial changes in conductivity, as illustrated in detail below. Note that there may be gradients in conductivity in the axial direction of a cable without generation of space charge, since the scalar product of the current density and the gradient vectors in Eq. (3.9) are equivalent [see Ohm's law (3.3)] to the scalar product of the electric field and the gradient vectors. Only when the conductivity gradient is in the same direction as the electric field will space charge be generated [17].

At zero load, the quasi-capacitive field is nonuniform in the cylindrical insulation layer of a cable. Thus, since the electrical conductivity has a field dependence in extruded polymers for HVDC cables, there will be a gradient in conductivity and space charge will be generated. The gradient is such that $1/\sigma$ increases with r; thus space charge will have the same sign as the voltage on the inner electrode, and is sometimes referred to as homocharge in the inner part of the insulation—where the charge has the same sign as the inner electrode—and heterocharge in the outer part of the insulation—where it has the opposite sign as compared to the outer electrode. The magnitude of the space charge is largest at the inner electrode since this is the location of the highest electric field [17] (see also Fig. 3.17).

From Eq. (3.9) it can be noted that the magnitude of space charge at steady state is not influenced by the magnitude of the conductivity: Indeed, if conductivity is multiplied by a factor, this factor appears outside the gradient operator as well as in the nominator of the term inside the gradient operator and will therefore cancel, so that the magnitude of space charge at a given location is preserved [17].

It must be emphasized that this space charge is not dependent upon the operation of any other space-charge generating mechanism. In particular, it is not localized charge stored at precise trapping centers, as essentially meant in Chapter 4, but rather delocalized charge that accumulates at disuniformities of permittivity and conductivity [12]. This space-charge distribution can be calculated using (3.9) in conjunction with (3.3) by means of numerical methods only, as shown in Section 3.2.3.3 [12].

The attainment of the space-charge distribution expressed by (3.9) requires a transient (i.e., time-varying) current in the external circuit, referred to as "polarization current," which lasts until the steady-state distribution of (3.9) is reached and both Eqs. (3.4) and (3.7) are satisfied [16, 17]. The current is thereafter assumed to attain a constant (steady-state) value, called the "leakage current": this is the current that appears in (3.9) [3, 12, 16, 18].

In the special case that both the permittivity and the conductivity are constant throughout the insulation, no polarization current will flow in the external circuit (except for a short-duration current caused by polarization of the particles constituting the insulating material). The current will then immediately attain the steady-state value.

In a homogenous medium the permittivity is usually constant. *This means that if a polarization current of some duration is measured, it is an indication of the conductivity not being constant.* On the other hand, the absence of a polarization current indicates that the conductivity is constant throughout the insulation. The polarization current may either be caused by a separation of charge formed inside the insulation, that is, without any supply of charge carriers from the electrodes, or by charge carriers injected from the electrodes (see Chapter 4). The longer distance charge carriers need to be transported, the longer will the transient current last. Thus a space charge close to one of the electrodes, formed by injection of charge carriers, may give rise to a polarization current of relatively short duration even when the conductivity is low [16].

Unfortunately, insulation conductivity of solid dielectrics tends to be highly dependent on temperature and, to a lesser extent, on electrical field. As no comprehensive theoretical model exists for conduction in polymeric materials, the following empirically based mathematical models are commonly used to describe the dependence of the (volume) electrical conductivity of solid dielectrics on electric field (electric stress) E and temperature T (see, e.g., [1, 3, 13, 14, 19, 20]):

$$\sigma(T,E) = \sigma_0 \exp[a(T - T_0) + b(E - E_0)] \tag{3.10}$$

where

T_0 = reference temperature (K)
E_0 = reference electric stress (kV/mm or MV/m), often considered equal to zero
σ_0 = insulation conductivity at reference temperature T_0 and electric stress E_0 (S/m)
a = temperature coefficient of electrical conductivity (1/K or 1/°C)
b = stress coefficient of electrical conductivity (mm/kV, or m/MV)

Alternative relationships that can be used for expressing the variation in the electric conductivity of the material when changes in temperature and electric field strength are as follows (see, e.g., [13, 16, 18, 20, 21]):

$$\sigma(T,E) = \sigma_0 \exp\left[a(T - T_0)\left(\frac{E}{E_0}\right)^{\gamma}\right] \tag{3.11}$$

$$\sigma(T,E) = \frac{\sigma_0 \exp(-G_a/k_B T)\sin h(cE)}{E} \tag{3.12}$$

where γ in relationship (3.11) is a material constant playing a role similar to b in (3.10), while c in expression (3.12) plays a role similar to b in (3.10), G_a is the thermal activation energy governing the temperature dependence of the conductivity, $k_B = 1.38 \times 10^{-23}$ J/K is the Boltzmann constant. Note that, because of the different functional dependences of σ on E and T according to Eq. (3.12) with respect to Eq. (3.10), no direct comparison can be made about the values of G_a, c in Eq. (3.12) and a, b in Eq. (3.10); moreover, constant σ_0 in Eq. (3.12) is not a conductivity.

As a consequence, by assuming as valid Eq. (3.10), electrical resistivity ρ_E can be trivially expressed as follows:

$$\rho_E(T,E) = \rho_0 \exp\{-[a(T - T_0) + b(E - E_0)]\} \tag{3.13}$$

where ρ_0 ($\Omega\cdot$m) represents insulation resistivity at reference temperature T_0 and electric stress E_0.

Table 3.1, taken from the milestone work [3], shows the resistivity of a number of HV insulating materials for both AC and DC operation at selected values of temperature and electric field (stress).[5] In turn these figures are derived from a broad series of studies all referenced in [3] and omitted here for the sake of brevity.

[5] In all figures and tables from [3] discussed here the electric field in the insulation is called "electric stress" or simply "stress."

Table 3.1 Typical Values of Resistivity ($\times 10^{12}$ Ω·m) of Various Solid Dielectrics at Selected Temperatures and Electrical Stresses[a]

Sample Details		Stress	Temperature		
		kV/mm	20°C	60°C	80°C
		Lapped			
Paper	*1*	10	300	20	3
		20	300	10	4
PPL	*1*	20	50000	1600	300
		30	40000	1000	200
		Polymeric			
Thermoset	*2*	—	8000	400	50
	3	—	200	2	1
	4	5	500	60	5
	4	7	40	20	3
Thermoplastic	*5*	60	30	0.3	0.03
	6	20	140	1.8	—
	6	30	28	1.3	—

[a]Key: For numbers in italic in table:

1. Low-viscosity impregnant.
2. Electrical, extrusion and mechanical properties designed and optimized for EHVDC and HVDC operation.
3. Standard material used for HVAC operation.
4. Flexible material designed for both MVAC and HVDC operation.
5. Ethylene-based polymer, compound stabilized for extrusion and long life.
6. Propylene-based polymer, compound stabilized for extrusion and long life.

Source: After [3].

The resistivity values of Table 3.1 relevant to paper, polypropylene paper laminate (PPL), and thermoplastic insulation at 20 kV/mm,[6] when plotted against temperature T in the temperature range of 20–80°C, show that the resistivity can change by 2–3 orders of magnitude (see Fig. 3.6).

Focusing on expression (3.13), parameters a and b in such a relationship are used to describe the variation in the electric resistivity of the material with changes in temperature and electric field strength, as they were used in Eq. (3.10) to describe the variation in the electric conductivity of the material with changes in temperature and electric field strength. Tables 3.2 and 3.3 show the values of a and b obtained using the best fit of (3.10) to the data listed in Table 3.1 [3].

Typical published values of the coefficients for oil-impregnated paper insulated cables that are reported in [1, 12] are $a = 0.1°\text{C}^{-1}$ and $b = 0.03$ mm/kV, in substantial agreement with Tables 3.2 and 3.3. Other studies with data on typical values of a and b for extruded cable systems are [14, 22], and such data have been used—in conjunction with space-charge studies—to screen potential candidates as dielectrics for HVDC applications.

[6] This is actually a quite interesting field value because it is of the same order of magnitude of the maximum electric field in current top-rated HVDC cable extruded insulation [1, 13].

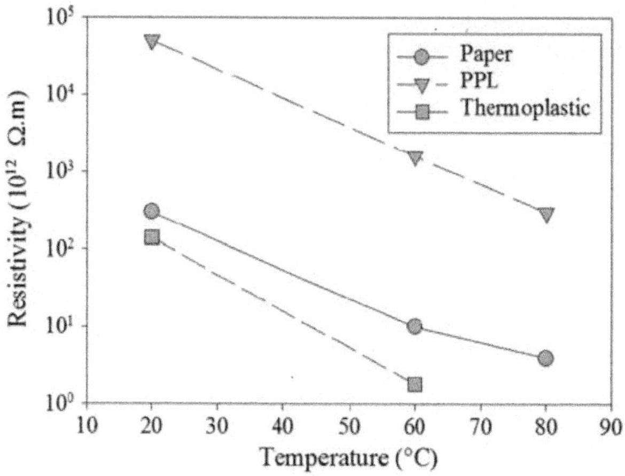

Figure 3.6 Resistivity of selected insulations (from Table 3.1) at 20 kV/mm and at different temperatures (after [3]).

An important observation that can be made when considering the values of a and b in Tables 3.2 and 3.3 is that in practice neither a is constant with electrical field nor b is constant with temperature; this observation will be developed further later on.

The consequences of the field and temperature dependencies described by (3.13) are that any thermal gradient will modify the electrical field distribution of the HVDC cable significantly with respect to the zero load distribution (similar to that of

Table 3.2 Values of Parameter *a* That Expresses the Dependence of Resistivity upon Temperature of Various Solid Dielectrics at Selected Values of Electric Field (Stress)[a]

Sample Details		Stress (kV/mm)	Temperature Coefficient a (1/°C)
Paper	*1*	10	0.074
PPL	*1*	20	0.086
		30	0.093
Thermoset	*2*	—	0.084
	3	—	0.072
	4	5	0.101
Thermoplastic	*5*	60	0.115
	6	20	0.104
	6	30	0.114

[a]The figures in italics refer to the descriptions in the key of Table 3.1.

Source: After [3].

Table 3.3 Values of Parameter *b* That Expresses the Dependence of Resistivity upon Electric Field (Stress) of Various Solid Dielectrics at Selected Temperatures[a]

		Stress Coefficient *b* (mm/kV)		
		20°C	60°C	80°C
Paper	*1*	0.028	0.018	0.029
PPL	*1*	0.0238	0.0296	0.0416
Thermoset	*2*	—	—	0.0645
Thermoplastic	*6*	0.0128	0.06	0.034

[a]The figures in italics refer to the descriptions in the key of Table 3.1.

Source: After [3].

an HVAC cable, as argued above). The steady-state electric field distribution under working DC conditions can be calculated through knowledge of [3]:

1. Field and temperature dependence of resistivity
2. Insulation geometry
3. Temperature distribution

A well-known approximate expression of the steady-state electric field distribution under DC voltage [23] can be derived starting from the temperature profile in the insulation wall as a function of the generic radius r; such temperature profile, given the Joule losses on the conductor W_C in W/m, the thermal conductivity $\lambda_{T,d}$ of the insulation compound in W/K m, and the external radius of the insulation r_o in m, can be evaluated via the following formula [24]:

$$\Delta T(r) = T(r) - T(r_o) = \frac{W_C}{2\pi\lambda_T}\ln\left(\frac{r_o}{r}\right) \tag{3.14}$$

where $T(r)$ is the temperature value at r and $\Delta T(r)$ is the temperature drop between r and the outer insulation radius r_o.[7] Note that when $r = r_i$, $\Delta T(r)$ becomes equal to the temperature drop across the whole insulation thickness, $\Delta T(r_i)$, that is referred to as ΔT for the sake of simplicity from now on.

For a material without thresholds for both temperature and electric field, the electrical resistivity after Eq. (3.13) can be rewritten as

$$\rho_E(T, E) = \rho_0 \exp[-(aT + bE)] \tag{3.15}$$

If I is the current that flows through the insulation wall, the absolute value of the current density J can be evaluated as a function of the generic radius r:

$$J = \frac{I}{2\pi r} \tag{3.16}$$

[7] Relationship (3.14) is a simple application of the so-called "thermal Ohm's law" to power cable geometry [24].

By inserting Eqs. (3.15) and (3.16) into Eq. (3.3), one obtains

$$E(r) = \frac{I\rho}{2\pi r} = \frac{I\rho_0}{2\pi r}\exp\{-[aT(r) + bE(r)]\} \tag{3.17}$$

When considering Eq. (3.14), Eq. (3.17) changes as follows:

$$\ln\left[E(r)\frac{2\pi r}{I\rho_0}\right] = -aT(r_o) - \ln\left(\frac{r_o}{r}\right)^A - bE(r) \tag{3.18}$$

where A is defined as (see also eq. (3.14)):

$$A = \frac{aW_C}{2\pi\lambda_{T,d}} = \frac{a\Delta T}{\ln(r_o/r_i)} \tag{3.19}$$

Equation (3.18) is the relationship used for the estimation of the electric field in the insulation wall of a screened DC cable. Unfortunately, such an equation does not yield a closed-form expression for the electric field, hence the electric field can be estimated from (3.18) only numerically. A closed-form expression for the electric field was obtained from (3.18) by Eoll in [25] by approximating the field term of the exponential in Eq. (3.17) as follows:

$$\exp[-bE(r)] \approx \left[\frac{eE(r)}{E_m}\right]^{-bE_m} \tag{3.20}$$

in which e is the Neper number, E_m is the mean electric field in the insulation given the conductor to screen (or pole-to-ground) voltage U_0:

$$E_m = \frac{U_0}{r_o - r_i} \tag{3.21}$$

Equation (3.18) can then be rewritten in a closed form:

$$E(r) = \left(\frac{I\rho_0}{2\pi r_o}E_m^B\left(\frac{r}{r_o}\right)^{A-1}\exp\{-[aT(r_o) + B]\}\right)^{1/(1+B)} \tag{3.22}$$

where

$$B = bE_m = \frac{bU_0}{r_o - r_i} \tag{3.23}$$

It is interesting to evaluate the error introduced with the approximation (3.20) as a function of the ratio $E(r)/E_m$ and considering the influence of the b parameter. This can be done by resorting to the following percent error function, err%:

$$\text{err\%} = 100\frac{\exp[-bE(r)] - [eE(r)/E_m]^{-bE_m}}{\exp[-bE(r)]} \tag{3.24}$$

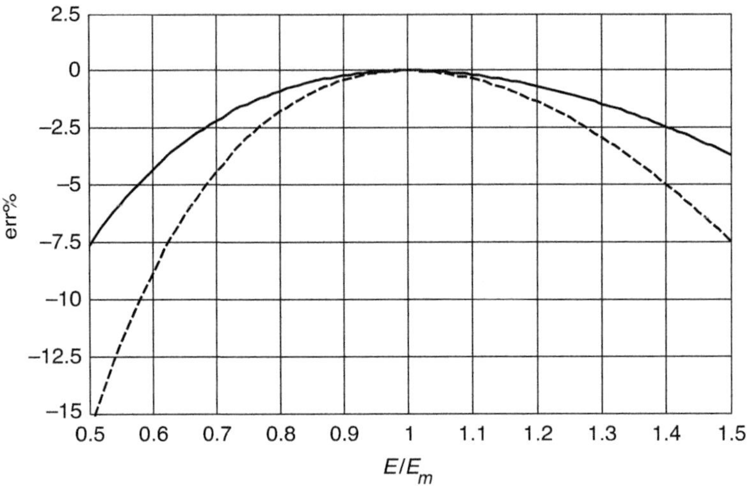

Figure 3.7 Percent error function, err% [see Eq. (3.24)], versus E/E_m for $b = 0.03$ mm/kV (solid line) and $b = 0.06$ mm/kV (dashed line).

Figure 3.7 shows that the larger the difference between E and E_m, the larger is the error introduced by the approximation (3.20). Furthermore, the larger the value of parameter b, the larger is the error. Consequently, it can be stated that the validity of Eq. (3.20) increases as the ratio E/E_m tends to 1, that is, as the field profile becomes more and more uniform within the insulation thickness.

Now, considering that U_0 can be evaluated as

$$U_0 = \int_{r_i}^{r_o} E(r)\, dr \tag{3.25}$$

inserting Eq. (3.22) into (3.25) yields

$$U_0 = \frac{k}{\delta}\left(\frac{I\rho_0}{2\pi r_o}\right)^{1/(1+B)} \left(r_o^{\delta} - r_i^{\delta}\right) \tag{3.26}$$

where [see (3.19) and (3.23)]

$$\delta = \frac{A+B}{1+B} = \frac{aW_c/(2\pi\lambda_{T,d}) + bE_m}{1 + bE_m} = \frac{aW_c/(2\pi\lambda_{T,d}) + bU_0/(r_o - r_i)}{1 + bU_0/(r_o - r_i)}$$

$$= \frac{\dfrac{a\Delta T}{\ln(r_o/r_i)} + \dfrac{bU_0}{(r_o - r_i)}}{1 + \dfrac{bU_0}{(r_o - r_i)}} \tag{3.27}$$

$$k = \left(E_m^B \left(\frac{1}{r_o} \right)^{A-1} \exp\{-[aT(r_o) + B]\} \right)^{1/(1+B)} \tag{3.28}$$

By extracting the expression of current I from (3.26) and introducing it in (3.22), one finally obtains

$$E(r) = \frac{\delta U_0 (r/r_o)^{\delta-1}}{r_o \left[1 - (r_i/r_o)^{\delta} \right]} \tag{3.29}$$

Expression (3.29) was actually developed for cables with paper insulation [25], however, it is equally applicable—and commonly applied—to extruded dielectric cables [1, 3]. A comparison of (3.29), with (3.1) shows that the DC expression is far more complex than the AC one, even under the approximation of (3.20).

In DC regime, the temperature drop across the insulation thickness can be ascribed to the losses on the conductor W_C only [see Eq. (3.14)], since the influence of ohmic losses due to the leakage current can be neglected in practice.[8] Then from Eqs. (3.29, 3.27), it can be readily seen that the *steady (or stable) HVDC field distribution does not depend on the absolute temperature at the various points within the insulation but on the temperature drop across the insulation thickness (ΔT) only* [12].

Therefore, it can be argued that, when the HVDC cable system is loaded, the resistive power loss in the conductor and the thermal dissipation through the cable give rise to a thermal gradient across the cable dielectric. The insulation close to the conductor is subjected to a higher temperature, which reduces the electrical resistivity with respect to the insulation close to the electric screen, thereby reducing the electric field in the insulation close to the conductor and raising it far from the conductor. For a high load, this may lead to an inversion of the electric field profile across the dielectric, so that—when loaded—and HVDC cable system may experience a higher electric field intensity at the insulation shield than at the conductor shield [1].

To be more quantitative, Table 3.4 after [3] details the electric field (stress) at the conductor and core calculated for a 100-kV cable geometry (500-mm^2, 6-mm insulation wall) in AC regime using Eq. (3.1) and in DC regime without and with load using Eq. (3.29) in conjunction with the values of parameters a and b for different materials given in Tables 3.2 and 3.3. Table 3.4 includes the DC leakage current I calculated for the material combinations.

The detailed stress distribution for the full-load case is shown in Figure 3.8. This figure shows that, in a cylindrical structure such as a cable, the highest field strength with no load is found near the conductor. The figure also shows that the

[8] Usually, the effect of the heating of the insulation by the leakage current may be disregarded, especially at zero load, as pointed out above. However, in certain cases, when the cable temperature and applied voltage are high enough, the field distribution is influenced by these insulation losses. They might even lead to an instability that causes breakdown of the cable. This holds especially for MIND cables, which suffer high temperatures and should be operated (also for other reasons) at a conductor temperature not higher than—typically—55°C [3,12].

Table 3.4 Electric Field (Stress) at the Conductor (kV/mm) and Insulation Shield Calculated for 100-kV Cable Geometry in AC Regime and in DC Regime without and with Load[a]

Stress (kV/mm)	Thermoset *2*	Thermoplastic *6*	Paper	PPL
AC				
Conductor	23.7	23.7	23.7	23.7
Insulation shield	17.7	17.7	17.7	17.7
DC no-load				
Conductor	21.8	21.8	22.7	22.4
Insulation shield	19.2	19.2	18.5	18.7
DC load				
Conductor	19.4	18.9	17.8	17.2
Insulation shield	21.3	21.8	23	23.6
***I* (μA/km)**	0.2	4	7	0.04

[a]The figures in italics refer to the descriptions in the key of Table 3.1. DC leakage currents are included.

Source: After [3].

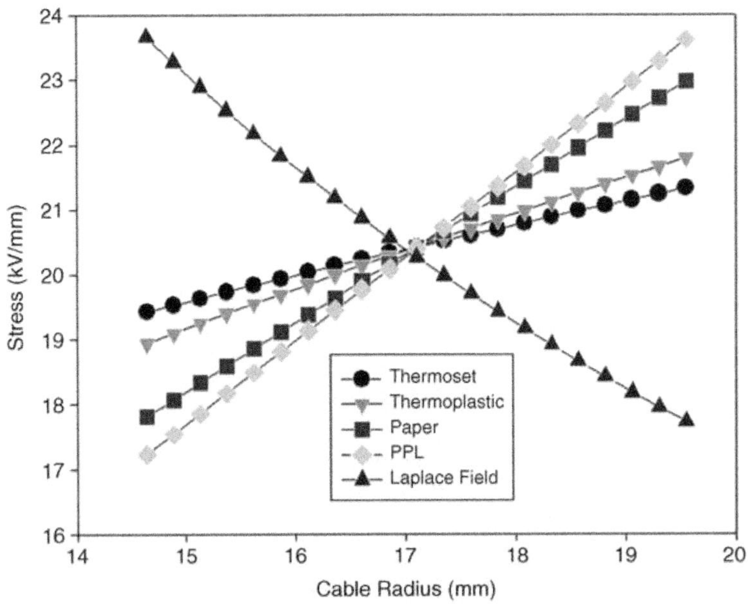

Figure 3.8 Electric field (stress) distribution, under load, for a 100-kV DC cable manufactured using different insulations. The power loss W_C is the same for all designs. The Laplace (AC) electrical stress distribution is also shown (after [3]).

effect of temperature dependency of the insulation resistivity *when the cable is loaded is an inversion of the field*: The highest field strength is now found near the insulation shield of the cable, so that *the highest electrical stress transfers from the conductor to the insulation shield.* Note that for this load and this set of cable and

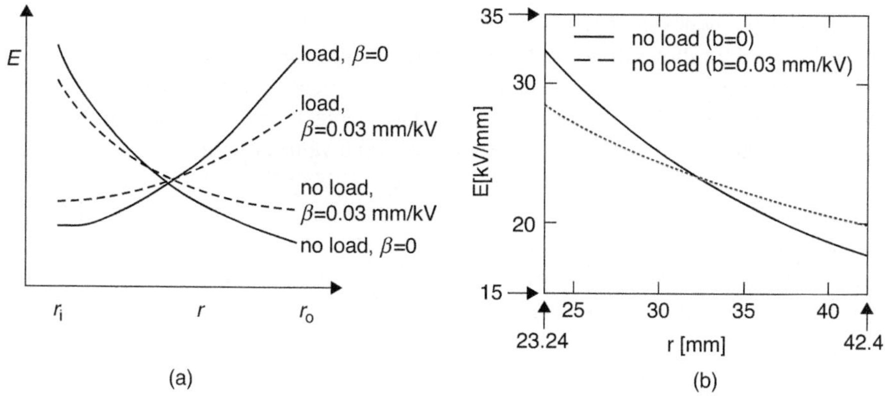

Figure 3.9 (*a*) Stable DC fields under load and no-load conditions for the standard cable of [12]. The solid lines represent the fields if the field dependency of ρ_E is disregarded, the dotted lines represent the fields when taking this effect into account. "β" is used instead of "*b*" (after [12]). (*b*) Stable DC fields under no-load conditions for the "standard cable" of [12] using Eq. (3.29).

material parameters, an inversion occurs for all materials, and that the no-load conductor stress and the loaded insulation–screen stress are of the same order of magnitude. Figure 3.8 clearly shows also that the magnitude of the inversion depends on the insulation, with the *polymeric insulations chosen here* (thermoplastic and thermoset) *showing a lower level of inversion* than the lapped (paper and PPL) ones; this is essentially *because of the lower thermal resistivity of the dielectric*, $\rho_{T,d} = 1/\lambda_{T,d}$, in the case of polymeric insulation, for example, 3.5 K·m/W for XLPE versus 6 K·m/W for mass-impregnated insulation [24].

From Table 3.4 one more consideration arises. When comparing the AC field values (referred to as Laplace field in Fig. 3.8, consistently with Fig. 3.5) with the DC no-load values (relevant to the situation when the cable is under DC voltage but not on load), it can be argued that even the DC no-load radial field distribution differs from the AC case. This difference between the AC and DC fields is normally small and arises essentially[9] from the electrical stress dependence of the conductivity. The radial DC field distribution is identical to the AC distribution only if the insulation conductivity is stress independent [i.e., $b = 0$ in (3.10), or $\gamma = 0$ in (3.11), or $c = 0$ in (3.12)] [3].

This is apparent from Figure 3.9*a*, where stable DC fields under load and no-load conditions are reported, relevant to a mass-impregnated paper cable referred to as "standard cable" (see below), with the temperature coefficient $a = 0.1\ °C^{-1}$ and

[9] Strictly speaking, one more difference arises from the heating effect of the leakage current, which sets a temperature gradient across the DC insulation, but this effect can be neglected at zero load, as pointed out above.

the field dependency coefficient $b = 0.03$ mm/kV [1,12]. For the sake of completeness, Figure 3.9*b* shows the same DC field profiles under no load as Figure 3.9*a* but in a more quantitative way, since it details also the field values calculated through Eq. (3.29) and the dimensions of the standard cable insulation. Interestingly, the value of a chosen for this cable is very close to the values in Table 3.2 for thermoplastic (e.g., LDPE) and particularly thermoset (e.g., XLPE) insulations for extruded cables over the whole range of electric stress considered, while the value of b is close only to that for thermoplastic insulation at 80°C. However, on the whole these values of a and b are not meaningless also for representing the situation of an extruded insulation, especially when considering that conductivity/resistivity is more sensitive to temperature (thus to the value of a) than to electric field (thus to the value of b). As can be seen from Figure 3.9, the effect of the field dependence of the insulation conductivity/resistivity is a leveling of the field strength.

A deeper insight into the field inversion problem comes from a careful observation of Eq. (3.29) and Tables 3.2 and 3.3. Indeed, as can be argued also from its derivation procedure [25], Eq. (3.29) assumes that the dielectric is homogeneous and considers single values of the parameters a and b. On the contrary, as observed previously, Tables 3.2 and 3.3 show that a and b are stress and temperature dependent, respectively. Thus, in normal operation the values of a and b will vary across the insulation. These stress and temperature effects should be considered to produce the best estimates of the stress distribution.

As a consequence, strictly speaking, an even more complex dependence of ρ_E on E and T than (3.13) should be considered, namely:

$$\rho_E(T,E) = \rho_0 \exp\{-[a(E)(T - T_0) + b(T)(E - E_0)]\} \tag{3.30}$$

where the dependence of a on electric field and b on temperature could be hypothesized as a first guess to be linear, namely:

$$a(E) = a_0 + a_1 E \tag{3.31}$$

$$b(T) = b_0 + b_1 T \tag{3.32}$$

where a_0, a_1 and b_0, b_1 are proper best fitting coefficients.

Sensitivity studies (see Table 3.5) based on the cable parameters used to obtain the results in Table 3.4 show that, with the temperature limited to 60°C, the temperature coefficient a plays a larger role in determining the electrical stresses than the stress coefficient b, to a first-order approximation. The uncertainty in the stress calculations due to the declared variations in a and b are of the order of 3% [3].

Resistivity inhomogeneities can occur in practice due to, for example, water ingress (decrease in resistivity) or removal of cross-linking by-products (increase in resistivity). The effects of these changes on the electric stress within a DC cable are shown in Figure 3.10. Such effects can be very significant and demonstrate the additional considerations that need to be made when designing, manufacturing, and operating DC systems.

Table 3.5 Sensitivity Study Showing Selected Electric Field (Stress) Calculations for PPL Using *a* and *b* Values for Different Stresses and Temperatures

Stress (kV/mm)	$a = 0.086\,°C^{-1}$ $b = 0.024$ mm/kV	$a = 0.093\,°C^{-1}$ $b = 0.024$ mm/kV	$a = 0.086\,°C^{-1}$ $b = 0.030$ mm/kV	$a = 0.093\,°C^{-1}$ $b = 0.030$ mm/kV
No load	22.6	22.6	22.4	22.4
Conductor shield	18.6	18.6	18.7	18.7
Insulation shield				
Full load	17.0	15.6	17.2	16.9
Conductor shield	23.9	24.3	23.6	24.0
Insulation shield				

Source: After [3].

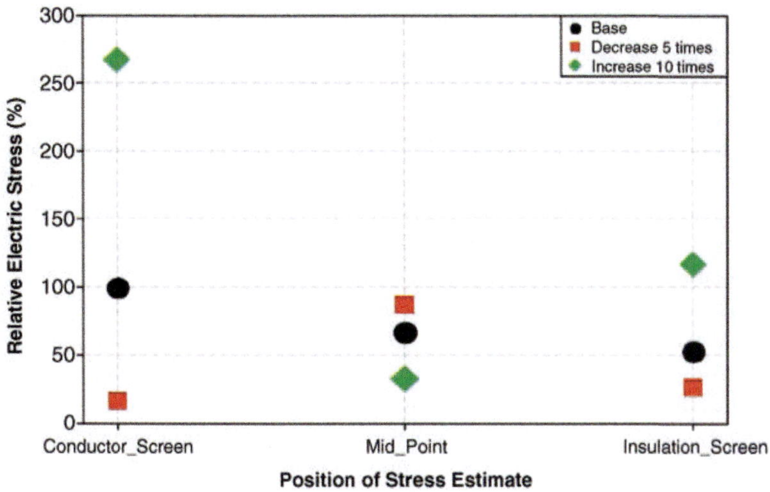

Figure 3.10 Modification of the equilibrium electric field (stress) (base case for 100% at the conductor shield) for selected increases and decreases in resistivity at both shields (after [3]).

3.2 TRANSIENT DC FIELD DISTRIBUTION

As pointed out at the beginning of Section 3.1.3.2, the expression (3.29) of electrical field distribution in an HVDC cable holds only *under steady-state* DC conditions. Indeed, immediately after DC voltage application there is no DC field distribution in the cable yet. The distribution is at first a "quasi-capacitive" one. Via an intermediate time-dependent field, the distribution changes from the quasi-capacitive to a purely resistive distribution [12]. To be exact: The purely resistive stage will be reached only after an infinite time, as pointed out above. After, for instance, a polarity reversal, yet another field distribution is found in the cable. Therefore, because the field distributions differ per operational stage, different computational schemes must be used for computing the transient DC field distribution that the cable experiences at every different operational stage [12]. This is the subject of the present section, that is mostly based on the treatment of [12]—actually a milestone paper for HVDC insulation.

3.2.1 Time to Reach Steady-State DC Field Distribution

First of all it is important to establish an order of magnitude of the time to reach steady-state DC field distribution after a change in the operation of the HVDC cable, for example, a change in the load current from zero to a non-zero value, which involves a given regime temperature of the cable conductor.[10]

For estimating the time to the steady-state field distribution within a DC power cable in operation, it must be remembered (see Section 3.1.3.2) that such distribution differs from the Laplace field ($\propto 1/r$) due to the fact that the conductivity of the insulation depends on the local temperature (which decreases from the conductor to the sheath) and electric field. These phenomena generate a divergence, that is, a nonuniformity over insulation thickness, which is in addition to that derived from the cable geometry [11].

The evolution of the additional divergence can be represented as a function of a dimensionless parameter t/τ, where t is time and τ is the "time constant": $\tau = \rho_E \varepsilon$, where ρ_E is the volume electrical resistivity and ε the permittivity of the dielectric.

Calculations show that a time equivalent to 10τ has to pass to approach the steady-state distribution of the divergence. Table 3.6 gives the range of times to stability (10τ) as a function of temperature for different materials that are likely to be used for DC extruded cables. Practically, this means that the time to achieve a stable electrical stress distribution will depend upon temperature. It is important also to select test times during load cycle tests (carried out in the framework of prequalification and type tests, see Section 5.9) that permit the insulation systems to be tested to reach a stable electrical stress distribution. These considerations provide the foundations for the times of the zero load, high load, and 24/48-h load cycle tests described in detail in CIGRE Brochure 496 [11].

Table 3.6 Range of Times to Stability (10 τ) as Function of Temperature for Different Materials Used for DC Extruded Cables

Temperature (°C)	ε (F/m)	ρ_E (Ω·m)	Time to Stability, 10 τ (h)
20	$2 \times 10^{-11} < \varepsilon < 3 \times 10^{-11}$	$10^{15} < \rho_E < 5 \times 10^{16}$	$55 < 10\,\tau < 4300$
60	$2 \times 10^{-11} < \varepsilon < 3 \times 10^{-11}$	$2 \times 10^{13} < \rho_E < 5 \times 10^{14}$	$1 < 10\,\tau < 43$
90	$2 \times 10^{-11} < \varepsilon < 3 \times 10^{-11}$	$10^{12} < \rho_E < 5 \times 10^{13}$	$0.06 < 10\,\tau < 4.3$

Source: After [11].

[10] As pointed out above, such temperature can be derived via the methods described in standard series of IEC 60287 [24]. More details about the thermal aspects of cable design are given in Chapter 5 at Section 5.5.

Note that injection of charges from the electrodes is also active; however, the 10τ time constant covers the time needed for these injection processes [11].

The origin of time constant $\tau = \rho_E \varepsilon = \varepsilon/\sigma$ can be understood by developing relationships (3.2–3.5) for the transient conditions between the initial application of the voltage and the long-term steady state treated in Section 3.1.3.2. By doing so, the transient form of Eq. (3.9) is obtained [17]:

$$\rho = -\frac{\varepsilon}{\sigma}\frac{\partial \rho}{\partial t} + \mathbf{J} \cdot \nabla\left(\frac{\varepsilon}{\sigma}\right) = -\frac{\varepsilon}{\sigma}\frac{\partial \rho}{\partial t} + \sigma \mathbf{E} \cdot \nabla\left(\frac{\varepsilon}{\sigma}\right) \tag{3.33}$$

Relationship (3.33) shows that the rate at which the long-term steady state is approached depends on the term $-(\varepsilon/\sigma)(\partial\rho/\partial t)$, which is governed by the dielectric time constant $\tau = \varepsilon/\sigma$. When the time derivative of space charge goes to zero, charge and field become steady and (3.33) reduces to (3.9). At that time, the condition for having a stable space charge within the dielectric is that τ – thus $1/\sigma$ in practice – is radially nonuniform, as already pointed out in Section 3.1.3.2 [17].

Relationship (3.33) shows also that during a transient situation the magnitude of the conductivity is very important. The dielectric time constant decreases when the conductivity increases; hence, the higher the conductivity, the more rapidly a steady-state resistive field is established: The time to reach a steady-state resistive field distribution will be dictated by the lowest conductivity in the insulation. Hence, when measurements of space charge are made (see Chapter 4), the time resolution of the measurements has to be adapted to the actual dielectric time constant, which can vary by several orders of magnitude depending on material characteristics, temperature and magnitude of electric field, as pointed out in Section 3.1.3.2 [17].

3.2.2 Definition of Operational Stages of HVDC Cables

Following the proposal after [26], the different operational stages of HVDC cables when switching on and off a DC voltage (top) and after a voltage polarity reversal (bottom) are described and defined with reference to Figure 3.11 (after [12], where actually paper insulated cables are treated, but most considerations in [12] hold also for extruded insulation cables, apart from the values of resistivity and permittivity, which are, however, of the same order of magnitude; see below).

Focusing first on the upper part of Figure 3.11 the relevant stages are as follows:

Stage I: The applied DC voltage U is raised from 0 up to its final value $+U_0$. At the beginning of this stage there is no space charge within and no temperature drop across the cable insulation. Thus, with a good degree of approximation, it can be argued that along the stage the electric field has a capacitive distribution determined by the geometry and the permittivity ε only.

Stage II: While the voltage has already reached its final value$+U_0$, the electric field is time dependent and its distribution is changing from a capacitive to a resistive one.

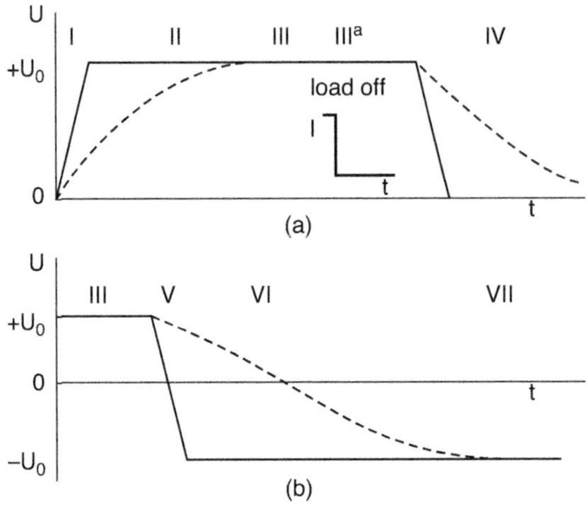

Figure 3.11 Different stages when switching on and off a DC voltage (*a*) and after a polarity reversal (*b*). The dotted lines represent the growth and decay of space charges in the insulation (after [12]).

Stage III: The electric field has become stable and its distribution is purely resistive.

Stage III[a]: It is a special case of stage III such that load current *I*—if any—is switched off.

Stage IV: The applied DC voltage is reduced to zero and kept to such a value. A field still exists after the voltage has been removed.

Consider now the lower part of Figure 3.11, which accounts for a polarity reversal subsequent to stage III above. The relevant stages are as follows:

Stage V: The voltage polarity is being reversed from its starting positive value, $+U_0$, to its final negative value, $-U_0$.

Stage VI: The voltage polarity has been reversed, but the electric field has not become stable yet.

Stage VII: The field after the polarity reversal has become stable.

In [12] a software is described that calculates the field distributions in each of the defined stages. The distributions are determined for different ambient temperatures, temperature drops, applied voltages, cable geometries, permittivities, and conductivities as a function of time where appropriate. The software calculates also space-charge density as a continuously varying function of space (and time where appropriate). Hence a software like this, though used in [12] for paper insulated cables (see above), it can be used for all HVDC cables since its theory is general, and it is sufficient to refer to the specific permittivities and conductivities of extruded insulation for applying it to HVDC extruded cables as well. Of course, extruded cables in particular can give rise to *localized trapped charge centers* related to charge traps associated with the highly nonuniform microstructure of the extruded insulation; actually, these charge centers give rise to a space-charge distribution that can be significantly different from the continuous and regular (i.e., fairly slow varying in space) distributions calculated in this software after [12] according to the conductivity model for homogeneous materials. As a consequence, also the actual field distribution can be fairly different from that calculated by such software, as treated in detail in Chapter 4. For this reason, other more refined softwares have been developed for calculating the Poissonian field associated with such "nonuniform" localized charge distributions [21,27]. Localized charge distributions can be evaluated experimentally via the various space-charge measurement techniques illustrated at Section 4.3. Such an approach is, however, skipped in this chapter and treated in Chapter 4, devoted to space charges in extruded HVDC insulation.

Even though the procedure described here neglects the effect of locally trapped space charges, in any case the software and the results described in [12] are reported in detail since they constitute a reference for all HVDC cables, and a sort of target for extruded HVDC cables that could be reached (and at which manufacturers are aiming through extruded compounds with low space-charge retention properties) if localized charge centers could be reduced or removed and only small diffused and delocalized charge storage takes place.

3.2.3 Field Distributions in the Various Stages

This section illustrates stage by stage the field distributions and the relevant computation method, as reported in [12] with reference to a so-called "standard" cable, featuring 450-kV rated voltage, 1600-mm^2 copper conductor cross section, inner insulation radius $r_i = 23.2$ mm, insulation thickness $d = 19.2$ mm,[11] outer insulation radius $r_o = 42.4$ mm, resistivity $\rho_0 = 1 \times 10^{16}\,\Omega\cdot$m at $T_0 = 293$ K ($\theta_0 = 20°$C[12]), $E_0 \sim 0$ kV/mm, temperature coefficient of resistivity $a = 0.1$/°C, field coefficient of resistivity $b = 0.03$ mm/kV, and relative permittivity $\varepsilon_r = 3.5$. It can be argued from Table 3.6 that the value of ρ_0 is in the very middle of the range of ρ_0 values for different materials that are likely to be used for DC extruded cables, so it can be used also for extruded insulation. Moreover, as discussed when commenting on Figure 3.9 (relevant to this same cable) in Section 3.1.3.2, the values of a and b chosen for this cable make some sense also for representing the situation of an extruded insulation, particularly of the thermoplastic type (e.g., LDPE) at fairly high temperatures. On the contrary, the permittivity is significantly higher, thus the calculations after [12] will be corrected here whenever possible, accounting for a permittivity value of $\varepsilon = 2.5$, in accordance with Table 3.6. In all calculations the quasi-static approximation will be used, that is, $\nabla \times \mathbf{E} = 0$.

[11] Actually the value of inner insulation radius r_i reported in [12] is 19.2 mm, but this is not consistent with a conductor cross section of 1600 mm^2. Since insulation thickness s is omitted in [12], it is assumed here that $s = 19.2$ mm and $r_i = 23.2$ mm, which descends from the conductor cross section plus a typical conductor stranding factor of ~0.95. This seems a good choice since it yields a perfect agreement with [12] as to the Laplace field, as shown in Figure 3.12.

[12] Note that in this book T denotes absolute thermodynamic temperature in K, θ temperature in °C.

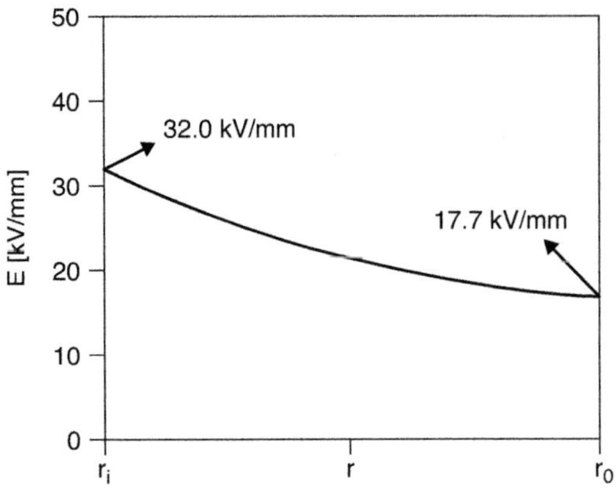

Figure 3.12 Capacitive field distribution at the end of stage I for the standard cable (after [12]).

3.2.3.1 Stage I: Raising the Voltage In stage I, the applied DC voltage is raised to its desired value ($U_0 = 450\,\text{kV}$ for the standard cable). This operation takes a relatively short time, usually of the order of some seconds, much shorter than the above time constant $\tau = \rho_E \varepsilon$ of the insulation (see Table 3.6). For this reason, in stage I the cable contains no space charge yet, the field distribution is purely capacitive, and the field may be calculated using Eq. (3.1), the same formula as used for HVAC cables [12].

The capacitive field distribution within the insulation wall at the end of stage I for the standard cable calculated with Eq. (3.1) is plotted in Figure 3.12. The figure shows that the field strength is the highest at the conductor semicon–insulation interface, as for AC cables (see Fig. 3.5).

3.2.3.2 Stage II: After Raising the Voltage After having raised the voltage, in stage II the field changes from a capacitive to a purely resistive distribution, as pointed out above. The field between these distributions, called an "intermediate field," cannot be calculated analytically but numerically on the basis of [12]:

1. Gauss law (3.4)
2. Continuity equation (3.5)
3. Ohm's law (3.3)
   thereby attaining eq. (3.33) as shown in Section 3.2.1.

   Proper initial and boundary conditions must be set, namely:

(a) The electric field distribution at $t = 0$ is the capacitive distribution of Figure 3.12.

(b) The voltage across the insulation is U_0 and the voltage at the outer semicon is 0.

(c) A function $T(r,t)$ describes the dependence of temperature on radius and time [needed due to the temperature dependence of the electrical conductivity in Eq. (3.3)].

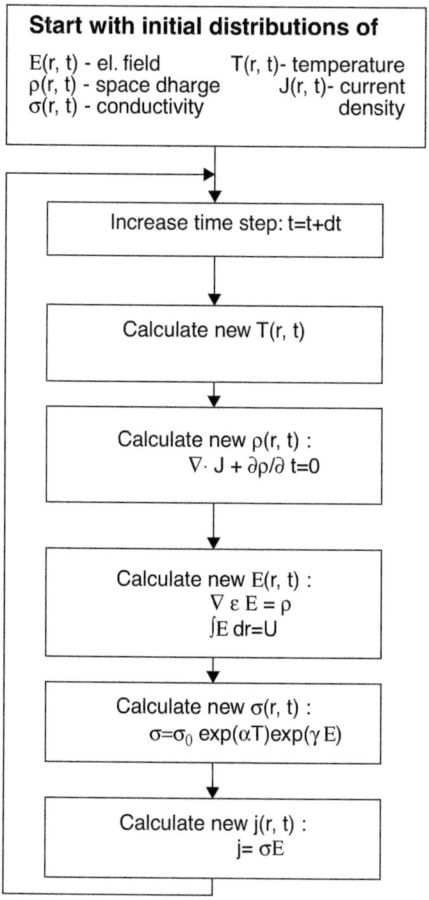

Figure 3.13 Flowchart of field-calculating software (after [12]).

The software after [12] calculates the electric field and the charge density by solving these equations in space and time, as shown in Figure 3.13. As far as the cable load is concerned, three different field distributions can be computed in stage II associated with three different time instants (see Fig. 3.14):

1. The intermediate field E as long as the load current of the cable is zero ($I = 0$)
2. The intermediate field E as soon as the load current of the cable is switched on
3. The intermediate field E of a steadily loaded cable (namely the load current of the cable has been switched on long before and the temperature at the various locations within the cable is stable)

In situation 1, temperature is constant across the insulation, hence field inversion does not occur and just a slight change in the field is observed due to the field dependence of the conductivity only (compare the lines "no load, $b = 0.03$ mm/kV"

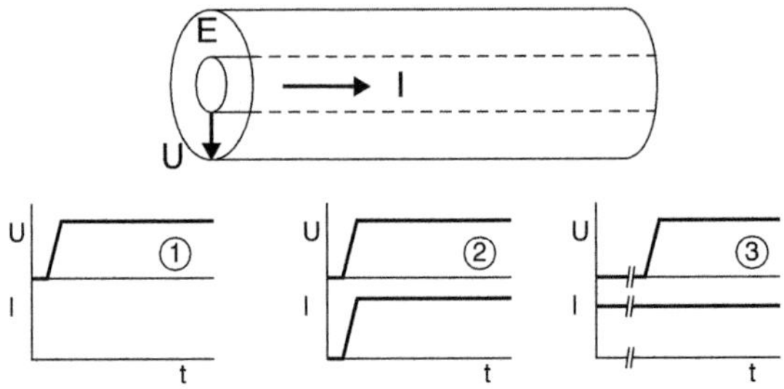

Figure 3.14 Different situations in stage II as explained in the text (after [12]).

and "no load, $b = 0$" in Figure 3.9*a*). In situations 2 and 3, the field inversion phenomenon eventually takes place. The intermediate fields in these two situations are almost the same and vary from a purely capacitive field to the inverted resistive field, differing solely as to the rate of variation [12].

For the sake of brevity, more details are given for situation 3 only. Figure 3.15 shows the field distribution in the standard cable at stage I ($t = 0$) and at different times during stage II, under the following hypotheses about the "thermal profile": The temperature drop across the insulation, $\Delta T = \Delta\theta$, is equal to 15°C; the stable temperature of the metallic screen, θ_s, is equal to 35°C[13]. At $t = 0$ the voltage is switched on, thereby yielding a capacitive field distribution (stage I). Thereafter, stage II starts: Intermediate fields at 10-min intervals are plotted in Figure 3.15, together with the purely resistive field at $t = \infty$. The figure shows that [12]:

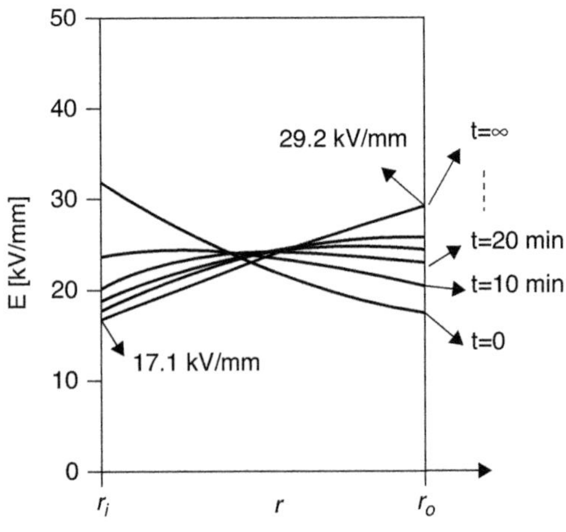

Figure 3.15 Field curves in stage II, situation 3 (see text) with $\theta(r_o) = 35°C$, $\theta(r_i) = 50°C$, $\Delta\theta = 15°C$ (after [12]).

[13] Note that for a HVDC extruded cable this means that the load current is significantly lower than the rated load current, that can be as high as to let the conductor temperature reach a value of 70°C or higher.

1. The field variation is rather small in the middle of cable insulation.
2. The field at the conductor drops faster than it rises at the metallic screen. Indeed, the insulation is 15°C warmer at the conductor side than at the metallic screen side, thus the electrical conductivity of the insulation close to the conductor is higher, given that the time constant $\tau = \rho_E \varepsilon = \varepsilon/\sigma$ is smaller.

3.2.3.3 Stage III: Steady Resistive Field and Relevant Space Charge

In stage III, the field does not vary with time anymore since the field distribution has become a steady resistive distribution. However, a gradient in the electrical conductivity exists because of its temperature and field dependence. Therefore, a space charge is present as already illustrated in Section 3.1.3.2 when deriving relationship (3.9) from the first Maxwell equation (3.4) and from the current continuity equation (3.5) [26]. The space charge gives rise to an electric field, usually referred to as "charge-induced" field, which—for the standard cable considered in [12]—gives rise to the phenomena of field inversion and field strength leveling illustrated in Figure 3.9*a*. In order to account for these phenomena, in [12] the following procedure is carried out: (1) calculation of the field, (2) calculation of the total insulation resistance of the cable, and (3) calculation of the space-charge distribution.

The exact electric field, including the effect of temperature and field dependence, is more complex than simplified equation (3.29); indeed, it can be shown to be given by [12,28]:

$$E(r) = U_0 \frac{r^{z-1}\exp(-bE)}{\int_{r_i}^{r_o} r^{z-1}\exp(-bE)\,dr} \qquad z = \frac{a\Delta T}{\ln(r_o/r_i)} \tag{3.34}$$

As in the case of simplified relationship (3.29), from Eq. (3.34) it can be readily seen that the field distribution does not depend on the absolute temperature but on the temperature drop ΔT only (once more if ohmic losses associated with the leakage current are neglected) [12]. Since electric field intensity appears at both sides of Eq. (3.34), this equation can be solved only numerically: The relevant electric field profiles obtained for different temperature drops ΔT are plotted in Figure 3.16*a*. The figure shows that:

1. When $\Delta T = 0$, the field is highest (29 kV/mm for the standard cable) at the inner semicon–insulation interface.
2. For large temperature drops (>15°C), the field at the insulation–outer semicon interface may overcome the highest possible field close to the conductor.
3. At a single point in the middle of the insulation, the temperature drop does not affect the field.

For the sake of comparison, Figure 3.16*b* displays the electric field profiles calculated via simplified Eq. (3.29) for the same temperature drops ΔT of Figure 3.16*a*. Very slight differences are found, in particular, the field at inner and outer semicons is a bit higher according to the simplified formula (especially at higher temperature drops), that can be thus deemed as conservative.

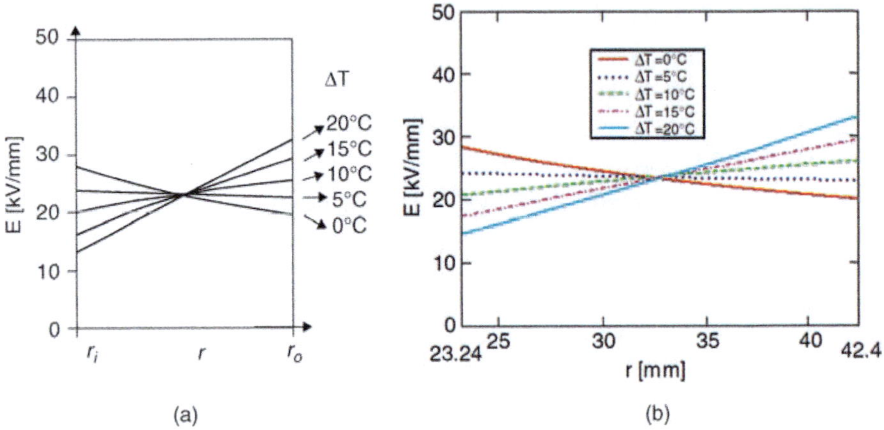

Figure 3.16 (*a*) Stable field distributions in stage III for the standard cable calculated for different temperature drops ΔT via Eq. (3.34) (after [12]). (*b*) Stable field distributions in stage III for the standard cable calculated for different temperature drops ΔT via Eq. (3.29).

As mentioned above, the temperature and field dependencies of electrical conductivity imply that in stage III space charge exists within the insulation: As pointed at Section 3.1.3.2, *this is not trapped charge but charge that accumulates at discontinuities of permittivity and conductivity* [12]. Then, the charge distribution in the cable insulation can be calculated using Eq. (3.3) plus the above-derived equation (3.9), reported here for the sake of convenience in a compact form:

$$\rho = \mathbf{J} \cdot \nabla\left(\frac{\varepsilon}{\sigma}\right) \tag{3.35}$$

This calculation can be done only numerically. The software illustrated when describing stage II was employed in [12], thereby attaining the charge distributions in the standard cable for different temperature drops ΔT and for $U_0 = 450\,\text{kV}$ reported in Figure 3.17 : for $\Delta T = 0°\text{C}$, the charge distribution arises just because of the field dependence of the conductivity. As proven by Figure 3.17:

1. For the lower temperature drops, the highest charge densities appear close to the conductor.
2. As the temperature drop rises, the charge moves progressively toward the metallic screen.
3. One single charge polarity is found at every temperature drop, thus at every load level.

This latter observation was already made at Section 3.1.3.2 with reference to zero load and was addressed to the field dependence of conductivity, which gives rise to a decreasing conductivity with r, thereby making the resistivity ($1/\sigma$) gradient increasing with r. Here it can be argued that the resistivity increases with r also at

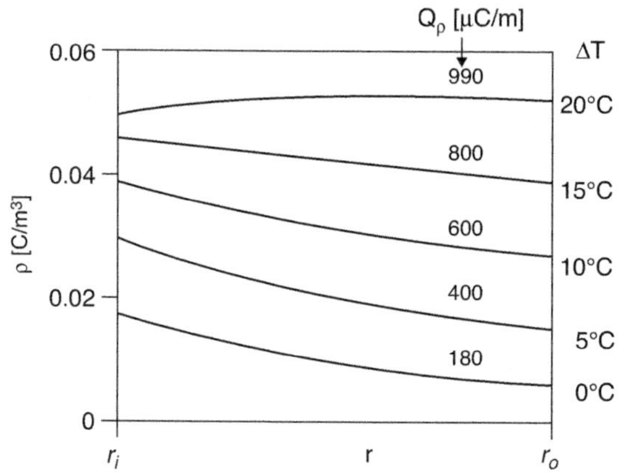

Figure 3.17 Charge distribution in the standard cable during stage III for different temperature drops ΔT at voltage $U_0 = 450\,\mathrm{kV}$ (after [12]).

nonzero load because of the temperature dependence of conductivity. Therefore, according to (3.35) space charge will have the same sign as the electric field vector, thus of the voltage on the inner electrode: positive if the field is directed outward from the cable, as happens when voltage at the conductor is positive, negative otherwise.

Figure 3.17 also shows the total charge per unit length (pul) of the cable, Q_ρ, thus showing that:

1. The accumulated charge can be as high as 600–1000 pC/m (a remarkable pul charge, indeed).
2. The higher the temperature drop, the higher Q_ρ [conversely it is proven in [12] that the higher the poling voltage, the higher Q_ρ (as quite obvious, see also Chapter 4].

One could conclude that the charge accumulation due to temperature and field dependencies in a HVDC cable may be calculated easily [12], but it must be remembered that this charge does not account for charge stored locally at particular trapping centers associated with insulation microstructure, which can only be measured with the techniques treated in Chapter 4.

3.2.3.4 Stage IIIa: After Switching off the Load Stage IIIa starts with switching off the load current, while keeping the voltage unchanged (see upper part of Fig. 3.11). As a consequence, the conductor temperature drops down to that of the room, the thermal gradient across the insulation disappears, and the field distribution is progressively reversed from the "inverted" field to the "quasi-Laplacian" field with maximum strength close to the conductor. Thus, the field in stage IIIa is a time-dependent "intermediate field" as during previous stage II and the same theory can be applied. Stage IIIa was introduced in [12] to emphasize that cooling down is critical especially for paper-insulated mass-impregnated cables.

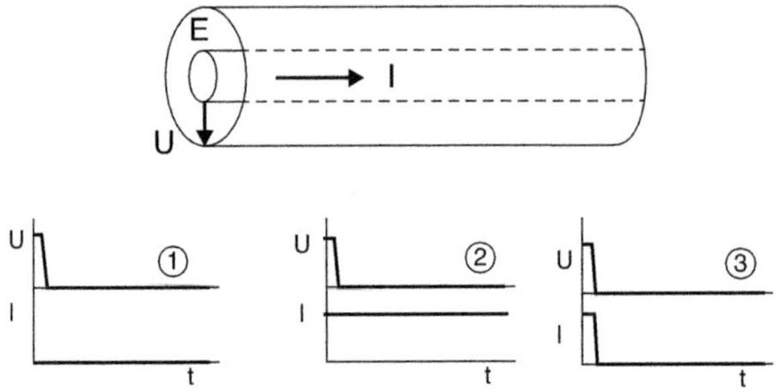

Figure 3.18 Different situations in stage IV as defined in the text (after [12]).

Indeed in this case, due to the shrinkage of the cable metallic layers during the cooling process, voids can arise in the mass-impregnated paper where partial discharges with an enhanced repetition rate could occur, with a possible harmful effect on the cable. For this reason, the operating temperature of mass-impregnated cables is limited to 55°C only [29]. Phenomena occurring during stage IIIa are carefully described in [28].

3.2.3.5 *Stage IV: After Switching off the Voltage* Stage IV begins with switching off the applied voltage, which goes to zero in a more or less short time, depending on the internal resistance of the voltage source and on cable capacitance. Nevertheless, the space charge decreases far slower, thus a "charge-induced" field may remain within cable insulation thickness for a far longer time. As to the load, three different cases may occur (see Fig. 3.18):

1. The cable load is zero before switching off the voltage.
2. The cable load is zero neither before nor after switching off the voltage.
3. The cable load is not zero before, but it is switched off together with the voltage.

In case 1, the remaining field is not so high (see Fig. 3.17 at $\Delta T = 0°C$). Hence this case is omitted in [12]. Cases 2 and 3 yield similar decays of the field, but case 3 features a slower field reduction rate (thus the field goes to zero in a longer time) than case 2, since in the former case the cable remains hot, while in the latter case the cable cools down due to switching off the load. This reduction rate can be made even slower—for a given temperature drop ΔT across the insulation—if the outer temperature of the insulation is very low. For example, in the case when the cable line follows a motorway path that includes viaducts, where the cable cannot be laid under the soil, but on the contrary is exposed to open air; then the outer temperature of the insulation can be very low, especially in winter conditions and/or in a mountain site.

In [12], case 2 only is considered, arguing that a previously loaded cable may keep a considerable charge-induced field. By assuming that the voltage before switching off was $U_0 = 450$ kV, while the cable current is constant and such that the temperature drop $\Delta T = 15°C$, the software introduced under stage II can be used. By indicating as $E(t=0^-)$ the field just before switching off the voltage and as $E(t=0+)$ the field just after switching off the voltage, $E(t=0+)$ can be calculated according to the following relationship [12]:

$$E(t=0+) = E(t=0^-) - E_{AC} \tag{3.36}$$

where E_{AC} is the capacitive field provided by Eq. (3.1). From (3.36) it can be deduced that the field $E(t=0+)$ is the field induced by the remaining space charge [12,28]. Thereafter, the charge-induced field distribution progressively decays until, at $t \to \infty$, the field is zero at every location in the cable.

3.2.3.6 Stages at Polarity Reversal Stages from V to VII in the lower part of Figure 3.11 are those related to a voltage polarity reversal. It is known that high stresses may occur at the conductor immediately after reversing the polarity of the applied voltage [12]. This is especially the case if the cable is loaded and then there is a temperature gradient across the insulation thickness.

This is caused by the space charge of the loaded cable. This trouble has been limiting the application of extruded HVDC cables to VSC half-bridge systems only, that is, with converters that reverse the current flow in order to reverse the power flow direction, without permitting any voltage reversal. A great research effort has tackled this problem and finally new XLPE-based compounds are being developed that show promising results in laboratory tests. Extruded HVDC cables able to withstand the polarity reversal at present have been developed and are available on the market (see Sections 5.6.5 and 7.2.8). The calculation of the field just after polarity reversal can be carried out in a similar way as during stage IV, with Eq. (3.36) where E_{AC} is now twice the field strength as calculated with Eq. (3.1) [12].

3.3 INFLUENCE OF THE ENVIRONMENT TEMPERATURE ON THE STEADY FIELD OF AN HVDC EXTRUDED CABLE

After the analysis of how the field changes during the various typical operational stages of an HVDC cable under various load conditions, it is stressed here how the temperature and the electric field profiles in the thickness of the insulation are affected by changes in the temperature of the laying environment with particular reference to extruded insulation. This can be important especially in the above-mentioned case of a cable line following a motorway path that includes viaducts; indeed, along viaducts the cable cannot be laid under the soil and is exposed to open air; then the outer temperature of the insulation can be very low, especially in winter conditions and/or in a mountain site.

A similar situation as to the low temperature can occur along buried sections of the cable line that are either at a high altitude, or in a very cold climatic area, or in both situations. Then, the soil may freeze and remain frozen for long periods over the year (the so-called "permafrost," encountered in high mountain regions and/or close to the polar circles). Hence the cable can be exposed to particularly low temperatures,

Table 3.7 Conductor Joule Losses vs. Soil Temperature for 300-kV DC Cable with 1400-mm^2 Copper Conductor Cross Section Carrying 1250 A

Soil Temperature θ_a (°C)	Conductor Temperature θ_C (°C)	Conductor Joule Losses W_C (W/m)
25	54.5	22.9
20	49	22.5
15	43.4	22.0
10	37.8	21.5
5	32.2	21.1
0	26.7	20.7
−5	21.1	20.2
−10	15.6	19.8

although being laid underground and protected from open air. The same type of cable line could be subjected to an opposite situation as to the ambient temperature if it enters tunnels passing under considerable mountain heights.

In order to understand the effect that the external temperature may have on the electric field distribution, let us consider an HVDC 300-kV 1400-mm^2 copper conductor cable buried at 1.3-m depth in a soil of uniform thermal resistivity equal to 1.3 K·m/W and with temperature ranging from 25°C in the hottest summer to −10°C in the coldest winter. Considering an insulation thickness of 21.1 mm, inner and outer semicon thicknesses of 2 and 1 mm, respectively, 1 mm of aluminum sheath, 4.5 mm of thermoplastic PE jacket, that result in an overall cable diameter of 101.5 mm and an ampacity I_C of 1250 A, the conductor temperature θ_C and Joule losses W_C with different soil temperature θ_a can be assessed through [24], as reported in Table 3.7.

The results reported in Table 3.7 lead to the conclusion that the variation of the conductor losses is negligible over a broad range of soil temperature, giving rise to corresponding negligible variations of the electric field distribution in the insulation wall when the cable is loaded at its full load. This is shown in Figure 3.19, which displays the electric field profiles within the insulation evaluated via Eq. (3.29) for the cable under exam. Note that the ordinate axis has been enlarged in order to emphasize the small differences between the various profiles.

The results here reported refer to a generic example. However, even if the parameters change, it is possible to show that the soil temperature has a negligible effect on the electric field distribution of a DC cable at full load.

3.4 IMPULSES SUPERIMPOSED ONTO A DC VOLTAGE

An HVDC cable in operation experiences switching overvoltages and—in the case of DC links with overhead line sections—lightning overvoltages, which are superimposed onto the DC service voltage. These events are considered in the following in the form of impulses of resulting peak voltage U_p superimposed on a DC service voltage U_{DC}; impulses can have both the same and the opposite polarity of DC

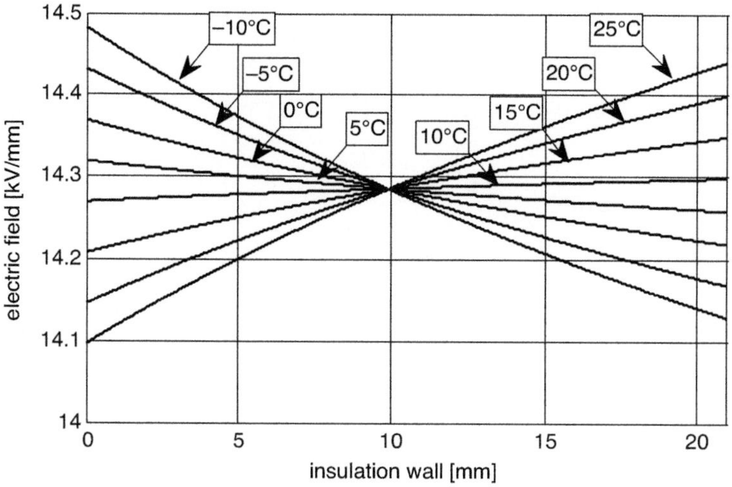

Figure 3.19 Electric field profiles—evaluated via Eq. (3.29)—in the insulation wall of a 300-kV DC cable (1400 mm^2 conductor cross section) carrying 1250 A for different soil temperatures. The ordinate axis is zoomed out in order to emphasize the small differences between the various profiles.

voltage [12,28] (see Fig. 3.20, consistent with [11]). The total electric field at the event of an impulse is expressed in [12] as follows:

$$E_t = E_{\mathrm{DC}} + E_{\mathrm{AC}} \tag{3.37}$$

where

$$E_{\mathrm{AC}} = \frac{U_p - U_{\mathrm{DC}}}{r \ln (r_o / r_i)} \tag{3.38}$$

Note that the capacitive field E_{AC} has to be meant as the "swing" of electric field from the DC to the AC distribution as a consequence of the resulting peak voltage U_p of the

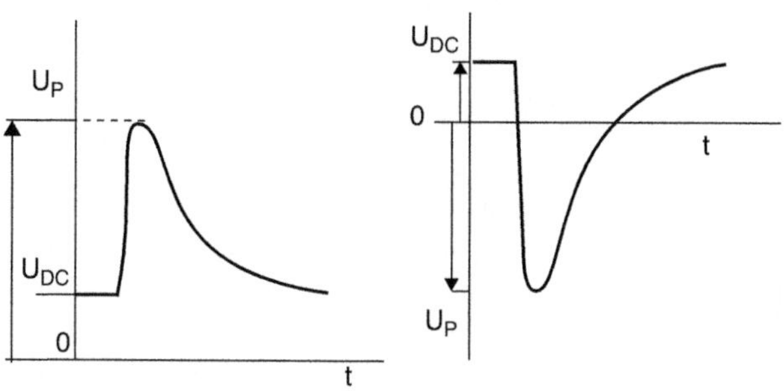

Figure 3.20 Definition of voltages U_{DC} and U_p (after [12]).

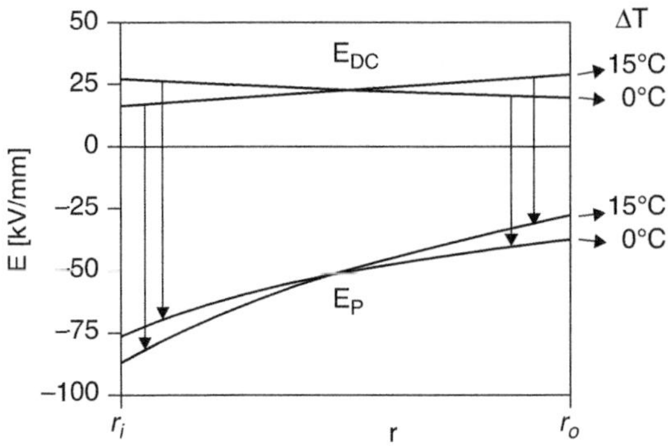

Figure 3.21 Total field distribution E_t in the event of an impulse superimposed on a DC voltage of opposite polarity. $U_{DC} = +450$ kV and $U_p = -1000$ kV. The DC field distribution E_{DC} is also shown, while the arrows highlight the capacitive field E_{AC} at some values of radius r (after [12]).

impulse superimposed onto the existing DC voltage; thus the absolute value of E_{AC} is larger for superimposed impulses of opposite polarity: $|-U_p - U_{\text{DC}}| > |U_p - U_{\text{DC}}|$ [12].

For the sake of illustration, in [12] relationships (3.37) and (3.38) are applied to the standard cable at $U_{\text{DC}} = 450$ kV with a superimposed impulse of opposite polarity such that $U_p = -1000$ kV, both in the case of unloaded cable ($\Delta T = 0$°C) and of loaded cable ($\Delta T = 15$°C). The results, displayed in Figure 3.21, prove that the total field E_t is maximum at the conductor, both when the cable is loaded and when it is not; but in the former case the cable experiences a higher total field stress. When the superimposed impulse has the same polarity, the field intensity at the conductor is always maximum for an unloaded cable, but it might become maximum at the metallic screen for a heavily loaded cable with high DC service voltage (this case is omitted in the figure).

As a further example, the values of the field at the conductor E_C for superimposed impulses of both the same and the opposite polarity of DC voltage, but all with the same absolute peak voltage $U_p = 1000$ kV, are plotted in Figure 3.22 against the DC prestressing voltage U_{DC} for both the unloaded ($\Delta T = 0$°C) and the loaded cable ($\Delta T = 15$°C). The figure shows that the field at the conductor decreases with the DC prestress for impulses of the same polarity, while it increases for impulses of the opposite polarity; the effect is enhanced by higher temperature drops.

By comparing the field values in Figure 3.22, it is observed in [12] that a test with an impulse superimposed on a DC voltage of opposite polarity (arrow *a*) leads to higher fields at the conductor than a test with an impulse of the same polarity (arrow *b*). It can be concluded—in agreement with the results of several studies relevant to both mass-impregnated cables and extruded cables [30,31]—that the breakdown field strength is lower for impulses of opposite polarity than for impulses of the same polarity as the DC prestressing voltage. According to [12], the reason of this behavior is twofold:

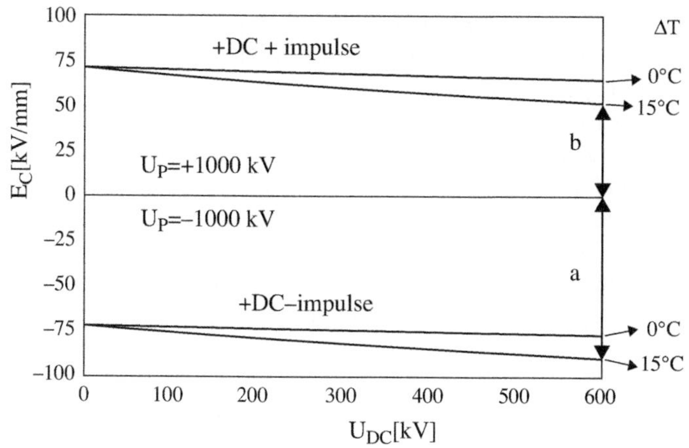

Figure 3.22 Field *at the conductor* in the event of an impulse superimposed on a DC voltage of the same (above) and of opposite (below) polarity (after [12]).

1. The capacitive field during the impulse, given by Eq. (3.38), is larger for impulses of opposite polarity than for impulses of the same polarity as the DC prestressing voltage; indeed, as previously highlighted, the numerator of (3.38) is larger in the former case than in the latter, since U_{DC} is added to U_p in the former case, subtracted from U_p in the latter. This is not compensated in the overall value of E_p by the DC field E_{DC} in (3.37), since for a same voltage the DC field at the conductor is lower than the AC field, both at zero load (see Fig. 3.9*b*) and especially at nonzero load (see Fig. 3.9*a*).
2. Mass-impregnated paper cables and extruded cables may contain space charge. In the case of a mass-impregnated cable, homocharge will be accumulated in the paper layers adjacent to the conductor and lead sheath as shown in [28]. At the event of an superimposed impulse of opposite polarity, these homocharges increase the stress in the butt gaps of a paper cable. Extruded cables also may contain homocharge [30, 31], but even heterocharge at various charge storage centers (see Chapter 4), that can be located both at the interfaces and in the bulk of the insulation. This makes the impulse behavior of an extruded cable much more complex and uncertain.

3.5 STATISTICAL APPROACH TO IMPULSE VOLTAGE TEST LEVELS FOR HVDC CABLES

The currently accepted test procedures for impulse voltage tests on DC cables are based on the superimposition of the DC rated voltage to the impulse wave. Only two reference testing procedures are universally accepted: a first one after [32] for oil–paper cables (i.e., both mass-impregnated and oil-filled cables) and a second one after [11] for extruded cables. They exhibit some minor differences in approaching impulse tests.

According to [32], both switching and lightning impulse tests should be performed by applying a lightning/switching impulse with polarity opposite to the DC superimposed voltage and the test should be repeated twice: first, with the DC superimposed voltage of one polarity and then with the other polarity. In more detail, a negative DC voltage equal to the rated cable voltage shall be applied between the conductor and the sheath, and then 10 positive switching/lightning impulses shall be superposed to the DC voltage. Later on, the same procedure is repeated with a positive DC voltage equal to the rated voltage applied between the conductor and the sheath, and then 10 negative switching/lightning impulses shall be superposed to the DC voltage. As to the peak value of the impulse, for lightning impulse tests reference [32] recommends a value at least 1.15 times the lightning impulse protective level (LIPL) of the surge arrestor connected to the cable and for switching impulse tests a value at least 1.15 times the defined switching surge level at a specific installation.

Coming now to extruded cables, reference [11] recommends two different procedures for switching impulse tests, depending on whether cables are qualified for polarity inversion under their operation or not. In the former case the procedure is the same as reported in [32], in the latter case it is as follows: A positive DC voltage equal to the rated voltage shall be applied between the conductor and the sheath, and then a series of 10 positive switching impulses and a series of 10 negative switching impulses shall be superposed in sequence to the DC voltage. Later on, the same procedure is repeated with a negative DC voltage equal to the rated voltage applied between the conductor and the sheath, namely 10 negative switching impulses followed by 10 positive switching impulses shall be superposed on the DC voltage. For the lightning impulse test, the procedure is the same as in [32]. Considering the peak value of the impulses, reference [11] states that for both lightning and switching tests the peak value shall be 1.15 times the maximum absolute value of the switching/lightning surge voltage that the cable system can experience.

In terms of stress absolute value, it can be argued that an impulse with the same polarity as the superposed DC voltage can give rise to higher electric field stresses in the insulation wall. On the contrary, the impulse with a polarity opposite to the superposed DC voltage can give rise to a very fast inversion that could be very stressing and hence harmful for the insulation. Whichever of the two has the higher probability to occur depends strictly on the system conditions, thus it should be evaluated case by case according to a procedure proposed in [33] and illustrated in what follows of this Section.

The situation of a DC line with a switching or a lightning impulsive overvoltage superimposed to the DC voltage is sketched in Figure 3.23, where a voltage impulse of amplitude U_i (to ground), the preexisting voltage U_{DC}, and the total voltage U_p (algebraic sum $U_i + U_{\mathrm{DC}}$) are reported: U_i has the same and the opposite polarity of U_{DC} in Figures 3.23*a* and 3.23*b*, respectively (symbols are fully consistent with previous Fig. 3.20).

In practice U_p is the overvoltage peak that is actually accessible, either via measurements performed through voltage measurement tools (e.g., voltage dividers) or via "worst-case" system overvoltage calculations that account for the

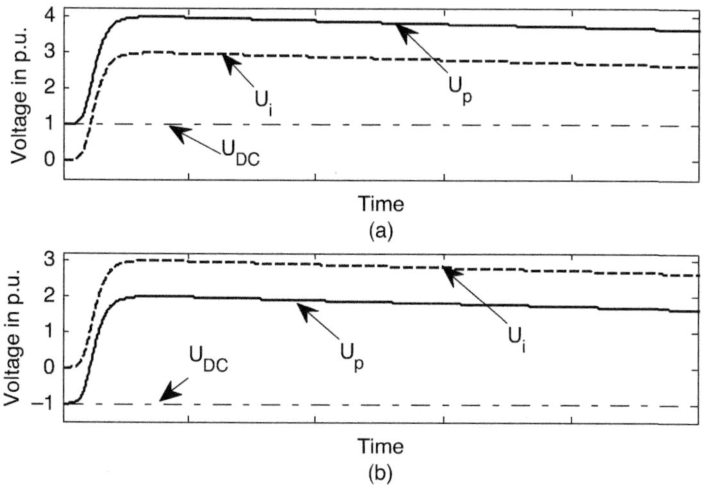

Figure 3.23 Voltage impulse U_i superimposed to DC voltage U_{DC} and their algebraic sum when U_i has (*a*) the same polarity and (*b*) the opposite polarity of U_{DC} (after [33]).

characteristics of the DC transmission link (e.g., presence/absence of overhead line sections, surge impedance of the cable/overhead line and of the busbars, surge protective level of surge arresters and spark gaps, direct lightning and/or back-flashover, and so on [33]). Commonly, U_p is expressed in pul of preexisting DC voltage U_{DC}.

As far as the electric field stress in the insulation wall is concerned, the voltage impulse of amplitude U_i (to ground) gives rise to a capacitive field component, whereas the steady DC voltage U_{DC} gives rise to a resistive field component. Therefore, in order to assess the total electric field E_t that stresses the insulation at the generic radial point r, the value of U_i has to be derived from the (measured or calculated) overvoltage peak U_p accounting for the polarity of U_p with respect to the preexisting U_{DC} voltage according to the scheme of Figure 3.23, namely:

Impulse of peak amplitude U_p has the same polarity as $U_{DC} \Rightarrow U_i = U_p - U_{DC}$.
Impulse of peak amplitude U_p has the opposite polarity to $U_{DC} \Rightarrow U_i = U_p + U_{DC}$.

Thereafter, the total electric field E_t that stresses the insulation at the generic radial point r has to be evaluated considering both the capacitive component of the electric field generated by the impulse voltage U_i and the resistive component generated by the preexisting DC voltage U_{DC} and performing the algebraic sum of these two components, whose sign once more depends on the polarity of U_p with respect to the preexisting U_{DC} voltage. Consequently, the total electric field E_t when the impulse has, respectively, the same ($E_{t,s}$) and the opposite ($E_{t,o}$) polarity of the DC voltage can be written as [34]:

$$E_{t,s} = g_C U_i + g_R U_{DC} = g_C(U_p - U_{DC}) + g_R U_{DC} = g_C U_p + (g_R - g_C)U_{DC} \tag{3.39}$$

$$E_{t,o} = g_C U_i - g_R U_{DC} = g_C(U_p + U_{DC}) - g_R U_{DC} = g_C U_p + (g_C - g_R)U_{DC} \tag{3.40}$$

Note that these relationships are in agreement with the observations already performed in Section 3.4 according to [12]; in particular, in $E_{t,o}$ the capacitive component of the electric field is due to the swing voltage in the insulation wall ($U_p + U_{DC}$, absolute values), consistently with [34]. In both (3.39) and (3.40) the voltage variables will be considered as absolute values; g_C and g_R are the geometric capacitive and resistive field factors, having the following expressions respectively:

$$g_C = \frac{1}{r \ln (r_o/r_i)} \tag{3.41}$$

$$g_R = \frac{\delta (r/r_o)^{\delta - 1}}{r_o[1 - (r_i/r_o)^{\delta}]} \tag{3.42}$$

where r_i and r_o are inner and outer radii of insulation, r is the generic radial coordinate in the insulation wall, and δ is given by Eq. (3.27).

It can be easily shown that both g_C and g_R are monotonic functions of r but, while g_C only decreases with r, g_R can either increase or decrease with r depending on cable load. Indeed, δ increases with cable current and may eventually give rise to the well-known "electric field inversion." In fact such a parameter gives rise to the highest electric field at the inner semiconductive–insulation interface when lower than 1 and to the highest electric field at the insulation–outer semiconductive interface when higher than 1 [33].

It must be pointed out that the inversion of electric field in the insulation wall is strictly associated with δ and consequently with insulation geometry (r_i, r_o), voltage between conductor and screen (U_{DC}), Joule losses (W_C), and insulation compound characteristics (a, b, $\lambda_{T,d}$). It can be demonstrated that the electric field inversion in the insulation wall occurs if $\delta > 1$: Hence, δ can be named as "field inversion coefficient."

Due to the monotonic behavior of both g_C and g_R, the maximum electric field value E_t ($E_{t,s}$ or $E_{t,o}$) that can arise in a DC cable is the highest among the following values [33]:

1. E_t at the inner semicon–insulation interface under no load: g_C and g_R with $r = r_i$ and δ with $W_C = 0$ ($\delta = \delta_{min}$)
2. E_t at the inner semicon–insulation interface under full load: g_C and g_R with $r = r_i$ and δ with $W_C = W_{C,max}$ ($\delta = \delta_{max}$)
3. E_t at the insulation–outer semicon interface under no load: g_C and g_R with $r = r_o$ and δ with $W_C = 0$ ($\delta = \delta_{min}$)

4. E_t at the insulation–outer semicon interface under full load: g_C and g_R with $r = r_o$ and δ with $W_C = W_{C,\max}$ ($\delta = \delta_{\max}$)

Dealing now with a cable installed in the field, these considerations enable the estimation of the maximum value of $E_{t,s}$ and $E_{t,o}$ for a specific value of U_p, which will be estimated through the maximum overvoltage U_h that impinges to the cable entrance:

$$U_p = h_c U_h \tag{3.43}$$

where h_c is the overvoltage reduction factor due to the cable line [35,36]. DC cable lines are usually longer than 50 km, and consequently h_c can be equal to twice the voltage refraction factor γ_r for a surge coming from an overhead line and impinging to the cable line [35,36]. Therefore,

$$U_p = 2\gamma_r U_h = 2\frac{2Z_2}{Z_1 + Z_2} U_h \tag{3.44}$$

where Z_1 and Z_2 are the surge impedances of the overhead line and the cable, respectively. Z_1 and Z_2 can be estimated with the following well-known formulas [37]:

$$Z_1 = 60 \ln\left(\frac{2h_o}{r_h}\right) \tag{3.45}$$

$$Z_2 = \left(\frac{60}{\sqrt{\varepsilon_r}}\right) \ln\left(\frac{r_o}{r_i}\right) \tag{3.46}$$

where h_o is the height to ground of the overhead conductor, r_h is the radius of the overhead conductor, and ε_r the relative permittivity of the cable insulation.

The value of U_h shall be evaluated properly at the cable entrances. For DC links without overhead line sections, U_h is the peak value of the overvoltage at the exit of the converter, that is, on the DC bay before the cable termination. Then Z_1 is the bay surge impedance.

Furthermore, a safety coefficient $k_s > 1$[14] can be considered in order to cover all the uncertainties in the estimation of E_t that can even take into account the effect of aging. Thus the electric field that the cable installed in field will be able to withstand, $E_{w,2}$[15], can be assessed as follows [33]:

$$E_{w,2} = k_s E_t \tag{3.47}$$

Since usually an insulation dielectric strength fits a two-parameter Weibull distribution well [38], the breakdown probability $P_{b,2}$ at the electric field value $E_{w,2}$ can be assessed as [33]:

[14] A typical value of k_s is 1.05.

[15] From now on subscript 2 will refer to the cable installed in the field, while subscript 1 will refer to the cable to be tested in the laboratory.

$$P_{b,2} = 1 - \exp\left[-\left(\frac{E_{w,2}}{\alpha_2}\right)^{\beta_E}\right] \tag{3.48}$$

With α_2 and β_E being the scale parameter and the shape parameter, respectively, of the Weibull distribution of dielectric strength of the installed cable.

Hence, for a given value of β_E (which depends on the insulation material only [38,39]), the value of $E_{w,2}$ associated with a specific value of $P_{b,2}$—for example, 1 or 0.1%—can be estimated as follows:

$$\alpha_2 = \frac{E_{w,2}}{\left[\ln\left(\frac{1}{1-P_{b,2}}\right)\right]^{1/\beta_E}} \tag{3.49}$$

Coming now to the cable tested in the laboratory, the relevant scale parameter of the Weibull distribution of dielectric strength, α_1, can be guessed at using the enlargement law [39–41], which will be treated extensively in Chapter 6; for an enlargement process concerning the cable length only, that is, when the cable tested in the laboratory has the same radial geometry as the cable installed in the field, this law reduces to:

$$\alpha_1 = \alpha_2\left(\frac{l_2}{l_1}\right)^{1/\beta_E} \tag{3.50}$$

where l_1 and l_2 are the lengths of the cable tested in the laboratory and of that installed in the field, respectively. As shown in [39–41], the shape parameter β_E does not change in the enlargement process.[16]

Now, the minimum electric field $E_{w,1}$ that the cable in the laboratory will withstand in order to guarantee the electric field that the cable installed in the field will withstand can be evaluated with a backward procedure. Indeed, by assuming a breakdown probability $P_{b,1}$ in the test performed at a maximum electric field in the cable insulation equal to $E_{w,1}$, the following expression of $E_{w,1}$ can be obtained from the two-parameter Weibull distribution of dielectric strength [33]:

$$E_{w,1} = \alpha_1\left[\ln\left(\frac{1}{1-P_{b,1}}\right)\right]^{1/\beta_E} \tag{3.51}$$

$E_{w,1}$ can be evaluated for both impulsive voltage with the same ($E_{w,1,s}$) and with the opposite ($E_{w,1,o}$) polarity of the superposed DC voltage. From previous relationships (3.39) and (3.40) it holds:

$$E_{w,1,s} = g_C U_{w,s} + (g_R - g_C)U_{\mathrm{DC}} \tag{3.52}$$

[16] It must be highlighted that strictly speaking β might depend on the test voltage waveform and temperature of the insulation [53]. These effects are neglected here for the sake of simplicity.

$$E_{w,1,o} = g_C U_{w,o} + (g_C - g_R) U_{DC} \tag{3.53}$$

where $U_{w,s}$ and $U_{w,o}$ are considered as the maximum absolute peak value of the withstand impulse voltage to be applied during the test performed, respectively, with the same and the opposite polarity of the superimposed DC voltage U_{DC}. Thus [33]:

$$U_{w,s} = \frac{E_{w,1,s} + (g_C - g_R) U_{DC}}{g_C} \tag{3.54}$$

$$U_{w,o} = \frac{E_{w,1,o} - (g_C - g_R) U_{DC}}{g_C} \tag{3.55}$$

It is important to remember that, for both the impulsive test with the same and the opposite polarity of the superposed DC voltage, the test condition will be selected considering the load condition that for the cable installed in the field leads to the highest electric field in the insulation wall, that is, $W_C = 0$ or $W_C = W_{C,max}$.

As an applicative case study of the assessment procedure of impulse voltage levels proposed in [33], a 320-kV DC extruded cable with 2500-mm^2 aluminum conductor is treated, to be operated without the need of fast polarity reversal (i.e., to be used in a VSC HVDC intertie). The link is 200 km long without overhead line sections. The input values are reported in Table 3.8. In this example, the switching

Table 3.8 Selected Input Values for the Example

U_{DC}	kV	±320
r_i	mm	30
r_o	mm	48
a	1/°C	0.084 [3]
b	mm/kV	0.0645 [3]
β_E	—	10
$W_{c,max}$	W/m	12
k_s	—	1.05
$P_{b,1}$	%	1
$P_{b,2}$	%	1
l_1	km	0.1
l_2	km	200
Z_1	Ω	280 (bay)
Z_2	Ω	18.5
U_h	kV	1.15∗1.9∗320
γ_r	—	0.1333
U_p	kV	192
A	m/W	0.0446
B	mm	20.8
δ_{min}	—	0.5361
δ_{max}	—	0.7842

Table 3.9 Output Values of the Example

$E_{t,s}$	kV/mm	31.6
$E_{t,o}$	kV/mm	15.6
$E_{w,2,s}$	kV/mm	33.2
$E_{w,2,o}$	kV/mm	16.4
$\alpha_{2,s}$	kV/mm	52.6
$\alpha_{2,o}$	kV/mm	25.9
$\alpha_{1,s}$	kV/mm	112.4
$\alpha_{1,o}$	kV/mm	55.4
$E_{w,1,s}$	kV/mm	71.0
$E_{w,1,o}$	kV/mm	35.0
$U_{w,s}$	kV	719
$U_{w,o}$	kV	438

surge only is considered for the sake of brevity. According to [11], such cable will be subjected to switching impulse tests using impulses with both the same and the opposite polarity to the rated DC voltage.

The values of $E_{w,2}$ for impulses with the same and the opposite polarity of the DC voltage are 31.6 kV/mm (inner field without load) and 15.6 kV/mm (inner field at full load, $W_C = 12$ W/m). The corresponding capacitive and resistive field factors are $g_{C,s} = 0.0709$ and $g_{R,s} = 0.0624$ for impulses with the same polarity as the DC voltage, and $g_{C,o} = 0.0709$ and $g_{R,o} = 0.0587$ for impulses with the opposite polarity to the DC voltage. In Table 3.9 the outputs of the example are reported. Note that the voltage impulse values $U_{w,s}$ and $U_{w,o}$ are 719 and 438 kV, which can be rounded to 720 and 440 kV, respectively.

It can be argued that the voltage impulse levels derived in this applicative example are in good agreement with those chosen for similar projects ongoing in the world. This supports the soundness of the procedure proposed in [33] and recalled here, which has the advantage of being based on a thorough analytical treatment - that includes the impulse withstand behavior with cable length via the enlargement law and the effect of overvoltage attenuation for surges that travel along the cable - rather than on rules of thumb based on experience.

3.6 MODIFICATION OF THE STRESS DISTRIBUTION BY TRAPPED SPACE CHARGE EFFECTS

The retention of electrical charge within a dielectric is a key issue when considering an insulation material for use in an HVDC cable system. The high resistivity of the materials combined with the temperature and field variations described above, as well as with material microstructural nonuniformities, can result in electric charge becoming trapped within the body of the dielectric, and this is particularly true in the case of dielectrics of the extruded-polymeric type.

Up to now, space charge has been considered as a charge that results as a consequence of Maxwell equations and temperature dependence of electrical conductivity/resistivity. From now on—and particularly in Chapter 4—space charge is the term given to electrical charges that become localized—that is, trapped—within the bulk of a dielectric. These trapped charges can occur in power cables as a result of an electric potential applied across the insulation [3,42–44], and their distribution within the insulation thickness can be more complex and less regular than the distributions derived in Sections 3.1 and 3.2, with both the magnitude and even the sign of charges changing abruptly within narrow layers in the dielectric thickness.

Things are even more complicated and critical in cable accessories (i.e. joints and terminations), due to the additional geometrical and structural complexity, the combination of different insulating and field-grading materials, the relevant presence of multiple interfaces, the high values of electro-thermal and thermo-mechanical stresses; such factors may all interact with each other and with space charges.

A key feature of the space charge present in a dielectric is that it will modify any electrical stress that may be applied across it. At high voltages the trapped space charge interacts with the applied electric field, such that the total electric stress may be increased in one area and decreased in another. The increased stress leads to increased rates of electrical aging and accelerated failure [42–45]. For instance (see Section 3.4), the superposition of this trapped space-charge field with, say, an opposite polarity field, may enhance the deleterious voltage polarity effects already known with oil–paper insulated HVDC cables [12], thereby reducing further the effective thickness of the dielectric and leading to insulation failure at lower than expected voltages [1]. The premature failures of the early extruded HVDC trial cables (see Chapter 1), where standard XLPE compounds developed for AC applications were used at DC, are believed to have been caused by space-charge phenomena. Space charge is not found to be an issue with the normal operation of HVAC cable systems, as the oscillating polarity reversal of the electric field does not allow sufficient time for charge entrapment to occur. Space charge is also the reason why HVAC XLPE cable systems are not recommended to be tested with high DC voltage in after laying tests [3].

As a result, strictly speaking, the HVDC electric stress within an insulation of the extruded-polymeric type cannot be predicted by standard methods such as those illustrated in Sections 3.1 and 3.2, based on Maxwell equations only [12], since these methods consider uniformly delocalized space charges only, a condition that is never faced in practice, and thus there is added complexity in the design of such cable systems. In particular, space-charge measurement becomes fundamental and space-charge measuring techniques of proven sensitivity, accuracy, and spatial resolution are required [3].

The understanding and measurement of space charge has therefore become a key issue in the progression of HVDC extruded dielectric technology. Many studies have been published on the topic of space-charge measurement with the aim to finding an extruded dielectric material suitable for HVDC applications. Particularly over the last two decades, there has been considerable interest in the measurement of the space-charge distributions within dielectrics under both AC and DC stress

Table 3.10 Typical Electrical Stress Modifications Due to Space Charge at Long Times

Material[a]	Stress modification (% of theoretical)
Thermoset *3*	200 – 800
Thermoset *2*	20 – 50
Thermoplastic *5 & 6*	0 – 10
Paper & PPL *1*	25 – 100

[a]The figures in italics refer to the descriptions in the key of Table 3.1.

Source: After [3].

conditions. Studies have shown that space charges can modify the electrical stresses over two time ranges [3,45,46]:

1. Short term: Charges modify the electrical field but decay quickly on removal of the voltage.
2. *Long term: Charges modify the electrical field but decay very slowly on removal of the voltage. These charges can influence the material strength and aging.*

Thus, it is the charges that develop following long-term electrification that have the largest impact upon the electrical stress distribution.

A number of studies have shown that the space-charge buildup under both short and long times are highly dependent upon the materials [43,47] and the process history [48,49]. Table 3.10 indicates typical magnitudes of the stress modification that have been reported as resulting from long-term space-charge buildup.

Space-charge-related problems, measurement techniques, and solutions, with a particular focus on HVDC cables, are treated extensively in Chapter 4.

3.7 DIELECTRICS FOR HVDC EXTRUDED CABLES

On the basis of the considerations made in this chapter—as well as of those made in Chapter 4, devoted to space charges—a dielectric to be chosen for realizing the extruded insulation of HVDC cables should have the following properties [1,50]:

1. Stable insulation resistivity (i.e., insensitive to variations of temperature, electric stress, polarity reversals, and constant during the time of electrification), as far as possible
2. Low thermal resistivity
3. Low space-charge retention properties
4. High DC breakdown strength, particularly under superimposed impulse conditions, and insensitive to temperature rise and polarity reversals

It is not easy to single out a material that can answer all the manufacturing requirements of HVDC extruded insulation cables, and an Electric Power Research Institute (EPRI) report dating back to December 2004 concluded that "so far no single material has been identified as the answer to all the manufacturers HVDC requirements" [1]. At the beginning of the era of extruded HVDC insulation, standard HVAC materials—which were readily available—were considered and tested, all based on polyethylene (PE); they were mainly cross-linked polyethylene (XLPE) and low-density polyethylene (LDPE), and even currently the materials adopted by extruded HVDC cable manufacturers are in fact all based on PE, though subjected to different thermochemical treatments—such as cross-linking—and/or mixed with proper additives [18]. Probably the attention has been focusing on PE-based material since cable manufacturers have a very deep knowledge in the processing of such materials, even if different polymer-based materials for HVDC application are being developed. Therefore, it is worthwhile achieving a deeper insight in the main characteristics of PE and other PE-based polymers considered as base materials for HVDC cables.

Polyethylene is the simplest hydrocarbonic polymer and is a typical example of a synthetic polymer. Indeed, it is obtained from the polymerization of ethylene, as illustrated in Figure 3.24. Polymers such as PE are classified into linear, branched, and cross-linked polymers (see Fig. 3.25).

The linear polymers are constituted of long chains, tied up through weak interactions (van Der Waals bonds), and feature the highest crystallinity degree. In branched polymers there are new chains that develop from intermediate points of the native chain and weak van Der Waals interactions are established since the chains are less packed: the degree of crystallinity is now lower than linear polymers. Cross-linked polymers are three-dimensional structures with primary bonds in all directions. Such materials, referred to as thermosets, are hard, mechanically resistant, nonfusible, and insoluble in all the reagents and feature a vast field of applications; on the other hand, because of these same characteristics, they have to be cast in the definitive form before or contemporarily with the attainment of their final structure.

In the cable industry, due to the thickness of the insulation wall, the cross-linking (also called "vulcanization") is usually realized chemically through peroxides. Such peroxides are activated at elevated temperatures, giving rise to reactions among residual functional groups of different molecules, with the formation of more complex molecules and the definitive hardening of the product. Cross-linking is an irreversible process for the material.

On the contrary, linear and branched polymers can be formed at high temperatures in order to exploit their higher plasticity at such temperatures, without involving further modifications of their molecular structure. The high temperature

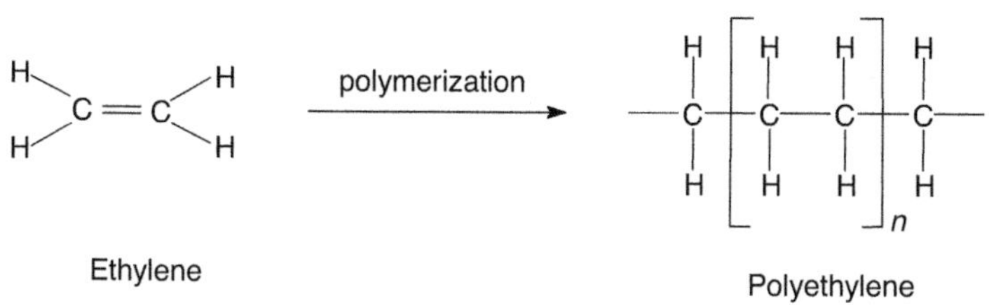

Figure 3.24 Scheme of the polimerization of ethylene, that gives rise to PE.

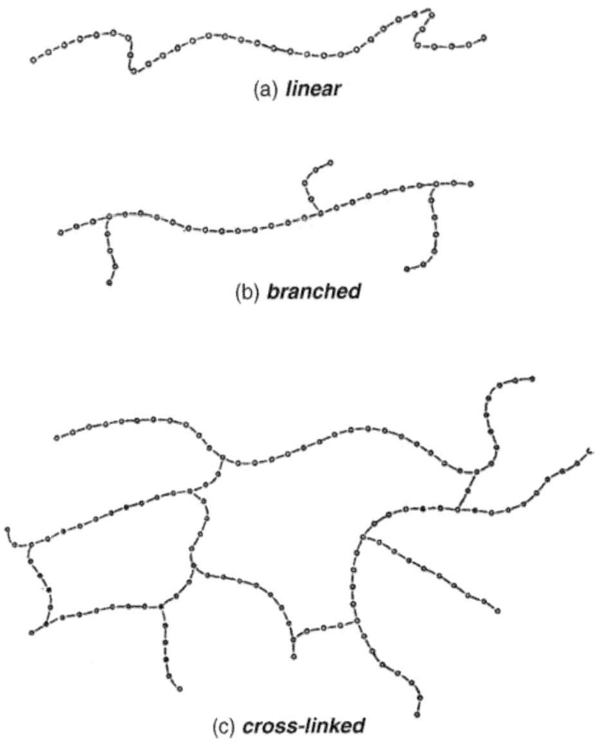

Figure 3.25 Polymers such as PE are classified into (*a*) linear, (*b*) branched and (*c*) cross-linked polymers.

weakens the van Der Waals bonds, so that the material resembles a paste and can be shaped. Therefore, linear and branched polymers are denoted as thermoplastic resins and, contrary to the thermosetting resins, their form is reversible in the sense that they can be softened again by heating and again formed.

Polyethylenes vary in molecular weight and degree of branching, with the highest branching occurring in LDPE. Other PE families are unbranched or high-density PE (HDPE) and short-branched linear low-density PE (LLDPE). The polyethylene most commonly used for HV insulation purposes is XLPE, which is LDPE cross-linked with an organic peroxide, typically dicumyl peroxide [18].

XLPE and EPR (ethylene–propylene rubber), a copolymer of ethylene and propylene, have been popular extruded dielectric materials because they offer good electrical characteristics, ease of processing, and acceptable cost. When properly formulated and compounded, EPR is in some ways superior to XLPE, but it is generally more expensive and definitely more lossy (see Table 3.11) which makes it impractical in EHV applications. XLPE and EPR have been used for almost 40 years, and their performance advantages and disadvantages are well known. Other nonpolar polyolefins and blends of LDPE, LLDPE, and HDPE have been used from time to time. Syndiotactic polypropylene also has excellent insulating properties [18], but it is prohibitively expensive at present.

Table 3.11 Comparison of the Properties of Main Dielectrics Used for Extruded Cable Insulation

Material	PVC	EPR	Impregnated Paper	PE	XLPE
Density (kg/m^3)	1400[a]	1200[a]	1335[c]	923[c]	923[c]
Relative permittivity	8	2.5–2.6[b]	3.6[d]	2.3	2.3
tan δ ($\times 10^{-3}$)	100	2[b]	2	0.4	0.4
Thermal resistivity (K m/W)	6	5	4.75 ÷ 5[c]	3.5[e]	3.5

[a]These values are strongly influenced by the compound receipt.
[b]Values for HV cables taken from [53] for replacing those after [50], which were actually relevant to MV cables.
[c]As to the authors' best knowledge.
[d]After [4].
[e]After [24].

Source: After [50] (unless otherwise specified).

In the production of XLPE, methane, acetophenone, and cumyl alcohol are among the by-products of the decomposition of dicumyl peroxide. Since methane is explosive, the XLPE insulation of cables is vented, that is, heated to 60–70°C in a fan-forced oven, to remove most of the methane (and other by-products) from it before the next stage of manufacture is commenced [51]. The amount of dicumyl peroxide initially added to the base polyethylene is also kept to a minimum, for the same reason [18].

The mechanical properties of XLPE are such that, when used as insulation in a power cable, it can tolerate a conductor temperature of 90°C indefinitely, 105°C for up to 36 h during emergencies (differently from EPR, which thanks to its superior thermal performances can tolerate a conductor temperature of 140°C for up to 36 h during emergencies [52,53]), and 250°C for a few seconds during short circuit. It undergoes a phase change from a ductile semicrystalline solid to a sort of non-crystalline elastomer in the temperature range 90–110°C, and *the associated expansion is an important factor in the design of cable joints and terminations* [18].

The reason for choosing polyethylene as base material is simple: It is known to have unique characteristics that make it an ideal cable insulating material. It is readily available in a sufficiently pure and uniform state, and it has a very high intrinsic dielectric strength (about 800 MV/m). Compared with oil-impregnated paper, it has a lower dielectric loss factor, lower relative permittivity, and lower thermal resistivity. Moreover, PE has satisfactory physical and chemical characteristics with good resistance to a wide variety of chemicals. Table 3.11 provides a comparison of the properties of the main dielectrics used for extruded cable insulation [50].

Polyethylene has been used for manufacturing high-voltage AC cables for about four decades. In France, for example, the first 225-kV low-density polyethylene (LDPE) cable was installed in 1969. Later on, significant improvements were achieved in PE cable manufacturing processes and materials [50], such as the use of *ultra-clean PE*, the *perfectly dry curing technology*, and the establishment of a manufacturing technology for *simultaneous three-layer extrusion* (triple co-extrusion process) of:

1. Semiconductive conductor shield
2. Insulation
3. Semiconductive insulation shield

These developments have led to the latest achievement in using cross-linked polyethylene (XLPE) for extra-high-voltage (EHV) AC cables. Today XLPE is used for manufacturing 500-kV AC cables [54].

Unfortunately, and despite the success of using PE as an insulating material for high-voltage and extra-high-voltage AC cables, the use of PE as an insulating material for HVDC cables was not so successful as for AC cables, and this was addressed to the accumulation of space charges. Space-charge measurement techniques and space-charge-related effects are treated extensively in Chapter 4, as already pointed out above.

Different attempts have been made to use this excellent insulating material for manufacturing HVDC cables over the past 35 years or so and, until a decade ago, many of these attempts were aborted and most of the existing HVDC cables were paper-impregnated or oil-filled cables [50]. However, since the mid 1990's, the development of HVDC cables with polymeric insulation received renewed attention, and, as a result, development projects were launched by cable manufacturers starting from Europe and Japan [50]. Variable results were achieved and different companies adopted different solutions.

Both semiconductive and insulation materials required modification to minimize space-charge effects under HVDC fields. Some companies opted for peroxide cross-linkable polymer systems based on high-pressure polyethylene resins. Others opted for low-density polyethylene compounds. Additives for improving space-charge behavior were also introduced into standard XLPE formulations with a degree of success [1]. Since all these solutions aimed at reducing space-charge retention properties of the used dielectrics, the reasons behind and the effects of these solutions can be better understood after reading Chapter 4, dedicated to space charges, thus more details about such solutions are reported later on in Chapter 5. These details come mostly from the archival papers [50] - dating back to 1997 - and [18] - dating back to 2003. These milestone papers are not "brand-new", but they are still actual as to the polymeric insulation for use in HVDC cables.

In the next Section, some hints at the morphology of polyethylene and its influence on electrical properties, taken mostly from [18], are given.

3.8 MORPHOLOGY OF POLYETHYLENE AND ITS INFLUENCE ON ELECTRICAL PROPERTIES

The improvement of PE characteristics from the viewpoint of HVDC cable insulation performances is strictly related to the morphology of polyethylene and its influence on electrical properties. Polyethylene (typically 15–40% crystalline) consists of spherulites (see Fig. 3.26), best described as a collection of lamella crystallites, surrounded by amorphous regions (see Fig. 3.27).

The spherulite size, which can influence the mechanical and electrical properties of the material significantly, depends on the processing temperatures and the cooling rates. Quenching gives smaller spherulites and lower crystallinity, while annealing gives larger spherulites and higher crystallinity. The heating and cooling of extruded layers of high-voltage cables is very carefully controlled and

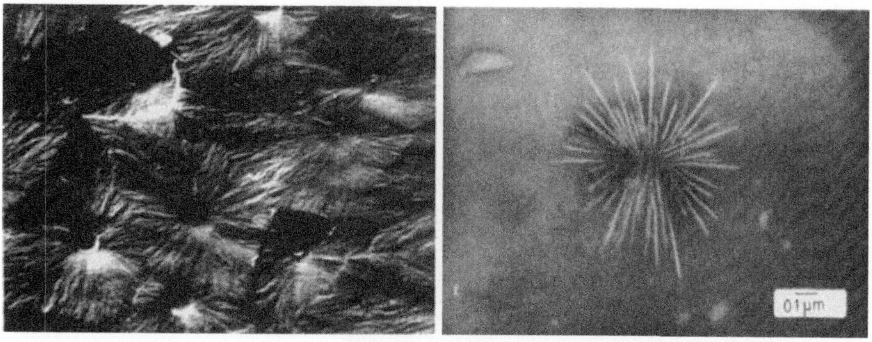

Figure 3.26 Typical spherulitic structures in polymers (after [55,56]).

takes place under high pressure, usually in an inert atmosphere so that the possibility of void formation is minimized. Additional cooling is usually performed, in water, under the same pressure as the previous step [18].

The addition of nucleating agents and the use of slower cooling rates gives higher crystallinity and smaller spherulites. This method is preferable because the end products have smaller void volumes and greater ductility. Nucleating agents are usually added to HDPE and polypropylene but not to LDPE and PE. However, particulate inorganic additives can act as nucleating agents in PE [18].

In Figure 3.28 a high magnification image of ruthenium-oxide-stained XLPE from a DC cable is shown. Although well-defined spherulites are not present, some aggregation of ruthenium stain is resolvable, suggesting poorly associated areas of low crystallinity. The electron diffraction pattern provides evidence for lamella spacing, although this is not obvious in the electron image [18].

Polymer structure and degree of crystallinity can influence dielectric strength and space-charge formation. *It has been noted in many studies that the amorphous*

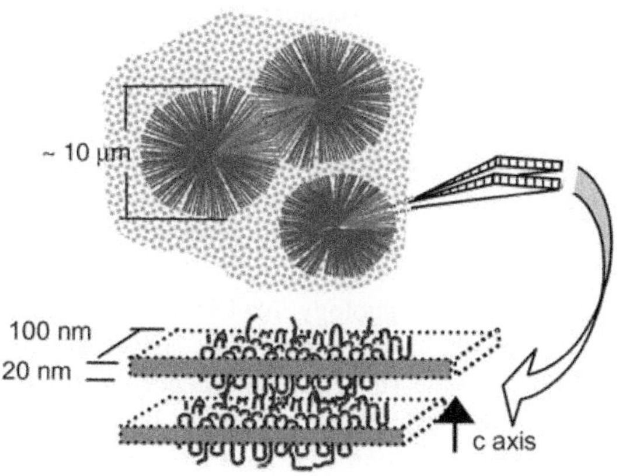

Figure 3.27 Scheme of the spherulitic structure of a crystalline polymer and of its lamellae (after [57]).

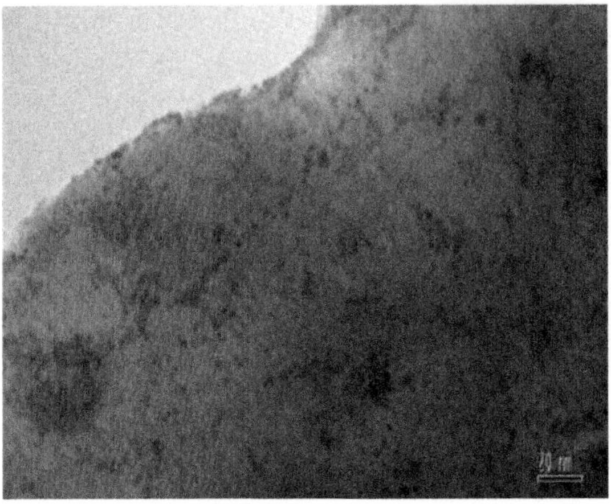

Figure 3.28 Transmission electron micrograph (TEM) of XLPE (after [18]).

interspherulite regions in polyethylene are the preferred trapping locations of charge-carrying species. Water plays an important role in the formation and transport of space charge in polyethylene, as also do defects such as microcavities. The accumulation of space charge at defect centers, for example, carbonyl groups formed by oxidation, can initiate local heating in a small region around the defect center, which in turn can lead to breakdown [18].

The morphology of polyethylene has been extensively studied in the scientific literature and is now well understood. *The amorphous regions play an important role in determining the electrical properties of polyethylene, and it is thought that charge transport occurs mostly within them.* However, the present understanding of the transport mechanisms involved is very incomplete. Impurities, uncrystallizable polymer molecules, additives, and voids are also thought to be concentrated within the amorphous regions [18]. More about charge injection and transport in insulating polymers can be found in Chapter 4.

The introduction of capillary voids into LDPE was found to limit the lifetime of the insulation for DC application. *All observed failure sites were located on the introduced defects, with the damage being caused by partial discharge activity within the voids, observed by microscopy after DC application and breakdown* [58]. Similar partial discharge phenomena from voids and subsequent breakdown are well known in HVAC applications [18].

Uncertainty about the structure and morphology of LDPE and XLPE necessitates further in-depth study, for example, the presence of spherulites, or only lamella as evident in Figure 3.28, and the subsequent influence of this on dielectric behavior. Not only is the initial morphology important, but previous work on LDPE has shown that thermal cycling combined with DC conductivity had the ability to change the morphology and chemistry of LDPE, resulting in a change in conductivity. In an impedance spectroscopy study of XLPE, Caron et al. [59] found that the spherulite structure of thin films (less than 250 μm thick) differed from that of thicker samples, as did the dielectric behavior (below 250 μm only crystalline lamellae were formed). A further study by Li et al. showed that the addition of a nucleating agent to LDPE changed the spherulitic structure and, hence, the electrical behavior [18,60].

The breakdown properties of polyethylene have been correlated with many factors, including morphology, chemical composition, additives, and impurities. In this context, morphology includes degree of crystallinity, size and shape of crystalline regions, and orientation. The DC electrical breakdown properties of polyethylene and copolymer blends are strongly affected by the orientation of the polymer chains in both the amorphous and crystalline regions. A further study of LDPE found that *the size of crystallites in the 020 plane had a strong correlation with impulse breakdown strength* [18,61].

REFERENCES

1. EPRI Report No. 1008720. "DC Cable Systems with Extruded Dielectrics." Palo Alto, California, December 2004.
2. J.C. Fothergill and R.N. Hampton, "Polymer insulated power cable", pp. 477–510, from *Advances in high voltage engineering* (Edited by A. Haddad and D. Warne), IET Power and Energy Series, published by the Institution of Engineering and Technology, London (UK), 2004.
3. R.N. Hampton. "Some of the Considerations for Materials Operating under High-Voltage, Direct-Current Stresses." *IEEE Electr. Insul. Mag.*, Vol. 24, No. 1, pp. 5–13, January–February 2008.
4. R. Bartnikas, Ed. *Power and Communication Cables*. IEEE Press Series on Power Engineering, New York, 2000.
5. J. Tanaka and A. Greenwood. *Advanced Power Cable Technology*, Vol. 1, CRC Press, Boca Raton, FL, 1983.
6. J. Castellon, P. Notingher, S. Agnel, A. Toureille, F. Brame, P. Mirebeau, and J. Matallana. "Electric Field and Space Charge Measurements in Thick Power Cable Insulation." *IEEE Electr. Insul. Mag.*, Vol. 25, No. 3, pp. 30–42, May–June 2009.
7. E. Ildstad and J. Sletbak, Eds. *Notes of the COMETT Course on Ageing and Diagnostic Testing of Polymeric High Voltage Cables*. Norwegian Institute of Technology (NTH), Trondheim, Norway, October 25–29, 1994.
8. Brochure CIGRE 398. "Third-party damage to underground and submarine cables." CIGRE WG B1.21, December 2009.
9. HVDC Light®: Technical Background. Prepared by: Sea Breeze Olympic Converter, LP, 2007.
10. C.A. Arkell and B.G. Gregory."Design of Self-Contained Oil-Filled Cables for UHV D.C. Operation." Paper No. 21-301, CIGRE Session 1984.
11. Brochure CIGRE 496. "Recommendations for Testing DC Extruded Cable Systems for Power Transmission at a Rated Voltage up to 500 kV." CIGRE Working Group B1.32, April 2012, also published on *Electra*, No. 261, pp. 88–95, April 2012.
12. M.J.P. Jeroense and P.H.F. Morshuis. "Electric Fields in HVDC Paper-Insulated Cables." *IEEE Trans. Dielectr. Electr. Insul.*, Vol. 5, No. 2, pp. 225–236, April 1998.

13. M.J.P. Jeroense. "HVDC, the Next Generation of Transmission. Highlights with Focus on Extruded Cable Systems." Proc. 2008 International Symposium on Electrical Insulating Materials (ISEIM 2008), pp. 10–15, Yokkaichi, Japan, Sept. 7–11, 2008.
14. B. Aladenize, R. Coelho, J.C. Assier, H. Janah, and P. Mirebeau."Field Distribution in HVDC Cables: Dependence on Insulating Materials." 5th International Conference on Insulated Power Cables (JiCable'99), Paper C2.13, June 20–24, 1999.
15. S. Boggs, D.H. Damon, J. Hjerrild, J.T. Holboll, and M. Henriksen. "Effect of Insulation Properties on the Field Grading of Solid Dielectric DC Cable." *IEEE Trans. Power Del.*, Vol. 16, No. 4, pp. 456–461, October 2001.
16. E. Ildstad, J. Sletbak, B.R. Nyberg, and J.E. Larsen."Factors Affecting the Choice of Insulation System for Extruded HVDC Power Cables." Paper D1-203, CIGRE Session 2004.
17. C.O. Olsson and M. Jeroense. "Evolution of the Distributions of Electric Field and of Space Charge in an Extruded HVDC Cable." Paper A.2.6, Proc. 8th International Conference on Insulated Power Cables (Jicable'11), Versailles—France, June 19–23, 2011.
18. T.L. Hanley, R.P. Burford, R.J. Fleming, and K.W. Barber. "A General Review of Polymeric Insulation for Use in HVDC Cables." *IEEE Electr. Insul. Mag.*, Vol. 19, No. 1, pp. 14–24, January/February 2003.
19. U. Nilsson, A. Campus, R.N. Hampton, M. Bergkvist, A. Farkas, and M. Jeronse."Material Solutions for Extruded HVDC Cables." Jicable, Paper C.10.1.2, pp. 753–758, 2003.
20. T. Worzyk. *Submarine Power Cables.* Spinger, Berlin, 2009.
21. W. Choo and G. Chen. "Electric Field Determination in DC Polymeric Power Cable in the Presence of Space Charge." Proceedings of 2007 Conference on Electrical Insulation and Dielectric Phenomena (CEIDP 2007) pp. 489–492, October 14–17, 2007.
22. J.O. Boström, A. Campus, U. Nilsson, P. Carstensen, A. Gustafsson, A. Farkas, and K. Johannesson. "Cross-Linked Polyethylene Materials for DC Power Cables." 5th International Conference on Insulated Power Cables (JiCable'99) Paper C2.12, June 20–24, 1999.
23. C.K. Eoll. "Theory of Stress Distribution in Insulation of High Voltage d.c. Cables. Part II," *IEEE Trans. Electr. Insul.*, Vol. EI-10, No. 2, pp. 49–54, June 1975.
24. IEC 60287 (Series). "Electric Cables—Calculation of the Current Rating." 2006.
25. C.K. Eoll. "Theory of Stress Distribution in Insulation of High Voltage d.c. Cables. Part I," *IEEE Trans. Electr. Insul.*, Vol. EI-10, No. 1, pp. 27–35, March 1975.
26. E.H. Kreuger. *Industrial High DC Voltage.* Delft University Press, Delft, The Netherlands, 1995.
27. W. Choo and G. Chen. "Electric Field Determination in DC Polymeric Power Cable in the Presence of Space Charge and Temperature Gradient under dc Conditions." International Conference on Condition Monitoring and Diagnosis, pp. 321–324, April 21–24, 2008.
28. M.J.P. Jeroense. *Charges and Discharges in HVDC Cables—In Particular in Impregnated HVDC Cables.* Delft University of Technology, Delft University Press, Delft, 1997.
29. M. Marelli, A. Orini and E. Zaccone. "Cavi sottomarini in corrente continua: soluzioni per il Mediterraneo." Proceedings of the 2009 AEIT Conference, Catania (Italy), September 27–29, 2009 (in Italian).
30. M. Pays, M. Louis, J. Perret, C. Alquie, and J. Lewiner."Behavior of Extruded HVDC Power Cables." CIGRE 21-07, 1988.

31. Y. Maekawa, A. Yamaguchi, C. Ikeda, and M. Hara."Research and development of DC XLPE cables." Jicable, B9 Cables and Accessories for HVDC Links—Submarine Cables (l), pp. 562–569, 1991.

32. GIGRE Electra 189. "Recommendations for Tests of Power Transmission DC Cables for a Rated Voltage up to 800 kV." WG 21.02, April 2000.

33. M. Marzinotto, G. Mazzanti, and C. Mazzetti. "Lightning and Switching Impulse Levels Selection for Long DC Cable Lines." Proceedings of the 2011 IEEE Conference on Electrical Insulation and Dielectric Phenomena (IEEE–CEIDP 2011), pp. 510–513, Cancun, Mexico, October 16–19, 2011.

34. CIGRE Technical Brochure 86. "Overvoltages on HVDC Cables—Final Report." WG 33/21/14.16, August 1994.

35. M. Marzinotto. "Relationship between Impinging and Stressing Overvoltages Statistical Distributions in Power Cable Lines." IEEE Power Tech 2007, Lausanne, Switzerland, July 1–5, 2007.

36. M. Marzinotto and C. Mazzetti. "Overvoltage Attenuation in Power Cable Lines—A Simplified Estimation Method." *Electri. Power Syst. Res. (EPSR)*, Vol. 80, Issue 5, pp. 506–513, Elsevier Editions, May 2010.

37. A.R. Hileman. *Insulation Coordination in Power Systems*. Marcel Dekker, New York, 1999.

38. W. Hauschild and W. Mosch. *Statistical Techniques for High Voltage Engineering*, IEE Power Series, Peter Peregrinus, London 1992.

39. M. Marzinotto, G. Mazzanti, and C. Mazzetti. "A New Approach to the Statistical Enlargement Law for Comparing the Breakdown Performances of Power Cables—Part 1: Theory." *IEEE Trans. Dielectr. Electr. Insul.*, Vol. 14, No. 5, October 2007.

40. M. Marzinotto and G. Mazzanti. "The Application of the Enlargement Law to HVDC Cable Lines." Proceedings of 2010 IEEE Conference on Electrical Insulation and Dielectric Phenomena (IEEE CEIDP), West Lafayette, Indiana, October 17–20, 2010.

41. M. Marzinotto. "On the Application of the Enlargement Law to Cable Lines." Proceedings of IEEE Power Tech 2005, St. Petersburg, Russia, June 27–30, 2005.

42. CIGRE SC15 Task Force 3. "Space Charge Measurement Techniques: A Review." *Electra*, Vol. 187, pp. 74–89, Dec. 1999.

43. M. Abou Dakka, A. Bulinski, and S. Bamji."Correlation between Space Charge Development and Breakdown in Polymeric Insulation under DC Fields." *IEEE ICSD*, Toulouse, pp. 166–169, 2004.

44. G.C. Montanari and G. Mazzanti. "Recent Developments in Insulation Life Modeling." *Revue Electri. Electron.*, Vol. 3, pp. 66–72, 1998.

45. T. Takada, K. Murata, and Y. Tanaka. "Dynamic Observation of Space Charge Formation in XLPE Using Pulsed Electro Acoustic Method." *IEEE CEIDP* pp. 161–164, 1996.

46. Y. Zhang, J. Lewiner, C. Alquie, and R.N. Hampton. "Space Charges of Different Mobilities: a Novel Data Processing Method." *J. Appl. Phys.*, Vol. 77, No. 10, pp. 5195–5200, 1995.

47. A Mendelsohn and J. Jow. "Novel Polymer Insulation Compounds for HVDC Applications." *Jicable*, Paper C.10.1.1, pp. 747–752, 2003.

48. U. Nilsson, A. Campus, R.N. Hampton, M. Bergkvist, A. Farkas, and M. Jeroense. "Material Solutions for Extruded HVDC Cable." *Jicable*, Paper C.10.1.2, pp. 753–758, 2003.

49. Y. Zhang, J. Lewiner, C. Alquie, and R.N. Hampton. "Evidence of a Strong Correlation between Space Charge Build up and Breakdown in Cable Insulators under DC Stres." *Jicable*, pp. 697–700, 1995.

50. M. Salah Khalil. "International Research and Development Trends and Problems of HVDC Cables with Polymeric Insulation." *IEEE Electr. Insul. Mag.*, Vol. 13, No. 6, pp. 35–47, November–December 1997.
51. T. Andrews, R.N. Hampton, A. Smedberg, D. Wald, V. Waschk, and W. Weissenberg. "The Role of Degassing in XLPE Power Cable Manufacture." *IEEE Electr. Insul. Mag.*, Vol. 22, No. 6, pp. 5–16, 2006.
52. H. Yamada, S. Nakagawa, S. Katakai, K. Kishi, T. Nakanishi, and Y. Murata. "Development of Heat-Resistant XLPE Cable and Accessories." Proceedings of the 7th International Conference on Properties and Applications of Dielectric Materials, 2003.
53. M. Marzinotto. *Insulation Coordination of Extruded Power Cable Lines*. PhD Thesis. Padis Editions, Rome, 2006.
54. T. Kubota, Y. Takahashi, S. Sakuma, M. Watanabe, M. Kanaoka, and H. Yamanouchi. "Development of 500-kV XLPE Cables and Accessories for Long Distance Underground Transmission Line. Part 1. Insulation Design of Cables." *IEEE Trans. Power Del.*, Vol. 9, pp. 1738–1749, 1994.
55. H. Wagner. "Pseudo-spherulite Structures in Cross-Linked Low-Density Polyethylene." *IEEE Trans. Electr. Insul.*, Vol. EI-13 No. 2, pp. 81–86, April 1978.
56. P.J. Phillips. "Morphology—Electrical Property Relations in Polymers." *IEEE Trans. Electr. Insul.*, Vol. EI-13 No. 2, pp. 69–81, April 1978.
57. J.P. Jones, J.P. Llewellyn, and T.J. Lewis. "The Contribution of Field-Induced Morphological Change to the Electrical Aging and Breakdown of Polyethylene." *IEEE Trans. Dielectr. Electr. Insul.*, Vol. 12, No. 5, pp. 951–966, October 2005.
58. G. Chen and A.E. Davies. "The Influence of Defects on the Short-Term Breakdown Characteristics and Long-Term dc Performance of LDPE Insulation." *IEEE Trans. Dielectr. Electr. Insul.*, Vol. 7, pp. 401–407, 2000.
59. L.E. Caron, C.M. Garcia, P.C.N. Scarpa, and A. Ruvolo-Filho. "Impedance Spectroscopy Behavior of Cross-Linked Polyethylene." *J. Macromol. Sci. Phys.*, Vol. B39, pp. 27–38, 2000.
60. X. Li, Y. Cao, Q. Du, Y. Yin, and D. Tu. "Charge Distribution and Crystalline Structure in Polyethylene Nucleated with Sorbitol." *J. Appl. Polym. Sci.*, Vol. 82, pp. 611–619, 2001.
61. D.C. Cho, T. Mori, T. Mizutani, and M. Ishioka. "Effect of Blending on Electrical Conduction and Breakdown of Low-Density Polyethylene Films with Different Densities." *Jpn. J. Appl. Phys.*, Vol. 40, pp. 206–212, 2001.

CHAPTER 4

SPACE CHARGE IN HVDC EXTRUDED INSULATION: STORAGE, EFFECTS, AND MEASUREMENT METHODS

This chapter tackles the key topic of space-charge storage in HVDC extruded insulation, the relevant manifold effects, and the techniques developed by researchers in order to measure the amount and distribution of space charge stored in the insulation, as well as the electric field profile associated with such distribution. Due to the significance of these effects on the performances of HVDC extruded cable systems, this chapter is quite massive and the various sections are divided in turn into several subsections. This requires some extra care in detailing the main contents of the chapter, which are as follows:

- An introduction (Section 4.1) outlining the objectives of the chapter, that is, singling out the effects and problems associated with space charge in HVDC extruded cables, the relevant mechanisms of space-charge storage, and the importance of the space-charge measurement methods developed over the years, with focus on the methods well suited for the insulation of HVDC extruded cables
- A review of the theoretical background of charge injection and transport in insulating polymers (Section 4.2), considering, first, low-field conduction mechanisms—both ohmic and ionic conduction; see Section 4.2.1—and, second, high-field conduction mechanisms—charge injection from electrodes and bulk-controlled high-field conduction mechanisms; see Section 4.2.2
- An illustration of the processes that result in space-charge accumulation (Section 4.3), namely charge generation—via electronic charge injection, field-assisted thermal ionization of impurities, spatially inhomogeneous electric polarization, steady DC current coupled with a spatially varying ratio of permittivity to conductivity; see Section 4.3.1—and charge trapping via physical defects and chemical defects; see Section 4.3.2

Extruded Cables for High-Voltage Direct-Current Transmission: Advances in Research and Development, First Edition. Giovanni Mazzanti and Massimo Marzinotto.

- An extensive, historical, and comparative review of the various space-charge measurement methods for HVDC extruded insulation introduced over the years (Section 4.4), classified into two categories, namely,
  1. Thermal methods, that is, thermal pulse method, laser intensity modulation method (LIMM), and thermal step method (TSM); see Section 4.4.1
  2. Pressure pulse methods, that is, pressure wave propagation (PWP) method, laser-induced pressure pulse (LIPP) method, and pulsed electroacoustic (PEA) method; see Section 4.4.2
- An illustration of the up-to-date developments of the best-suited techniques for measuring space charges in extruded insulation for HVDC cables (Section 4.5), namely the TSM (see Section 4.5.1), the PEA technique (see Section 4.5.2), and the PWP technique (see Section 4.5.3)
- A final synthetic comparison between the best space-charge measurement methods for power cables: PEA versus TSM (Section 4.6). This very last section is intended as a final commentary on the selection of the space-charge measurement method that best fits the intended goals of the involved subjects in light of the broad collection of results from many sources reported in the rest of the chapter

4.1 SPACE CHARGE IN HVDC CABLE INSULATION

The design of HVDC extruded cables is one of the most challenging issues in the cable industry. It is well established that the electric field distribution over the insulation thickness is strongly affected by space charge, which can control the cable system behavior, in particular, its long-term reliability and life expectancy. Indeed, when a DC voltage is applied across a solid extruded dielectric, space charge accumulates in it at a more or less rapid rate, depending primarily on the voltage level and the properties of the insulation compound and of the electrodes. If the space-charge density becomes sufficiently high, the local field strength may exceed the breakdown strength of the dielectric, leading to insulation failure.

Several calculations of the field and space-charge distributions within cable insulation in the steady state have been published (see, e.g., [1–5]) and have been extensively treated in Chapter 3. They assume that volume conductivity/resistivity of the insulation depends on temperature and field. Such approaches are commonly used by engineers to design HVDC cables and serve as an important reference.

However, from an R&D standpoint they are unsatisfactory for two reasons. First, the conduction mechanisms operating in the insulation under service conditions are very complex, involving charge injection from the electrodes—that is, through the semiconductive screens—charge accumulation in the proximity of the electrodes, charge transport, trapping, and detrapping in the insulation bulk. Second, it should be taken into account that there are no homogeneous materials in practice and consequently local conditions may affect the conduction processes strongly. In fact, material dislocations, impurities, and chemical and physical defects alter locally the insulation material, making the analytical treatment of conduction mechanism very difficult and requiring complex numerical simulations that take into account random defects. On the other hand laboratory measurements represent the alternative and usually the only way to assess the behavior of a specific material under different stresses. Thus, measurements on cables being tested are usually performed to assess the various conduction mechanisms that have been proposed. This experimental approach has been widely studied by manufacturers and researchers and is the main topic of this chapter.

As pointed out in CIGRÉ Brochure 219 [6]—which from a historical viewpoint is still a milestone document for testing DC extruded cable systems, although it has been replaced recently by CIGRÉ Brochure 496 [7]—the potential of polyethylene-insulated cables for DC underwater high-voltage applications was recognized and studied in the 1950s following the successful implementation of thousands of kilometers of polyethylene underwater communication cables. Experimental polyethylene DC cables were produced in the 1960s for operation at 200 kV to ground. The commercial application of extruded DC cables was hampered later on by a tendency for the cables to fail during polarity reversal tests. This was generally attributed to the presence of space charges trapped within the cable insulation [6], as discussed in more detail in this chapter. Indeed, one of the inherent characteristics of HVDC cables is the buildup of space charges under the influence of the DC field [8].

The concept of space charges was originally introduced to identify charges other than the surface charges, which are the only ones present on perfect dielectrics or on perfect conductors [8]. Considerable work has been done on space charges in solid dielectrics during the past five decades, and a huge amount of literature exists. Comprehensive theoretical as well as experimental surveys on the space charges in solid dielectrics have been published [9, 10]. Using electromagnetic field theory, it was shown that the buildup of space charge in solid insulators is a consequence of the dependence of the insulation conductivity (resistivity) on temperature and electric field [11, 12] (see Fig. 4.1*a*), which leads to the well-known field inversion or stress inversion phenomenon (already illustrated in Chapter 3) when a load rising from zero to high load is applied to the cable [1–5, 8] (see Fig. 4.1*b*; more details can be found in Chapter 3). More quantitatively, Poisson's equation indicates that a space-charge density of 1 $\mu C/cm^3$ in a planar XLPE sample will cause the field to change by 50 kV/mm over a distance of 1 mm [13].

Investigations of conductivity and space charge in polymers by several authors using different techniques indicate that space-charge formation in such materials is closely connected with conduction, charge transport, and trapping/de-trapping mechanisms in these materials. Space charge is formed due to [8]:

1. *Injected charge carriers* at the electrodes, whether mobile or trapped
2. *Charge already present in the bulk polymer* due to the presence of *ionic dissociable additives, impurities, dislocations, and chemical or physical defects*

The injection of charges from the electrodes involves the transfer of electrons (and holes) through the electrode–polymer interface. This process is highly dependent on the conditions of the interface, including the electrode material, the surface defects, the impurities, and the oxidation level.

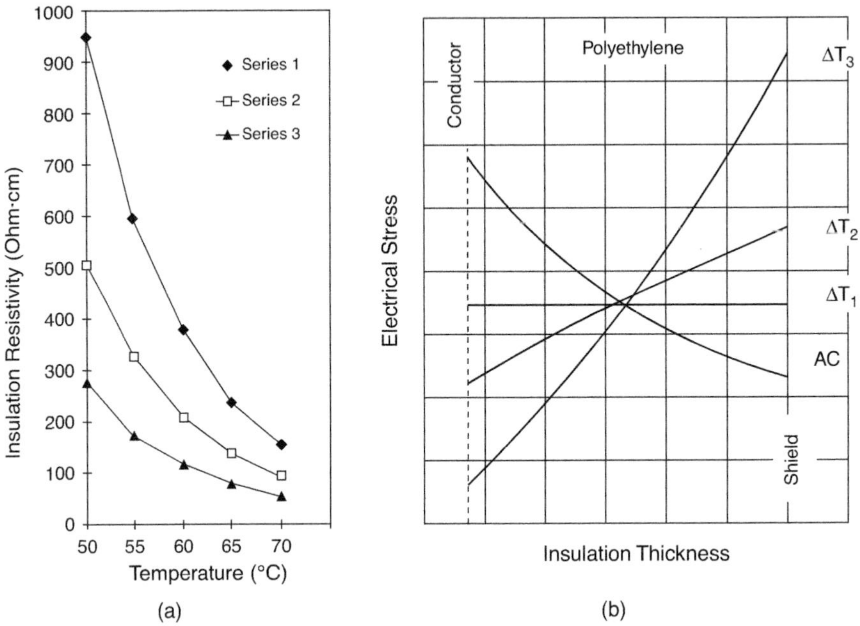

Figure 4.1 (*a*) Dependence of DC insulation resistivity of LDPE on temperature and electric field. Series 1: $E = 17$ kV/mm. Series 2: $E = 22$ kV/mm. Series 3: $E = 28$ kV/mm (after [8]); and (*b*) qualitative sketch of the phenomenon of stress inversion in a polyethylene DC cable insulation ($\Delta T_3 > \Delta T_2 > \Delta T_1$) vs. the AC stress (after [8]).

On the contrary, as mentioned above, the formation of space charge in a polymer is usually associated with ionization of some chemical species. These species can be introduced during manufacturing of the material, such as antioxidants, or cross-linking by-products, or other impurities. For space-charge accumulation to take place, the injected charges and/or the charges already present in the polymer must be trapped in the material; thus the phenomenon depends on the availability and nature of traps. The residence time of a charge carrier in a trap depends on the trap depth, the energy required to extract the carrier, the temperature, and the applied electric field. In general, polymers exhibit both "shallow traps" and "deep traps"—energetically speaking. In polyethylene a depth of traps ranging between 0.1 and 1.4 eV has been found [12].

Many characteristics of the material have an influence on the trapping process, for example, the degree of crystallinity, oxidation, impurities, lattice structure defects (both chemical and physical), and by-products from the cross-linking reactions.

With reference to the charge trapped in the vicinity of the electrodes, two types of space-charge distributions can be observed [8, 12, 13] (see Fig. 4.2): (i) homocharge and (ii) heterocharge.

The *homocharge* has the same polarity of the neighboring electrode (see Fig. 4.2*a*) and is usually due to trapping of injected charges. This condition occurs

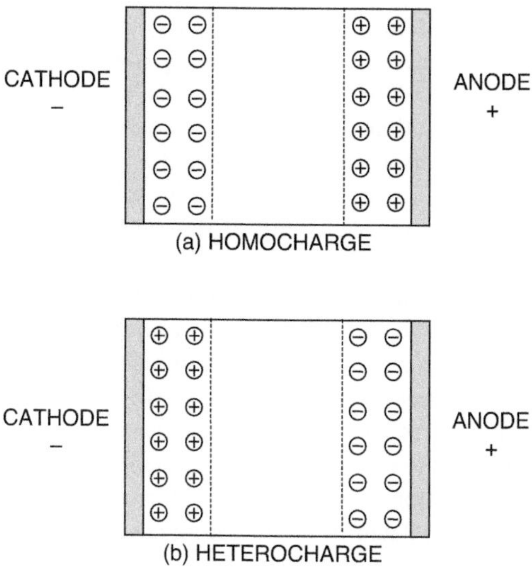

Figure 4.2 Qualitative sketch of (*a*) homocharge and (*b*) heterocharge at the electrodes.

when charge injection prevails over charge transportation. Assuming homocharges are injected from both electrodes, as shown in Fig. 4.3*a*, the electric field at the interfaces of the electrodes becomes lower than the uniform field E_0, whereas the field in the bulk becomes higher [13].

On the contrary, the *heterocharge* has the opposite polarity to the electrode near which it accumulates (Fig. 4.2*b*) and can be generated by ionization of the

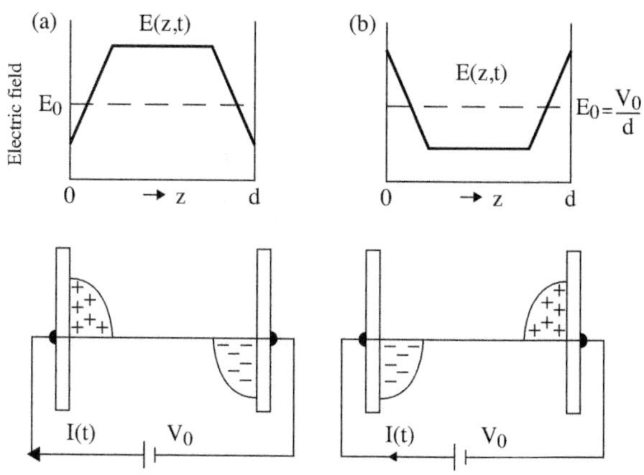

Figure 4.3 Space-charge distortion of the electric field distribution between parallel electrodes with spacing d and voltage V_0 so that the average electric field is $E_0 = V_0/d$. (*a*) Bipolar homocharge injection so that the field is reduced at both electrodes and enhanced in the central region. (*b*) Bipolar heterocharge distribution where the field is enhanced at both electrodes and depressed in the central region (after [13]).

dissociable chemical species present in the polymer. Indeed, when an electric field is applied, the electrons and positive ions generated by dissociation, respectively, migrate toward the electrode of opposite polarity, where they can be trapped. Consequently, in this case charge transportation prevails over charge injection. The effect is an intensification of the field near the electrodes and a consequent reduction in the middle of the insulation thickness, as shown in Fig. 4.3*b*.

Therefore, it can be argued that, from the viewpoint of the endurance and reliability properties of an extruded HVDC cable, a main effect of space charge that accumulates in the material is to deform the field by creating zones of high overstress. If the space-charge density is sufficiently high, the local field strength may exceed the breakdown strength of the dielectric, leading it to failure [14]. Even if in the overstressed zones created locally by trapped space charge the field is not so high as to exceed the breakdown strength of the dielectric, the overstress can, however, ease the degradation reactions by the electrostatic energy associated with the accumulated charge and cause the formation of weak points within the material that become the precursors of aging, as speculated in the literature (see [15] for a comprehensive review).

Dealing with the deformation of the field, it should be noted that often in the case of the formation of homocharge the worst condition is that of a possible voltage polarity reversal. In this case, the homocharge becomes heterocharge and under-stressed points are suddenly subjected to much higher fields. This fact is confirmed by experimental data that show that polarity reversal causes a reduction in the life of the material [16]. On the other hand, for some materials the presence of hetero-charges can give rise to problems during a polarity reversal for the instantaneous transition from heterocharges to homocharges. Indeed, as explained above, the buildup of heterocharges means that charge transportation prevails over charge injection. Consequently, most charge injected is transported through the material to the other electrode, where it tends to accumulate.[1] When a fast polarity reversal occurs, heterocharge accumulated near the electrode is now homocharge, and due to the fact that charge transportation prevails over charge injection/extraction, the charges from each electrode cross the material more or less quickly, giving rise not only to recombination but possibly also to damage of the material structure [17].

Cables in service can be subjected to polarity reversal, also depending on converter type: in HVDC systems with current source converters it is the only way to invert the power flow. As to HVDC systems with voltage source converters, at present the industry can offer VSC converters that can perform the required characteristics of power inversion (switching from rectifier to inverter and vice versa) using current reversal (VSC-CR) only or both voltage and current reversal (VSC-CVR); actually the latter are still in a very infant stage of the development process[2]. It is worthwhile pointing out that in HVDC systems the power reversal function, when provided, can be:

[1] It is supposed here that charge extraction can be assimilated to charge injection.

[2] This topic is extensively treated in the paper M.R. Guarniere, A. Iliceto, M. Marzinotto and R. Niccolai, "HVDC link SA.CO.I.: operational experience and future developments", 8th Mediterranean Conference on Power Generation, Transmission, Distribution and Energy Conversion, MEDPOWER 2012.

1. A "slow polarity reversal," scheduled by the transmission system operator. Typically a slow polarity reversal starts with a first period lasting a few minutes whereby the voltage is ramped down slowly, followed by a second period lasting some tens of seconds whereby the voltage is ramped down at a faster rate till to zero. Thereafter the cable conductor is left grounded for a period ranging from some minutes to some tens of minutes. Afterwards the voltage is ramped up to the opposite polarity with the same procedure.
2. A "fast polarity reversal" when the HVDC system is called to support the grid quickly (due to the occurrence of an unexpected contingency) through an immediate power reversal in the link. Typically it is requested that a fast polarity reversal is completed in less than one second without earthing. Thus a fast polarity reversal is more stressful than a slow polarity reversal.

Moreover, polarity reversals can be associated also with electromagnetic transients on the DC link. These transients are essentially switching and lightning overvoltages, the latter to be considered only for mixed DC transmission interties consisting of both overhead lines and underground cable lines.[3] Such events are explicitly mentioned in CIGRÉ Brochure 86, where only current source converters are mentioned since at that time voltage source converters were not on the market yet [18]. Ad hoc tests relevant to voltage source converters are included in the more recent CIGRÉ Brochure 219 [6] and in the updated CIGRÉ Brochure 496 [7].

In summary, it can be said that *charge injection processes are responsible for homocharge* distributions. *Heterocharge* distributions result *from bulk processes, that is, charge transportation processes or possible charge carrier partial blocking at one electrode of charges that were injected at the other electrode.* The net space-charge distribution will depend on competition between these processes. The density and distribution of space charge in an extruded insulating material may exhibit a complex pattern that is dependent on a multitude of factors, such as experimental conditions (temperature, level of poling field, and its duration), the geometry of the sample, the electrode type and material, the nature and properties of the electrode–polymer interface, and the polymer material itself, its chemical composition, degree of crystallinity, morphology, manufacturing history, and content of impurities and/or additives [8]. However, in materials such as PE, the nonhomogeneous structure and the excellent insulating properties cause space charge to persist for long periods of time at room temperature [12].

In order to get a deeper insight into the processes that lead to charge injection at the electrodes and to charge already present within the polymer, it must be realized that when an insulator is submitted to a HV gradient, several phenomena may occur [19]. Internal space charges move under the influence of the electric field, ionic

[3] Transmission interties consisting solely of underground cables are usually well protected from overvoltages due to direct lightning.

dissociation and recombination can take place, and, depending on the nature of the electrode–insulator interfaces, charges can be transferred at these interfaces [20]. The application of an external electric field to PE causes dielectric absorption, resulting in a polarization that persists even after the removal of the polarizing field. This persistent polarization in a nominally nonpolar semicrystalline film may arise from the injection of charge carriers from the electrodes and their subsequent localization in the polymer as space charges [21]. Commercially available PE also contains polar impurities such as phenanthrene, benzoic acid, and benzophenone, which may be oriented by the external field to give a net dipolar polarization. Concentrations of such impurities have been estimated to be as high as 4 mol% [22]. The presence of space charge at the trapping sites such as crystalline–amorphous interfaces, may produce regions of high electrical stress, leading to breakdown of the PE insulation used in HV applications [19]. This holds particularly under DC regime, due to the long poling time at constant DC voltage, as well as to voltage polarity reversals—when they occur—as pointed out above.

Electrical stress is not the only parameter involved in the space-charge buildup inside the insulating materials. Investigations have shown that cross-linking, melting, softening, and permeation processes in polymers also can result in space-charge buildup. Although some measures are taken to reduce accumulated charge, some charge may remain even after shipment. Thus, when cables with such materials are used, the remaining charge may distort the normal field distribution profile [19].

A historical evidence of space-charge effects even on HVAC extruded cables follows. During an EPRI-sponsored project performed by Detroit Edison, DC testing on extruded aged cables was performed to check their integrity after a period of cable system operation. DC testing resulted in the loss of life for aged XLPE (cross-linked low-density polyethylene) cables [19, 23]. Space charges induced by the DC stress period are believed to have played a role in such deterioration. It has also been reported in other investigations that space charges can play a very important role in dielectric aging through the initiation and growth of electrical trees [24, 25].

After examining preliminary the space-charge effects and problems related to HVDC extruded cables, this chapter illustrates the mechanisms of space-charge storage in HVDC cable insulation and the space-charge measurement methods developed over the years, with focus on the polymeric insulation of HVDC cables. Thus, for a better understanding of how space charge is introduced and accumulated in the thickness of HVDC polymeric insulation, in Sections 4.2 and 4.3 the mechanisms of charge injection, transport, and accumulation in insulating polymers are examined in more detail—with a main focus on extruded insulation for HVDC cables. Thereafter, the various space-charge measurement techniques developed over the years are reviewed and compared extensively in Section 4.4. Later on, Section 4.5 illustrates the up-to-date developments of the best-suited techniques for measuring space charges in extruded insulation for HVDC cables, that is, the TSM, the PEA, and the PWP techniques. A final synthetic comparison between the best space-charge measurement methods for power cables, PEA versus. TSM, in Section 4.6 closes the chapter.

4.2 CHARGE INJECTION AND TRANSPORT IN INSULATING POLYMERS

This section treats the basic processes on which charge injection and transport are based, in particular the effects of low and high electric fields—as well as other factors such as temperature—on both charge transport through the bulk and the way in which charge is injected into insulators from the electrodes. Among the many references available for these subjects, Chapter 9 of the book by Dissado and Fothergill [12] is extensively quoted in the rest of this section: readers interested in a deeper knowledge of these matters should refer to this book [12].

4.2.1 Low-Field Conduction Mechanisms

The models for low-field conduction encompass ohmic conduction and ionic conduction: The former is based on the movement of electrons (or holes) in a crystal lattice (in the same way as in a metal or semiconductor), the latter on the displacement of a vacancy through an ionic lattice.

Ohmic conduction has been proposed for some polymers, ionic conduction for others, while for a few—for instance, PET (polyethylene therephthalate)—there is no universal agreement [12]. If temperature rises progressively, thereby overcoming the glass transition point first and the melting point later, ionic conduction is likely to rise as well, whereas electronic conduction drops.

4.2.1.1 Ohmic Conduction At low electric fields the dependence of voltage on current in insulating polymers can be expressed essentially in the form of Ohm's law. However, it must be emphasized that currents in polymeric insulation are usually very small and dielectric relaxation can give rise to fairly strong transients, thus the ohmic behavior is hard to verify in practice at low electric fields. As is well known (see Section 3.1.3.2), Ohm's law states that the current density J through a material can be expressed as [12]

$$J = \sigma E = \frac{E}{\rho_E} \tag{4.1}$$

where E is the applied electric field, σ is the electrical conductivity, and ρ_E is the electrical resistivity of the material.

"Ohmic" conduction in insulating polymers is usually explained in terms of two models, namely the trap-free and trap-limited ohmic conduction [12]. According to the *trap-free ohmic conduction* model, when no electric field is applied, the thermal energy of electrons in the conduction band of a solid makes them free to move in all directions. As a consequence, these electrons "collide" randomly with thermal vibrations of the lattice, crystal imperfections (e.g., those associated with impurities), and voids. Due to these random collisions, the movement of electrons is isotropic and with no prevailing speed. Therefore, their *net speed* is *zero*, but their average velocity is $\sim 3k_BT/m_{\text{eff}}$, where $k_B = 1.38 \times 10^{-23}$ J/K is Boltzmann constant, m_{eff} is the "effective mass," and m_e is the true or *free electron* mass: taking

$m_{\text{eff}} \sim m_e$ yields $3k_BT/m_e = 10^5\,\text{m}\,\text{s}^{-1}$ at room temperature. Hence, the mean kinetic energy of electrons is a function of the lattice temperature [12].

In the presence of an electric field, the electrons are accelerated between one collision and another in the direction opposite to the electric field by a force $F_e = -q_eE$, $q_e = 1.6 \times 10^{-19}C$ being the electric charge of an electron. Hence, a net movement of electrons takes place whose mean velocity in the direction opposite to the field is referred to as the *drift velocity*, denoted here as v_e. This charge motion gives rise to an electric current due to electrons, whose current density J_e is given by [12]

$$J_e = q_e n_e \mu_e E \tag{4.2}$$

where n_e is the number of electrons per unit volume and μ_e is their mobility, defined as their drift velocity per unit electric field, namely,

$$\mu_e = \frac{v_e}{E} \tag{4.3}$$

As a consequence, from relations (4.1) and (4.2) the conductivity of the insulating polymer due to electrons can be expressed as follows [12]:

$$\sigma_e = \frac{J_e}{E} = \frac{q_e n_e \mu_e E}{E} = e n_e \mu_e \tag{4.4}$$

In general, if N_c charge carrier species are present and the ith ($i = 1, \ldots, N_c$) charge carrier has electric charge q_i, volume density n_i, and mobility μ_i, it holds [12]:

$$\sigma = \sum_{i=1}^{N_c} |q_i n_i \mu_i| \tag{4.5}$$

It can be argued from (4.2) and (4.3) that charge carriers—commonly electrons or holes—gain a mean speed proportional to the field. In fact the trap-free ohmic conduction model always predicts fairly high mobility values (not far from those encountered in heavily doped semiconductors), which are not encountered in insulating polymers: For example, mean mobilities greater than $10^{-10}\,\text{m}^2\,\text{V}^{-1}\,\text{s}^{-1}$ are seldom observed in polymers even at high field, and maximum fields before intrinsic breakdown are usually around $10^9\,\text{V}\,\text{m}^{-1}$. This implies that mean thermal speeds ($\sim 10^5\,\text{m}\,\text{s}^{-1}$ at room temperature, see above) are 5–6 orders of magnitude higher than usual drift velocities, the latter being lower than $\sim 10^{-10}\,\text{m}^2\,\text{V}^{-1}\,\text{s}^{-1} \times 10^9\,\text{V}\,\text{m}^{-1} = 0.1\,\text{m}\,\text{s}^{-1}$. Thus even high applied electric fields cannot alter the distribution of electron speeds, the mean free path, and the mean free time between collisions significantly [12].

This is because while simple "band conduction has been thought to occur in some polymers, in general the situation is much more complicated. Electrons are unlikely to be free to move at all and those that do have sufficient energy to be in the conduction band within polymer molecules are likely to be scattered very quickly and fall back into a trapped state" (after [12], p. 211, reproduced by permission of the Institution of Engineering and Technology).

Moreover, a temperature increase should reduce the electrical conductivity in the framework of trap-free ohmic (also referred to as "true ohmic") conduction. Indeed, the collision rates between electrons and the lattice increase at high temperatures, which raise lattice vibrations (or phonon density), thereby tending to decrease the mean time between collisions and reduce the mean speed of electrons. Thus, according to trap-free ohmic conduction, one should expect that conductivity has a negative temperature coefficient; however—contrary to metals and intrinsic semiconductors—polymers do not exhibit such temperature dependence even at low fields. It can be concluded that a linear J versus E relationship in polymers at low fields should not be addressed to trap-free ohmic conduction [12].

An amendment to the "true ohmic" conduction is thus required for insulating polymers. It is the so-called *trap-limited ohmic conduction*, based on the presence of "traps"—that is, localized states accessible for charge carriers—in the forbidden band gap between the conduction band and the valence band, where carriers may stay for a certain time, thereby diminishing the effective number of carriers available or, equivalently, reducing their effective mobility. Let us consider here, for the sake of simplicity, that the traps have number density N_t, and are located all at the same energy level, G_t, lying by an amount ΔG_t (the so-called "trap depth") below the bottom of the conduction band, G_C, namely $G_C - G_t = \Delta G_t$ (see Figure 4.4).

Resorting to the Fermi–Dirac distribution and to the wide band-gap hypothesis (i.e., $G_C - G_F \gg k_B T$, G_F being the Fermi energy level, as well as $G_F - G_V \gg k_B T$, G_V being the top of the valence band) the number density of conduction band electrons, n_c, and the number density of *occupied* trap states, n_t, may be calculated, respectively, as follows [12]:

$$n_c \approx N_{\text{eff}} \exp\left(-\frac{G_C - G_F}{k_B T}\right) \tag{4.6}$$

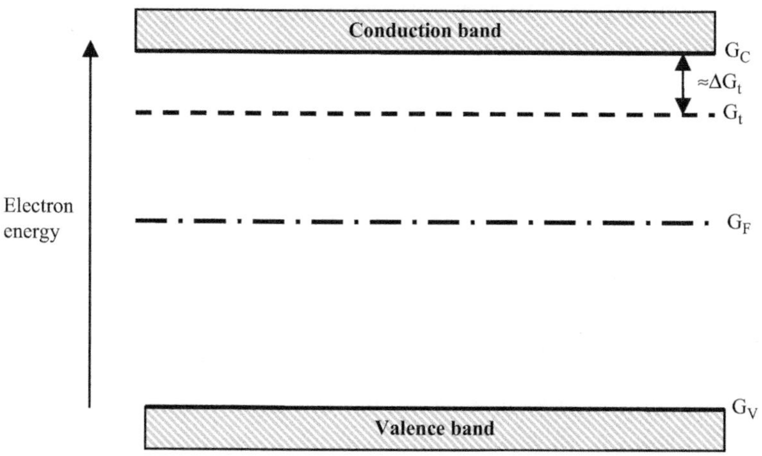

Figure 4.4 Energy-band diagram for dielectric with trap states in the band gap.

$$n_t = N_t\left[1+\exp\left(\frac{G_t - G_F}{k_B T}\right)\right]^{-1} \tag{4.7}$$

where N_{eff} is the density of states available for the conduction band.

For shallow traps—that is, those well above the Fermi level, so that $G_t - G_F \gg k_B T$—the latter relationship reduces to

$$n_t \approx N_t \exp\left(-\frac{G_t - G_F}{k_B T}\right) \tag{4.8}$$

The ratio of Eq. (4.6) to Eq. (4.8) yields the ratio of free charge carriers to trapped charge carriers, $\theta = n_c/n_t$, namely,

$$\theta = \frac{n_C}{n_t} = \frac{N_{\mathrm{eff}}}{N_t}\exp\left(-\frac{G_C - G_t}{k_B T}\right) = \frac{N_{\mathrm{eff}}}{N_t}\exp\left(-\frac{\Delta G_t}{k_B T}\right) \tag{4.9}$$

Therefore,

> The number of carriers available for conduction is reduced from n to $\theta\, n$. Historically this has been known as *"trap-limited mobility"* rather than "trap-limited carrier concentration", since θ *is also approximately the ratio of the time carriers spend in the conduction band to that spent in traps.* Thus, alternatively, in the presence of traps the carriers could be said to take longer to drift the same distance as without traps, so their mobility is apparently reduced while their number density can be said to be constant. Either way, the conductivity $\sigma = en\mu$ is reduced by a factor $\theta \sim 10^{-10}$–10^{-6}. [Thus,] *apparent mobilities* of $\sim 10^{-14}\,\mathrm{m^2\,V^{-1}\,s^{-1}}$ are reasonable in the light of this model. (After [12], p. 212, reproduced by permission of the Institution of Engineering and Technology).

4.2.1.2 Ionic Conduction

An Arrhenius-like relationship frequently rules low-field electric conduction in polymers, as follows:

$$\sigma(T) = \sigma_\infty \exp\left(-\frac{G_a}{k_B T}\right) \tag{4.10}$$

As to σ_∞ and G_a it can be speculated that σ_∞ is a reference value of conductivity at very high temperatures ($T \to \infty$), whereas G_a is the activation energy of conductivity in the Arrhenius-like relationship expressed by Eq. (4.10). Whatever their physical meaning, σ_∞ and G_a can be estimated in practice by ad hoc conductivity measurements. Whenever they are observed to be constant over a proper temperature range, this is usually addressed to an ionic conduction mechanism, although it must be pointed out that the carrier type can hardly be recognized.

For the sake of completeness and comparison with the ohmic conduction, a brief description of the ionic conduction theory follows. It was conceived for the

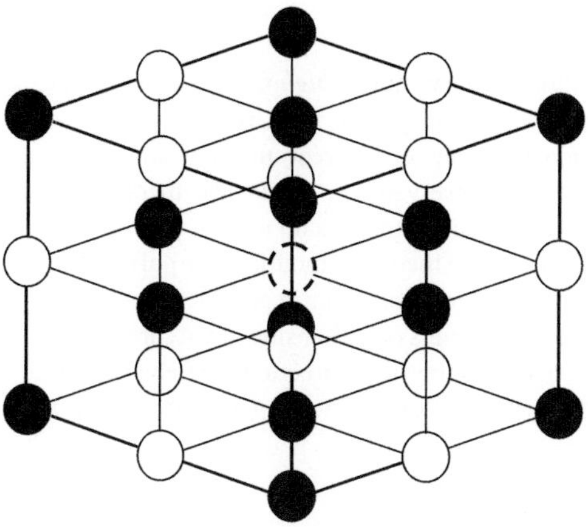

Figure 4.5 Cubic crystal structure containing anions (●) and cations (○) with a vacancy in the central position instead of a cation.

displacement of ions or vacancies through an ionic lattice and it is based on a few main assumptions [12]:

1. The concentration of charge carriers and their mobility depend on temperature;
2. The carriers are thermally activated;
3. The carriers are located in potential wells that limit their mobility.[4]

Let us focus on the cubic crystal structure displayed in Figure 4.5. If the central cation is missing, it can be replaced by anyone of 12 cations provided that its thermal energy can overcome the coulombic forces that keep it in place. The probability dP_{trans} that this "closest-neighbor" transition takes place in time dt can be calculated from classical statistical mechanics, as follows [12]:

$$dP_{\text{trans}} = \nu_0 \exp\left(-\frac{\Delta G}{k_B T}\right) dt \tag{4.11}$$

where

ΔG	=	Gibbs free energy of activation
$\nu_0 \approx \frac{k_B^3 T^3}{d_M h^3 \nu^2}$	=	attempt to escape frequency
$h = 6.62 \times 10^{-34}$ J/s	=	Planck constant
d_M	=	number of directions in which the defect can move
ν	=	vibrational frequency around the defect site in a plane orthogonal to the direction of movement ($\nu \sim 10^{12}$ Hz, thus $\nu_0 \sim \times 10^{13}$ Hz at room temperature)

[4] Under proper conditions, the two latter assumptions are meaningful for polymers, too. Also the dependence on electric field and temperature might be somewhat similar, notwithstanding that the scaling factors might differ [12].

By developing this model:

> at low fields a linear voltage-current relation is found, but *at high fields the behaviour becomes "super Ohmic", i.e. the current increases faster with voltage than that predicted by Ohm's law*. At high fields however a different mechanism usually dominates and this behavior is unlikely to be observed . . . Whilst the theory does not appear to be directly applicable to polymers, since there is no regular polar lattice through which the ions could move, the basis of the theory as modified for polymers is the *hopping of carriers over potential barriers* which can occur in polymers. Under an electric field the relative heights of the potential barriers on either side of the carrier may be changed and hopping becomes more likely in the field direction. (After [12], pp. 207–208, reproduced by permission of the Institution of Engineering and Technology).

4.2.2 High-Field Conduction Mechanisms

The conduction mechanisms that may be active at high fields can be classified as electrode-controlled mechanisms or bulk-controlled mechanisms [12]. The first category comprises the Schottky injection and the Fowler–Nordheim injection, while the second encompasses the so-called space-charge-limited conduction (SCLC), the Hopping conduction, the Poole–Frenkel mechanism, and the field-limited space-charge (FLSC) model [12].

4.2.2.1 Charge Injection From Electrodes A coulombic potential barrier usually limits the transfer of electrons between electrodes and polymeric insulation.[5] *This barrier can be reduced both in height and width by high electric fields* [12].

The *reduction in barrier height increases the probability that there will be electrons with sufficient thermal energy to overcome it*: this is known as *Schottky injection*, a process in which the electric field assists temperature in making electron injection from the electrode to the insulation easier. The other effect of high fields previously mentioned, that is, *the width reduction, makes electron "tunnelling" through the barrier possible*, a quantum-mechanical process called *Fowler–Nordheim injection* that has nothing similar in classical physics: For example, for extremely thin insulating films (say, $< \sim 10\,\text{nm}$) under proper conditions electrons can even pass through the film directly from one electrode to the other [12].

Both Fowler–Nordheim injection and Schottky injection are based on the very simple electron-energy band diagram of Figure 4.6.[6] The figure shows the above-mentioned coulombic potential barrier that electrons must overcome to leave the metal and enter the insulation, as well as the modifications of the barrier by an applied uniform electric field E as a function of the distance of the electron from the interface, x, in four simple stages: (a) total barrier height, (b) shape of barrier

[5] "Ions necessarily build up at metal electrode-insulator interfaces under DC conditions unless some electrolytic reaction occurs. Hole transfer is possible however since electron-hole recombination would then occur at the electrode" (after [12], p. 208, reproduced by permission of the Institution of Engineering and Technology).

[6] The nature of the electrode–insulation interface—in particular its energy band structure—has been strongly simplified here for staying within the scope of this book. For a more detailed description, the reader is referred to [12].

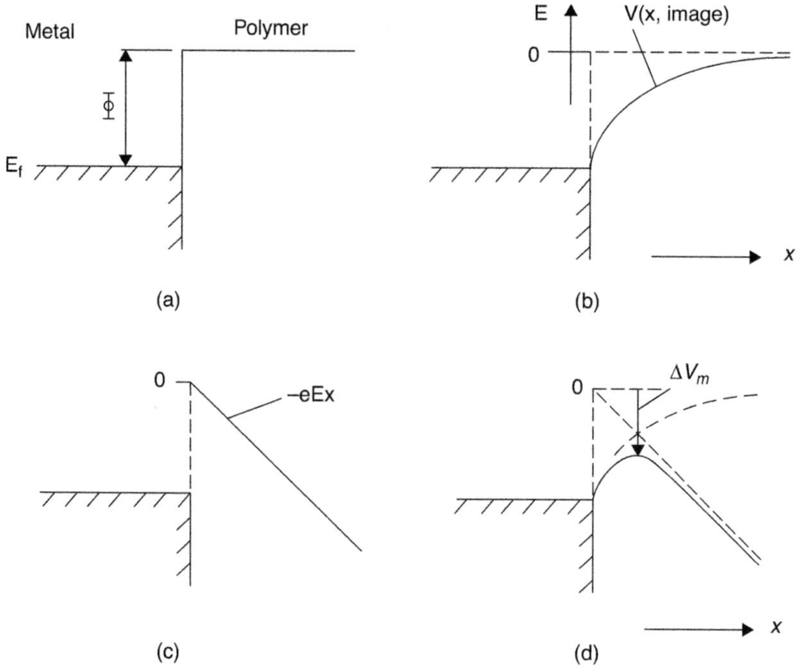

Figure 4.6 Modification of the coulombic potential barrier at a metal–polymer interface by an applied uniform electric field E as a function of the distance of the electron from the interface, x: (*a*) Total barrier height; (*b*) shape of barrier including effect of coulombic image force; (*c*) potential energy due to applied electric field; and (*d*) total barrier shape (after [12]; reproduced by permission of the Institution of Engineering and Technology).

including the effect of coulombic image force, (c) potential energy due to applied electric field, and (d) total barrier shaping that results from the combination of (b) and (c), where ΔV_m represents the overall barrier height reduction due to coulombic image force and applied field [12].

As far as the barrier height, ϕ, is concerned, the relevant theoretical developments began with electron injection from a metal into the vacuum: Then the barrier height coincides with the work function of the metal, ϕ_m. On the other hand, when dealing with the metal–polymer interface the barrier height and shape are modified by the local conditions and depend on the metal–insulation work function difference and local electrical polarization [12]. In an ideal case the electrons would have to be excited into a conduction band in the polymer: Then, the barrier height would be reduced by the electron affinity, GA, of the polymer.[7] However, *in PE the barrier height is increased since polyethylene has a negative electron affinity*, that is, only hot electrons with energy in excess of the vacuum level are delocalized [12].

[7] In a solid, the electron affinity (GA) is the energy released when an electron is placed at the bottom of the conduction band from the vacuum level [12], or conversely required to bring an electron from the vacuum into the conduction band.

4.2.2.2 *Bulk-Controlled High-Field Conduction Mechanisms* Various bulk-controlled conduction mechanisms occur at high fields. Consider the following [12].

Space-Charge-Limited Conduction (SCLC) In a thin insulation with good contacts, although the initial amount of free charge may be quite small, more free charge may be injected under high field, thereby leading possibly to a significant charge accumulation within the insulation if it has a high dielectric constant. Hence, more charge may enter the dielectric than leave it, at least in a first stage. In a subsequent stage, a further displacement of this charge injected from the electrodes through the bulk material might be difficult, thereby leading to the formation of space-charge distributions at the electrodes and within the bulk: such distributions limit further movements of the charge. This phenomenon is referred to as *space-charge limited conduction* (SCLC): The relevant theory is commonly based on an energy band diagram with traps.

The development of the theory is illustrated in two subsequent steps [12]:

1. The *trap-free dielectric*, that is, the ideal insulation with no thermally generated carriers, only negatively charged carriers (i.e., electrons), good injecting electrodes, and no traps, is treated first;
2. The dielectric with traps, that is, the interaction of injected charge with traps and thermally generated carriers, is considered later.

For the *trap-free dielectric*, since—for the reasons given above—the charge density in the dielectric, n_{ch}, is made up of two parts n_o and n_1, where n_o is the intrinsic carrier density and n_1 is an injection term, the current density can be thought of as being made up of two components as well. The details are omitted here for the sake of brevity but, as extensively discussed in [12], these two components of current density can be expressed as follows:

$$J = n_o e\mu \frac{V}{d} \quad \text{(ohmic contribution)} + \frac{9\varepsilon\mu V^2}{8d^3} \quad \text{(SCLC contribution)} \qquad (4.12)$$

where d is the thickness of the dielectric. The second term is referred to as the Mott and Gurney square law, or in some cases as Child's law for solids[8] [12].

Whenever $n_1 \gg n_o$, the space-charge-limited contribution is overwhelming. Let us define as "transition voltage," V_{tr}, the voltage above which the SCLC current overwhelms the ohmic current. By plotting the experimental values of J versus V in a log–log coordinate system, V_{tr} can be evaluated as the voltage at which the slope changes from unity to two (see Figure 4.7) [12].

On the contrary, for the *dielectric with traps* all electrons injected into the insulation contribute to the space charge, but only a fraction $\theta = n_C/n_t$ expressed by Eq. (4.9) does contribute to the current: It is the same fraction as for the effect of traps on ohmic conduction (see Section 4.2.1.1). θ may be very small ($\sim 10^{-6}$–10^{-10}) and has a major effect on the Mott and Gurney square law and the transition voltage. As a

[8] Child's law for the vacuum provides $J \propto V^{3/2}$ [12].

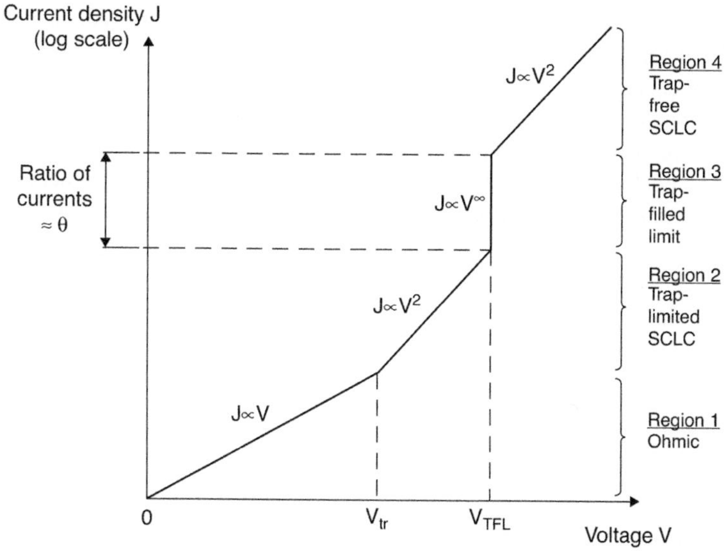

Figure 4.7 Schematic graph showing current density versus voltage for an ideal case of space-charge-limited current (after [12]; reproduced by permission of the Institution of Engineering and Technology).

consequence, J becomes [12]

$$J = \theta \frac{9\varepsilon\mu V^2}{8d^3} \tag{4.13}$$

Equation (4.13) holds for shallow traps, when $G_t - G_F \gg k_B T$ (compare with Fig. 4.4) and Eq. (4.7) can be approximated to (4.8). Anyway, as the voltage is raised to the so-called trap-filled limit, V_{TFL}, the number density of injected charge carriers becomes more or less the same as the number density of traps ($n \to N_t$) and the Fermi energy level overcomes the trap energy level; then, the traps should no more be regarded as shallow but as deep, and Eq. (4.7) reduces to $n_t \sim N_t$, that is, all the traps are filled. As a consequence, when $V = V_{\text{TFL}}$ the value of θ changes abruptly from its small value to unity, and the current density increases very quickly [see Eq. (4.13)]. Under the above assumption of one single energy level for all traps, when $V = V_{\text{TFL}}$, the current increases at an infinitely fast rate with voltage, and once the traps are filled and $\theta = 1$ the Mott and Gurney square law [$J \propto V^2$, see second term of Eq. (4.12)] holds again [12].

In summary, four relationships exist between voltage and current density, identified as regions 1–4 in the log(J) versus log(V) graph in Figure 4.7 [12]:

- *Region 1*, $0 < V < V_{\text{tr}}$, $J \propto V$, slope = 1: ohmic conduction associated with thermally generated carriers
- *Region 2*, $V_{\text{tr}} < V < V_{\text{TFL}}$, $J \propto V^2$, slope = 2: trap-limited SCLC [Eq. (4.13)]
- *Region 3*, $V = V_{\text{TFL}}$, slope = ∞: all traps at energy level G_t get filled
- *Region 4*, $V > V_{\text{TFL}}$, $J \propto V^2$ [Mott and Gurney square law, second term of Eq. (4.12)], slope = 2: trap-free SCLC

In the real world, the energy levels of traps are more or less largely distributed, thus the trap-filled limit is not well defined. By progressively raising the voltage, thereby increasing charge carrier injection, the Fermi level is displaced upward while more and more trap levels are filled. As a consequence the distinction between regions 2 and 3 is fuzzy in practice. Since the current increases rapidly before region 4 is encountered, breakdown usually takes place before the last region is reached. In general, *SCLC holds in thin insulating films and depends strongly on insulation thickness* [12].

Hopping Conduction

At high fields:

> The band structure may be less well defined, for example, states may be scarce around the Fermi level and gradually increase in concentration towards the "mobility edges" . . . In this case there are very few carriers in the higher-energy states which are sufficiently close for carriers to easily move from one to another and there are many carriers trapped in the lower-energy states which are too far apart for carriers to easily move between them. Conduction may however take place by thermally-activated hopping between states over potential barriers. (After [12], p. 208, reproduced by permission of the Institution of Engineering and Technology).

This can be explained as follows [12]:

> In insulators the Fermi energy is less than the mobility edge, G_C, and conduction is either by excitation to the mobility edge (cfr. trap-limited mobility and space-charge limited mechanisms) or by thermally-activated transfer between localized sites often named "hopping". Excitation to the mobility edge is unlikely in insulators with large mobility gaps in which states are likely to be filled to around the Fermi level but mostly empty a few k_BT above it. *Conduction is therefore more likely to be via a thermally-activated local transfer mechanism . . . in which an electron in an occupied state near the Fermi level acquires sufficient energy from the thermal vibrations of the lattice to move to a nearby unoccupied state of similar energy.* (After [12], pp. 232–233, reproduced by permission of the Institution of Engineering and Technology).

The displacement of an electron from one site to another can occur via the following mechanisms:

1. Thermal excitation over a potential barrier
2. Tunnelling through a potential barrier
3. A combination of mechanisms 1 and 2 [12]

Poole–Frenkel Mechanism Two main theories exist for modeling the electronic carrier density enhancement that may occur at high fields [12]:

1. The so-called Fròhlich amorphous-solid model: This theory infers a critical field above which electrons acquire energy at a faster rate than they lose it. Since this model may also be deemed as an electrical breakdown mechanism, it

could be better included in a discussion about electronic avalanche breakdown, which is beyond the scope of this book (for details, see [12]).

2. The so-called Poole–Frenkel mechanism: According to this theory, high electric fields are able to lower potential barriers that lock charge carriers within the dielectric bulk, thereby releasing localized carriers and assisting the thermally activated hopping conduction.

Focusing on the Poole–Frenkel mechanism for the bulk material, it can be argued that it resembles the Schottky effect for the electrode–insulation interface (see Section 4.2.2.1): In fact, according to the former, the field reduces the height of the barriers that localize charge carriers within the insulation; according to the latter, the field reduces the height of the barrier at the electrode–insulation interface. The Poole–Frenkel mechanism postulates that the wide band-gap insulation contains donors or acceptors, which—contrary to n-type or p-type semiconductors for electronic circuits—have commonly less energy available than that needed to be ionized. Resorting to the energy band diagram, the acceptor states lie several k_BT above the valence band, whereas the donor states lie several k_BT below the conduction band: thus, at room temperature generally neither free electrons are created by donating electrons to the conduction band nor free holes are created by accepting electrons from the valence band [12].

A detailed analysis of the Poole–Frenkel mechanism is carried out in [12], by considering for the sake of simplicity the case of an insulation with donors only, although a similar analysis can be performed for an insulation with acceptors or both donors and acceptors. Omitting further details here that can be found in [12], it can be concluded that the diagram of electron potential energy due to a coulombic field at no (or low) electric field near an ionized donor site, illustrated in Figure 4.8*a*, is modified by a high electric field as depicted in Figure 4.8*b*, thereby greatly easing the hopping mechanism. In Figure 4.8*b*, ΔV_m is the variation of the maximum height of

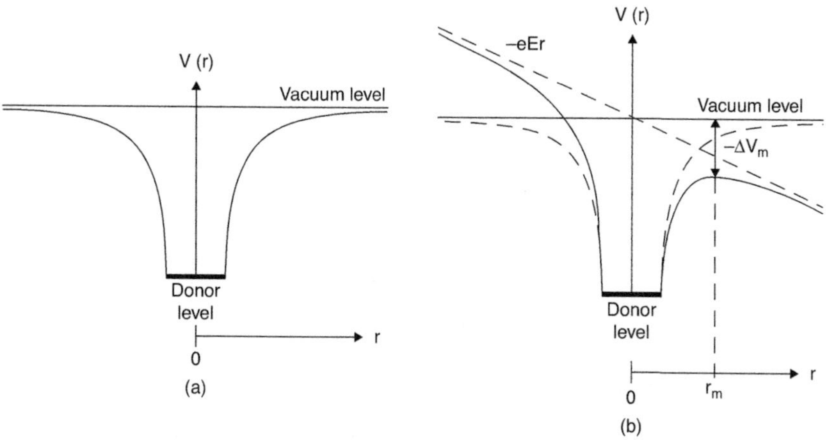

Figure 4.8 Diagrams of electron potential energy due to coulombic and electric field near an ionized donor site, (*a*) no (or low) field and (*b*) high electric field (after [12]; reproduced by permission of the Institution of Engineering and Technology).

the barrier in the field direction, and r_m is the position of the maximum in the potential surface after the application of a (high) electric field [12].

Field-Limiting Space-Charge (FLSC) Model It is well assessed that:

1. Whenever electric charge is injected by an electrode into an insulation, there is a chance that a space-charge cloud is formed adjacent to the electrode so as to lower the electric field at the electrode–insulation interface. For instance, this is the case of a "sharp point electrode–vacuum system" where a high negative voltage is applied to the electrode: The sharp point electrode injects electrons into the vacuum, thereby lowering the field at the point edge and contrasting breakdown occurrence [12].
2. The mobility of charge carriers rises with the electric field, as illustrated above when dealing with the Poole–Frenkel mechanism.

By developing these concepts, the *field-limiting space-charge (FLSC)* theory was suggested by Zeller and Schneider [26] and Schneider [27] for insulating polymers, namely a critical field, E_{mc}, would exist above which electrons gain enough kinetic energy to stay above the mobility limit of the σ-electron conduction band, and as a consequence charge carrier mobility would increase abruptly from usual low values ($<\sim 10^{-10}\,\mathrm{m^2\,V^{-1}\,s^{-1}}$) to those typical of the band transport ($>\sim 10^{-4}\,\mathrm{m^2\,V^{-1}\,s^{-1}}$) [12]. This mobility increase would be so sharp that it could be meaningfully approximated as follows:

$$\mu = 0 \qquad E < E_{\mathrm{mc}} \tag{4.14a}$$

$$\mu = \mu_\infty \qquad E > E_{\mathrm{mc}} \tag{4.14b}$$

where μ_∞ is the high-field limiting value of μ.

In some respects the FLSC theory resembles the abrupt current versus voltage increase at the trap-filled limit of the SCLC theory. In real polymers neither of these effects are observed though because of various sources of inhomogeneities in physical parameters, for example, the existence of amorphous regions and the anisotropies of crystalline regions. However, mobility might be expected to rise sharply at E_{mc}. Moreover, E_{mc} might exceed the breakdown strength itself in some cases. In practice, the treatment becomes rather complicated and goes beyond the scope of this book; thus readers who wish further information should refer to [12].

4.3 SPACE-CHARGE ACCUMULATION

In this section, the ways in which space charge is generated and trapped in polymeric insulation, thereby giving rise to charge accumulation, are illustrated with particular reference to measurements in practical cases. Comprehensive references about this subject are [14] and [28], which are extensively quoted in the rest of this section.

4.3.1 Charge Generation

4.3.1.1 Electronic Charge Injection Electron injection from the cathode (and hole injection from the anode) has been discussed in considerable detail in Section 4.2.2.1. It can be expected to occur at nearly every conductor–solid insulator interface, usually at electric field strengths exceeding 10 kV/mm, and is not readily suppressed. It generates homocharge, that is, charge of the same sign as the adjacent injecting electrode (see Section 4.1), and reduces the strength of the local electric field [14]. On the contrary—as already pointed out in Section 4.1—if charge transport prevails on charge injection, this may lead to distribution of charges having the opposite polarity of the adjacent electrode, so as to increase the field strength locally. In a study on LDPE dating back to 1998 [29] the space charge density was found to be dependent on the temperature. *At lower temperatures the density was considerable* and was attributed to injection of charges from the electrodes. Above 55°C it was much smaller, and *the authors attributed the decrease to increased mobility of the charges within the polymer matrix* [14, 29].

4.3.1.2 Field-Assisted Thermal Ionization of Impurities Field-assisted thermal ionization of impurity molecules would generate positive ions and electrons. Most of the latter would recombine within a very short time. The electrons that did not recombine would be extracted at the anode or deeply trapped in transit; the transport of these electrons would generate a decreasing current, which would eventually disappear, assuming no electron injection from the electrodes. The remaining ions, which would be expected to have very low mobility, would form a stable net positive space charge distributed fairly evenly through the sample volume [14].

If the field were initially applied with the opposite polarity, the same positive space-charge profile would be expected, assuming identical electrodes and a homogeneous sample. However, when significant space-charge density has been observed well away from the electrodes, it has usually shown a good approximation to mirror image behavior, that is, when the polarity of the applied voltage was reversed and equilibrium reestablished, the sign of the space charge reversed but its density was largely unchanged [14]. An example of the mirror image effect, observed in a length of power cable with XLPE insulation, appears in Figure 4.9 [30].

Thus, while ionization of impurities in the bulk is to be expected in most polymeric insulators, it is unlikely to be the only or the dominant space-charge generating mechanism. In this context it should be noted that water vapor, even at concentrations as low as 50 ppm, promotes the ionization of impurities and of cross-linking by-products in polyethylene [31]. Exclusion of water vapor is, therefore, very important in cable manufacturing. In the 1980s, steam pressure was used to cross-link XLPE, but today radiant heaters and an inert atmosphere have replaced steam, even in medium-voltage cable manufacture [14].

4.3.1.3 Spatially Inhomogeneous Electric Polarization Spatially inhomogeneous electric polarization gives rise to a bound (or apparent) volume space charge with density $-\nabla \cdot \mathbf{P}$, and a bound surface charge $\mathbf{P} \cdot \mathbf{a}_n$, where $\mathbf{P}$ is the local polarization and $\mathbf{a}_n$ is an outwardly directed unit vector normal to the surface [14].

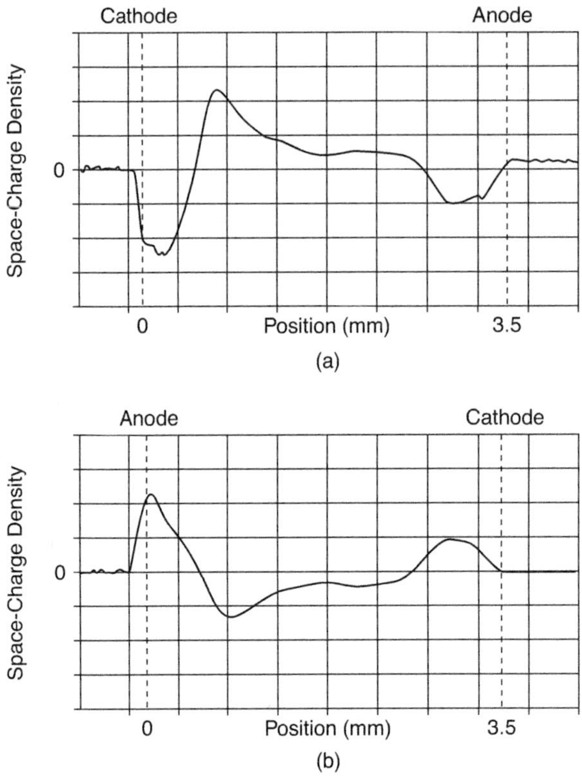

Figure 4.9 Radial space-charge profiles in the XLPE insulation of a length of cylindrical power cable, measured with grounded electrodes using the LIPP method (see Section 4.4.2.2). The insulation thickness was 3.5 mm. The sample had previously been held at 82°C for 2 h at atmospheric pressure (*a*) after +54 kV had been applied for 24 h and (*b*) after –54 kV had been applied for 80 h. Vertical scale 0.14 C/cm^3 (after [14]; see also [30]).

Such a polarization could be due to partial alignment of permanent dipole molecules distributed inhomogeneously through the volume of the insulation. Permanent dipole molecules with large dipole moments, for example, acetophenone and cumyl alcohol (approximately 2.9 and 1.7 Debye, respectively) are formed in XLPE by the cross-linking reactions and diffuse toward the insulation surfaces [32]. The bound charge in the volume is exactly compensated by the surface charge; however, it gives rise to a spatially varying electric field, which is formally equivalent to space charge unless the divergence of the field is zero. This mechanism is obviously consistent with the mirror image effect since the polarization vector will reverse direction but maintain the same magnitude when the applied field is reversed [14].

Figure 4.10 shows several space-charge profiles for 2-mm-thick planar XLPE specimens with semiconducting electrodes [32]. The pulsed-electroacoustic (PEA) technique (see Section 4.4.2.3) was used, and the measurements were made with the electrodes grounded. Figure 4.10*a* shows that, as the applied voltage increases, the positive peak close to the cathode increases in amplitude and eventually saturates.

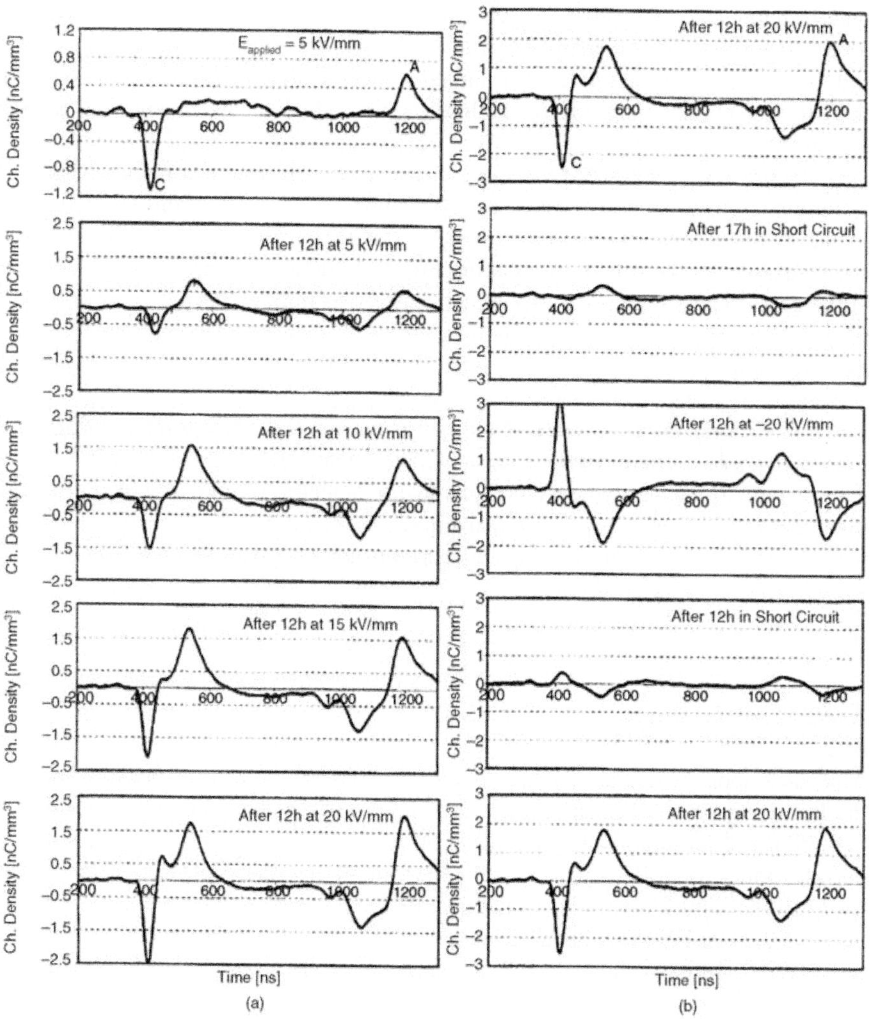

Figure 4.10 Space-charge profiles for a 2-mm-thick XLPE sample with semiconductor electrodes. (*a*) The top profile shows the charge induced on the cathode (C) and anode (A) by the applied field. The other profiles were obtained with the electrodes grounded after the field had been applied as specified. (*b*) All profiles were obtained with the electrodes grounded. The decay of the space charge in short circuits, and the inversion of the profile on reversing the applied voltage polarity, can clearly be seen (after [14]; see also [32]).

It was suggested that this peak is due to partial alignment of permanent dipole molecules in the applied field and that these molecules are concentrated near the electrodes during fabrication of the samples and/or diffuse outward from the center on to the sample surfaces. There is a corresponding negative peak adjacent to the anode, but it is attenuated and broadened relative to the cathode peak because the pressure pulses are broadened in transit. Note also the point of inflexion on the ascending side of the cathode peak, which becomes a minimum at higher fields. This feature was attributed to electron injection from the cathode and does not saturate at higher applied fields. Presumably, the electronic space charge associated with the injection lowers the field close to the cathode, thus causing the dipole alignment to saturate [14].

Figure 4.10*b* shows that the space-charge profile gradually collapses when the electrodes are grounded [32]. This is consistent with injected electrons trapped close to the cathode returning to the cathode under their self-field and with thermal randomization of the dipole orientation. Figure 4.10*b* also shows the mirror image effect very convincingly. The dipole concentrations implied by the amplitude and width of the associated space-charge peaks are shown to be consistent with the expected additive and impurity levels in XLPE [14].

4.3.1.4 *Steady Direct Current Coupled with a Spatially Varying Ratio of Permittivity to Conductivity*

As shown in Chapter 3 (and in particular in Section 3.1.3.2), electromagnetic theory predicts that a dielectric sample in which a steady direct current of density J is flowing, and in which the ratio permittivity/conductivity varies with position, acquires a space-charge density ρ given by [see Eq. (3.8)]:

$$\rho = \mathbf{J} \cdot \nabla\left(\frac{\varepsilon_0 \varepsilon_r}{\sigma}\right) \tag{4.15}$$

where ε_0 is the permittivity of free space, ε_r is the relative permittivity of the dielectric, and σ is its conductivity [14, 33]. This space-charge component is not dependent upon the operation of any other space-charge-generating mechanism. The marked dependence of σ on temperature and electric field in polymeric insulation [34, 35] was discussed in Chapter 3, Section 3.1.3.2, too, as well as the much smaller corresponding variation in ε_r [33]. Relationship (4.15) with typical parameter values for a 250-kV 500-MW XLPE insulated cable yields space-charge densities of the order of 10 nC/cm^3 [36], which are measurable using state-of-the-art equipment (see Sections 4.4, 4.5); Coelho et al. [37] arrive at a figure 10–15 times larger, based on a slightly different choice of cable parameters.

However, it is the change in electric field profile associated with this space charge, not the space charge itself, that is of interest to cable manufacturers. As to this field, it was pointed out in Chapter 3, Section 3.1.3.1, that the electric field profile $E_{AC}(r)$ within the insulation of an HVAC cable is completely determined by the inner and outer radii of the insulation (r_i and r_o) and the applied voltage U_0, assuming the relative permittivity ε_r of the insulation is constant [see Eq. (3.1)]; as load current flows through the conductor, heating it and creating a temperature gradient through the insulation, the temperature gradient will cause little change in $E_{AC}(r)$ because, although the conductivity $\sigma(T,E)$ of the insulation is strongly dependent on temperature and electric field, the associated leakage current is always far smaller than the capacitive charging current at normal power frequencies (50–60 Hz) [14]. On the contrary, as pointed out in Chapter 3, Section 3.1.3.2, in an HVDC cable, even if the insulation is isothermal, the field dependence of its conductivity will alter the electric field profile and as the load current flows through the conductor, the

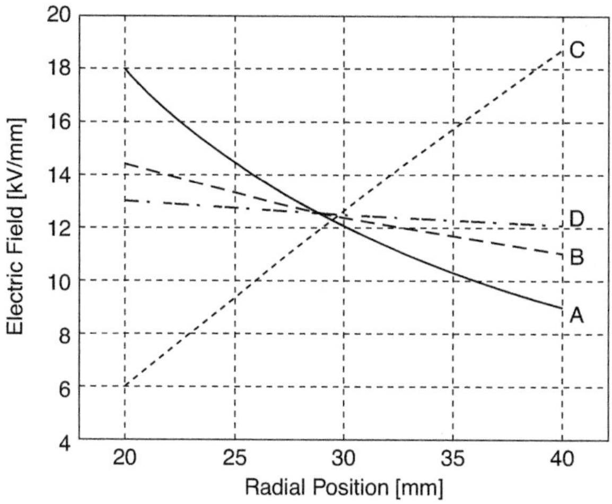

Figure 4.11 Electric field profiles within the insulation of a power cable with design voltage $U_0 = 250$ kV, $r_i = 20$ mm, $r_o = 40$ mm. Curve *A*: HVAC cable; curve *B*: HVDC cable with isothermal insulation, $c = 2 \times 10^{-7}$ m/V; curve C: HVDC cable with conductor temperature of 80°C, insulation temperature at r_o of 40°C, $G_a = 0.8$ eV, $c = 2 \times 10^{-7}$ m/V; curve *D*: HVDC cable with conductor temperature of 80°C, insulation temperature at r_o of 40°C, $G_a = 0.1$ eV, $c = 3 \times 10^{-7}$ m/V (after [14]).

temperature gradient through the insulation will alter much more of the field, possibly causing stress inversion, if the load current is high enough.

Examples of these situations have been given extensively in Chapter 3. As a further example that is useful for measuring the associated space charge and other features, let us consider Figure 4.11. Curve *A* of Figure 4.11 shows $E_{AC}(r)$ for a 250-kV HVAC cable with $r_i = 20$ mm and $r_o = 40$ mm. Let us assume, according to [35], the following dependence of σ on *E* and *T* [see also Eq. (3.12) in Chapter 3]:

$$\sigma(T, E) = \frac{\sigma_0 \exp(-G_a/k_B T)\sinh(cE)}{E} \tag{4.16}$$

where σ_0 is a constant, G_a is the thermal activation energy governing the temperature dependence of the conductivity, k_B is Boltzmann's constant, *T* is the absolute temperature, and the field dependence of the conductivity is given by $\sinh(cE)/E$, where *c* is a constant. Now, curve *B* of Figure 4.11 shows $E_{DC}(r)$ for an isothermal 250-kV HVDC cable with the same dimensions as for curve *A*, and $c = 2\times10^{-7}$ m/V. Curve *C* of Figure 4.11 shows *E*(*r*) for a 250-kV HVDC cable where the conductor temperature is 80°C, the temperature at the outer radius of the insulation is 40°C, $G_a = 0.8$ eV, and $c = 2\times10^{-7}$ m/V. In this case, the field *increases* monotonically from the inner to the outer radius: It is the above-mentioned field inversion [1–5, 11, 12].

It may also happen that, in some insulating materials under proper load conditions, the opposing influences of the temperature and field dependence of the conductivity gives rise to an almost constant field across the insulation, as once more shown in Sections 3.1.3.2 and 3.3; now, curve *D* of Figure 4.11 shows this situation for a 250-kV HVDC cable with $G_a = 0.1$ eV, $c = 3 \times 10^{-7}$ m/V, and the same temperatures as for curve *C*. In general, the smaller G_a and the larger *c*, the smaller the variation in *E*(*r*) with *r* [14].

The dependence of quasi-steady-state current, measured in the external circuit as a function of sample temperature and mean field (= applied voltage/sample thickness) suggests that, in many polymers, $G_a = 0.6-1.2$ eV and $c = 1-3 \times 10^{-7}$ m/V [35]. Note, however, that these values do not necessarily apply within the insulation itself, e.g., the externally measured activation energy G_a may be dominated by the energy required to inject electrons into the insulation from the conductor or semicon screen, the hopping transport of electrons through the insulation requiring much less energy. It is suggested below that this is the case in LDPE [14].

Bambery, Fleming, and Holbøll [38] measured steady-state space charge and current densities in 1.5-mm-thick plaques of high-purity LDPE at applied field strengths up to 15 kV/mm, with and without temperature differences up to 20°C across the thickness. The component of the total space charge due exclusively to the temperature gradient had density of the order of 10 nC/cm^3. However, setting $G_a = 1.1$ eV in accordance with their steady-state current density measurements, they calculated a value of 200 nC/cm^3 from Eq. (4.15), 20 times larger than their experimental value. (Although their calculation did not include the field dependence of the conductivity, it is unlikely that such inclusion would reduce the discrepancy to the experimental uncertainty level.) They suggested that the discrepancy arose because the activation energy G_a, derived from the temperature dependence of the current flowing in the external circuit, was much larger than the activation energy G_{hop} associated with the hopping transport of electrons through the sample volume. More specifically, G_a would be expected to consist of G_{hop} plus a component originating in the injection/extraction of electrons at the cathode/anode, respectively. Rewriting Eq. (4.16) in the form [14]

$$\frac{1}{\sigma(x)} \cdot \left[\frac{d\sigma(x)}{dT(x)}\right] = \frac{G_{\text{hop}}}{kT_{\text{av}}^2} \tag{4.17}$$

where $\sigma(x)$ is the "internal" conductivity at coordinate x within insulation sample thickness, and T_{av} is the average temperature of the sample containing the temperature gradient, they deduced $G_{\text{hop}} \approx 0.06$ eV [38]. This figure is in fairly good agreement with an investigation by Meunier and Quirke [39], who calculated an electron trap energy distribution for the amorphous volume of PE based on the electron affinities of an all-trans tridecane molecule n-$C_{13}H_{28}$ and the associated anion, thereby finding a trap depth range of 0–0.3 eV, with maximum concentration around 0.15 eV. They comment that, if they had taken the interaction with neighboring molecules into account, the trap depths would have been reduced by approximately 40%, that is, the most likely trap depth would be around 0.09 eV.

It would therefore appear that the hopping transport of electrons through LDPE requires much less thermal energy than would be expected from the activation energies deduced from steady-state current measurements. If this were also true for

XLPE, the density of space charge accumulating in XLPE cable insulation due to thermal gradients generated by Joule heating of the conductor would be much less than expected up to date. This would make designing an HVDC cable a much simpler task [38].

It should be noted that estimates of electron trap energies associated with impurities and structural defects are around 1 eV [14], with chemical defects such as hydroxyl or ketone functions or double bonds acting as deeper traps for electrons and holes up to 1.5 eV [40]. The residence time of carriers into such traps at room temperature is of the order of ≈500 s for a 1 eV trap, of the order of 10^{10} s for a 1.5 eV trap. This suggests that long-lived space charge in polymeric insulators is trapped at such sites [40], [41]. *It seems now that the space charge associated with the temperature gradient moves between the much shallower intrinsic traps.*

Therefore, nondestructive measurement of space-charge profiles are essential for evaluating the overall distribution of space charge within the thickness of extruded insulation for HVDC cables, that is, both the charge trapped in shallow traps (mainly associated with the intrinsic disorder in the supra-molecular organization of the polymer) and the charge trapped in deep traps (mainly associated with the strongly non-uniform molecular structure of the polymer).

4.3.2 Charge Trapping

Even if one can affirm that "an energy band structure does not apply to an insulating polymer" or that "the transfer of concepts derived from semiconductor physics is useless in understanding insulating polymer properties" ([28], page 859). Section 4.2 has shown that the energy band diagram is a good mean to illustrate how the concepts of quantum physics and quantum chemistry can be useful in giving a background for describing the insulating properties of polymers, even if they do not provide a complete picture still due to the inherent complexity of this kind of materials [28], [40]. Focus here is mainly on polyethylene (PE), essentially for two reasons: 1) all materials considered for the insulation of HVDC extruded cables are PE-based; 2) because of its simple chemical formulation PE was a subject of many theoretical studies. Calculations of its band structure have been done because one needs to describe the electronic structure to understand and interpret the experimental data on the valence and conduction band of polymers, derived from classical techniques such as XPS (X-ray photoelectron spectroscopy) or EELS (electron energy loss spectroscopy). An excellent review can be found in [42].

Polyethylene, as previously stated, is a wide band-gap material, that is, no accessible energy levels for charges exist between the top of the valence band and the bottom of the conduction band. Calculation or experimental measurements provided a value for the forbidden gap on the order of 8.8 eV [43]: This implies that PE would have an intrinsic conductivity of 10^{-45} S/m [44], but this level of conductivity has never been measured because the reality of PE is much more complex than the "idealized" band diagram, as discussed in many articles and books (see, e.g., [12] and [45]). Two main factors alter the "idealized" band diagram in PE.

First of all, PE molecules can have different conformations, which give rise to "disorder" in the supramolecular organization, for example, physical,

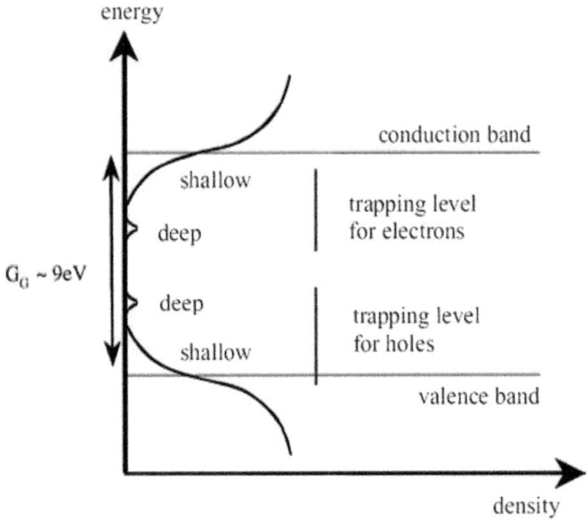

Figure 4.12 Schematic representation of state density in a disordered dielectric material. Shallow and deep traps are related to physical and chemical disorder, respectively (after [28]).

conformational, or topological disorders [28]. The local disorder of chain segments in the amorphous phase produces localized states (i.e., accessible states for carriers) on either side of the forbidden gap, i.e. at the edge of conduction and valence bands [40] (see Fig. 4.12, after [28]), all being less than 0.3 eV from the extended states of the bands with typical energies of ~0.15 eV [40]. These states, having high density (~10^{21} cm^{-3} [40]), are referred to as "shallow traps" (see above), as well as "conduction states" or "Anderson states" [28]; the residence time of carriers in shallow traps ranges typically from ~10^{-11} to ~10^{-13} s for typical energies <0.2 eV [40]. The way shallow traps assist charge transport is twofold [28]:

1. The localized states on either side of the forbidden gap are in thermal equilibrium with extended states of the conduction and/or valence bands: Transport can occur by thermal activation of carriers to a band, leading to the concept of conduction through multiple trapping/detrapping steps, as previously mentioned. This is where the term "shallow trap" comes from because charges move from trap to trap, experiencing each time a transition to the extended states of the band.
2. a large density of localized states in interaction exists, involving an overlap of the trap wavefunctions. Thus, carriers can tunnel from one level to another without undergoing a transition to the extended state. This is referred to as phonon-assisted tunneling (also called the "hopping" mechanism; see Section 4.2.2.2).

The energy G_C of conduction band minimum (CBM) for electrons – as well as the energy G_V of valence band maximum (VBM) for holes – is named the mobility edge because carrier mobility increases sharply from the effective mobility value due to transport in localized states—where carriers spend some time—to the "real" mobility associated with transport in a band, i.e. the conduction band for free electrons, the valence band for free holes (see Figs. 4.4, 4.12) [28]. From this viewpoint, some authors describe the behavior of electrons at very high field, assuming that carrier mobility is zero below G_C and infinite above, providing a support for the FLSC model proposed first by Zeller et al. [26], [46] (see Section 4.2.2.2) and further refined by Boggs [47].

The second factor that modifies the "idealized" band diagram in PE is the "chemical" disorder, due to one or more atoms that are not part of the repeat unit that constitutes the backbone of the molecular chain [28]. In PE, all sort of chemical disorders can occur, for example, side chains, additives, reaction products, impurities (from which the concept of impurity states [45]), and so on. Chemical disorder induces physical disorder in its surroundings too because of the difference in size and position of the foreign atoms with respect to the atoms constituting the polymer itself. The electronic properties of the foreign atom may differ from those of the atoms constituting the polymer backbone, thereby introducing additional energy levels in the forbidden gap, commonly at a deeper energy level (up to 1.5 eV, with typical energies of ≈1 eV [40]) and with a much lower density than shallow traps [28]. These energy levels are often called "deep traps" (see Fig. 4.12, after [28]) or "real traps" because their contribution to the conduction process is low, as opposed to shallow traps that assist conduction (as mentioned above). In summary, it can be said that chemical disorder controls the space charge. It must be also emphasized that, as mentioned above, other factors—for example, ions, electrochemical reactions, additives, by-products, and so on—can control the electrical behavior of organic insulators: In this respect, the formulation of the insulation might be so complex that discriminating between additive and matrix is impossible in practice [28].

As to the trap depth ΔG_t, introduced at Section 4.2.1.1, it can be defined also with reference to *GAd* and *GA*, that are respectively the electron affinity of the system considered with and without the defect, being it physical or chemical [39], as shown in Figure 4.13. A positive ΔG_t means a potential trap for the electron since in that case the electron has a greater affinity for the trap than for the "pure" material, and conversely for the hole [28].

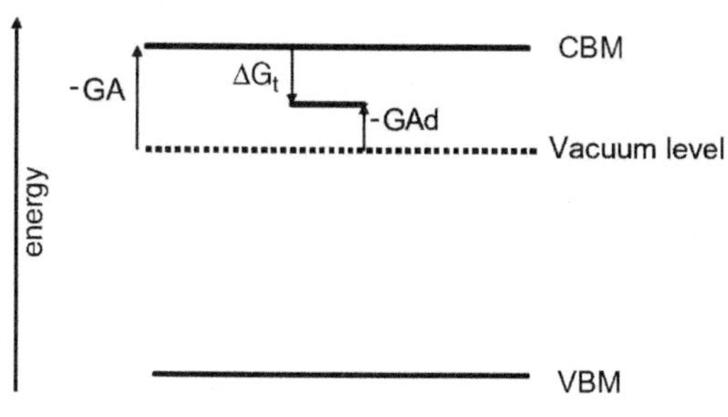

Figure 4.13 Definition of the trap depth for electron (ΔG_t = GAd - GA), where GAd is the electron affinity of the defect and GA the electron affinity of the matrix without the defect. CBM and VBM stand for conduction band minimum and valence band maximum, respectively (after [28]).

4.3.2.1 Physical Defects Numerical models were used in order to simulate a polymer matrix consisting of the n-$C_{13}H_{28}$ (tridecane) molecule in its minimum energy, that is, in its all-trans configuration [28]. It provided the reference for the electron affinity [48, 49]. The conformational disorder present in PE was simulated by molecular dynamics simulations and conformational defects in crystalline phase only. Simulation outputs highlighted that such defects produce shallow traps with energies below 0.3 eV [39], with a density of about 10^{20} cm^{-3}, leading to an average distance between traps of 15 Å. The residence time of the electrons in these states was estimated to be between 10^{-11} and 10^{-13} s, consistent with the fact that physical traps in PE assist the conduction process and do not provide sites for the long-lasting space charge, which is measured using conventional space-charge imaging techniques (which explains the use of "localized conduction states" instead of "traps" to name these states) [28]. A more refined approach in order to take into account the amorphous phase properties was developed in [50], where electronic states and the mobility of excess electrons have been determined in amorphous polyethylene. The minimum energy of electronic state obtained in the simulations is located below the vacuum level at $G_t = -0.32$ eV and the mobility edge G_C about the vacuum level, implying a trap depth energy $\Delta G_t \approx 0.3$ eV due to local disorder (which is the same value as the calculation in [39]). This leads to the following expression for the mobility [28]:

$$\mu = \mu_0 \exp\left(-\frac{\Delta G_t}{k_B T}\right) \tag{4.18}$$

where the preexponential factor has been estimated as $\mu_0 = 7.5 \times 10^{-2}$ m^2/V · s, leading to a mobility value of about 2×10^{-7} m^2/V · s, with an uncertainty of one order of magnitude. Note that (4.18) is valid in the limit that $\Delta G_t \gg k_B T$ and results from an approximation of the Kubo–Greenwood equation [50].

4.3.2.2 Chemical Defects Two groups of chemical defects were modeled using the n-$C_{13}H_{28}$ molecule in its all-trans configuration as a reference for electron affinity [28]:

1. *In-chain or side-chain imperfections*, that is, hydroxyl and ketone functions, double or triple carbon–carbon bonds, branching, and so on, simulated as a chemical modification of an alkane chain [41, 51]. The chain is modified through the creation of a chemical defect, thus the minimum energy conformation may change: The calculations accounted for this. In conclusion it was found that chemical defects containing carbonyl or conjugated double bonds are the deepest electron traps, whereas nonconjugated double bonds are shallower.
2. *By-products of cross-linking reactions* in the case of chemically cross-linked PE. Trap depths for electrons were derived for dicumyl peroxide, acetophenone, α-methylstyrene, cumene, and methane [48]. Among those, α-methylstyrene and acetophenone were identified as deep electron traps. Hole affinities were also derived for some of these chemical species in order to estimate their potential role in trapping positive charge [52]. Table 4.1 reports a summary of the simulation results on the trap depth relevant to some chemicals: Most values in the table were attained neglecting the polarization induced by

Table 4.1 Summary of Trap Depths Calculated Using ab Initio Methods for Different Chemical Defects in PE

Nature of Chemical Defects		Molecule with:	Depth of Trap (eV) For Electrons	For Holes (when available)
6-Tridecene[a] ($C_{13}H_{26}$)		In-chain C=C	0.16	0.57
5,7-Tridecene[a] ($C_{13}H_{24}$)		In-chain conjugated C=C	0.51	1.35
6-Tridecanone[a] ($C_{13}H_{26}O$)		Side-chain C=O	0.49	0.43
5-Decanone[b] ($C_{10}H_{20}O$)		Side-chain C=O	0.453	—
5-Decene[b] ($C_{10}H_{20}$)		In-chain nonconjugated C=C	0.122	—
4, 6-Decene[b] ($C_{10}H_{18}$)		In-chain conjugated C=C	0.443	—
5-Decyne[b] ($C_{10}H_{18}$)		In-chain nonconjugated triple bond	0.041	—
5-Vinyl nonane[b] ($C_{10}H_{20}$)		Nonconjugated C=C	0.157	—
5-Decanol[b] ($C_{10}H_{21}O$)		Hydroxy!	0.186	—
5-Decanal[b] ($C_{10}H_{20}O$)		Carbonyl group	0.445	—
4-Propyl heptane[b] ($C_{10}H_{22}$)		Saturated	0.121	—
By-products of cross-linking reactions when using decumyl-peroxide as a cross-linking agent	α-Methylstyrene[a] (C_9H_{10})	Conjugated double bond	1.53	0.79
	Cumylalcohol[a] ($C_9H_{12}O$)	Hydroxyl	0.28	0.36
	Acetophenone[a] (C_8H_8O)	Carbonyl group	0.9	0.04
	Cumene[b] (C_9H_{12})		0.04	

[a]After [52].
[b]After [41].
Source: After [28].

neighboring molecules on the model molecule [28]. To estimate this effect, a cluster model consisting of alkane chains surrounding the chemical defect was used in the calculations [41], but the relevant change in trap depth was found to be small, although in general polarization effects may be significant, particularly with a polar matrix [53]. Other calculations of trap depths related to chemical species present in semiconductive screens of cables and capable of spreading into the PE have been issued in the early 2000's [54].

4.4 REVIEW OF SPACE-CHARGE MEASUREMENT METHODS FOR HVDC EXTRUDED INSULATION

In the early 1970s, direct methods for space-charge observation in solid insulators were not available. *Indirect evidence of the existence of space charge in PE was obtained from the effect of prestressing on its impulse breakdown strength.* Thermally stimulated currents (TSC, see Section 4.4.3) were also used for the same purpose [8].

Early direct methods of investigating space-charge distributions in polymers employed destructive techniques such as the dust figure method and the field probe technique. These techniques were used in investigating space-charge distributions in thick samples (~2 mm) and cable sections made of plain LDPE and XLPE [55, 56]. Destructive methods have usually involved difficulties of execution and of interpretation.

From the late 1970s to the mid-1990s, a number of new nondestructive methods for investigating space charge in solid polymeric materials have been developed. Nondestructive measurement of space-charge profiles was first reported in the early 1980s and is now routinely performed in many laboratories [14]. An exhaustive and detailed review of experimental approaches to the determination of spatial charge and field distributions in the thickness of the dielectrics was carried out in 1997 by Ahmed and Srinivas in their fundamental study [19]. This work emphasizes nondestructive methods of direct probing and provides detailed information on the principles used in each method, typical resolutions, limitations, and advantages, which are summarized in Table 4.2. The methods treated in [19] are as follows:

1. Thermal pulse method
2. Laser intensity modulation method (LIMM)
3. Laser-induced pressure pulse method (LIPP)
4. Thermoelastically generated LIPP
5. Pressure wave propagation [or pressure wave pulse [8]) (PWP) method]
6. Nonstructured acoustic pulse method
7. Laser-generated acoustic pulse method
8. Acoustic probe method
9. Piezoelectrically generated pressure step method
10. Thermal step method (TSM)
11. Electroacoustic stress pulse method, or pulsed electroacoustic (PEA) method
12. Photoconductivity method

Table 4.2 Overview of Methods and Capabilities

Method	Disturbance	Scan Mechanism	Detection Process	r^a (μm)	t^b (μm)	Comments
Thermal pulse method	Absorption of short light pulse in front electrode	Diffusion according to heat conduction equations	Voltage change across sample	≥ 2	~ 200	High-resolution requires deconvolution.
Laser intensity modulation method	Absorption of modulated light in front electrode	Frequency-dependent steady-state heat profile	Current between sample electrodes	≥ 2	~ 25	Numerical deconvolution required
Laser-induced pressure pulse method	Absorption of short laser light pulse in front electrode	Propagation with longitudinal sound velocity	Current between sample electrodes	1	100–1000	No deconvolution required
Thermoelastically generated LIPP	Absorption of short laser light pulse in thin buried layer	Propagation with longitudinal sound velocity	Current or voltage between sample electrodes	1	50–70	Deconvolution required
Pressure wave propagation method	Absorption of short laser light pulse in metal target	Propagation with longitudinal sound velocity	Voltage or current between sample electrode	10	5–200	Resolution improved with deconvolution. Also used for surface charge measurements
Nonstructured acoustic pulse method	HV spark between conductor and metal diaphragm	Propagation with longitudinal sound velocity	Voltage between sample electrode	1000	≤ 10000	Used for solid and liquid dielectric. Higher resolution with deconvolution
Laser-generated acoustic pulse method	Absorption of short laser light pulse in thin paper target	Propagation with longitudinal sound velocity	Voltage between sample electrodes	50	≤ 3000	Deconvolution is required. Target and sample immersed in dielectric liquid

(continued)

Table 4.2 *Continued*

Method	Disturbance	Scan Mechanism	Detection Process	r^a (μm)	t^b (μm)	Comments
Acoustic probe method	Absorption of laser light pulse in front electrode	Propagation with longitudinal sound velocity	Voltage between sample electrodes	200	2000–6000	
Piezoelectrically generated pressure step method	Electrical excitation of piezoelectric quartz plate	Propagation with longitudinal sound velocity	Current between sample electrodes	1	25	Deconvolution required
Thermal step method	Applying two isothermal sources across sample	Thermal expansion of the sample	Current between sample electrodes	150	2000–20,000	Deconvolution required
Electroacoustic stress pulse method	Force of modulated electric field on charges in sample	Propagation with longitudinal sound velocity	Piezoelectric transducer at sample electrode	100	$\leq$ 10,000	Deconvolution required. Also used for surface charge measurements.
Photoconductivity method	Absorption of narrow light beam in sample	External movement of light beam	Current between sample electrodes	$\geq$ 1.5	—	Nondestructive for short illumination time
Space-charge mapping	Interaction of polarized light with field	Parallel illumination of sample volume or movement of light beam or sample	Photographic record	200	—	Mostly used on transparent dielectric liquids
Spectroscopy	Absorption of exciting radiation in sample	External movement of radiation source or sample	Relative change in the observed spectrum	$\geq$ 50	—	Few applications
Field probe	None	Capacitive coupling to the field	Current	1000	$\leq$ 20,000	Destructive

[a] r is the method's resolution.
[b] t is the sample thickness.
Source: After [19].

13. Space-charge mapping
14. Spectroscopy
15. Field probe

However, according to [8]—which dates back to the same year 1997 and constitutes a good review of international research and development trends of HVDC polymeric cables—only some of the above-listed methods reported in [19] are well-established techniques used in investigating space charge in solid polymeric materials for HVDC cable insulation (thereby giving rise to a considerable number of articles published about space-charge measurements in LDPE, HDPE, and XLPE samples and cables made of the same materials), namely:

1. The thermal pulse method
2. The thermal step method (TSM)
3. The laser intensity modulation method (LIMM)
4. The pressure wave propagation (PWP) method
5. The laser-induced pressure pulse (LIPP) method
6. The pulsed electroacoustic (PEA) method

Methods 1–3 (i.e., thermal pulse, thermal step, and LIMM) can be classified as "thermal methods" [19, 57], namely methods based on thermal heating applied to the sample, whereas methods 4–6 (i.e., PWP, LIPP, and PEA) can be classified as "wave propagation methods," that is, methods that use waves that propagate through the charged sample. These methods are reviewed in detail in this section, which also gives some hints about the main techniques for the estimation of the mobility and the energy levels occupied by the trapped charge carriers constituting the space charge.

In reviewing the methods for space-charge measurement, this section follows the description made in [19]—whose theoretical concepts still hold—plus integrations when needed. Indeed, a further evolution of space-charge measurement techniques since 1997 took place: For example, it was argued in [14]—dating back to 2003—that "the . . . space charge measurement techniques . . . in use today have been extensively documented by Ahmed and Srinivas [19]. Of these, four are much more widely used than the others are. They are the laser-intensity-modulation-method (LIMM), the thermal step method (TSM), the pressure-wave-propagation (PWP) or laser-induce-pressure-pulse (LIPP) method, and the pulsed-electro-acoustic (PEA) method ([14], page 18)."

This observation still holds today, with only LIMM being less popular, particularly as far as cable investigations are concerned. Hence, Section 4.5 discusses the latest improvements and measurement results by these main methods employed up-to-date, with focus on the TSM, PWP, and PEA methods and particular reference to space-charge measurements on HVDC extruded cable materials and insulation.

4.4.1 Thermal Methods

The thermal methods can be divided into three groups according to the signal used for excitation of the sample. While in the thermal pulse method and in the LIMM (referred to as "thermal wave method" in [57] in a more general sense) a short

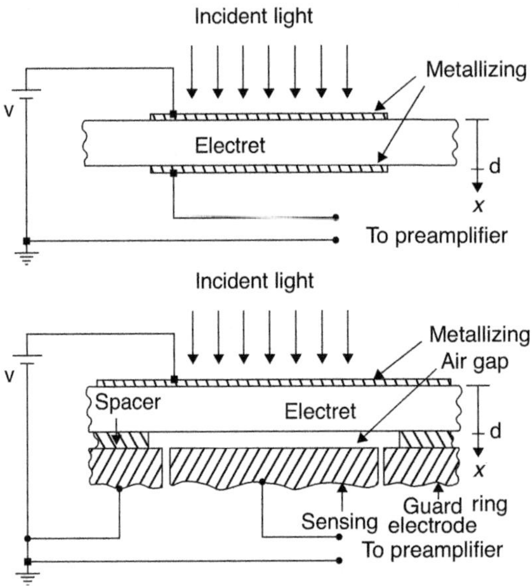

Figure 4.14 Schematic diagram of the apparatus for the thermal pulsing experiment in the double-metalizing and single-metalizing configurations (after [19]; see also [57]).

($< 100\ \mu s$) light pulse or a modulated light beam, respectively, is used to illuminate (and thus heat) the absorbing sample electrode, the thermal step method generally utilizes heating of one sample side relative to the other by application of a constant temperature difference between the two electrodes. In all cases, the absorbed energy diffuses into the sample and causes thermal expansion as well as changes in the dielectric permittivity. This results in a variation of the surface potential in an open-circuited (e.g., a one-side metalized) sample or in electrode currents in a short-circuited sample [57].

4.4.1.1 *Thermal Pulse Method* The thermal pulse method was first proposed by Collins [58] and later modified by others [59, 60]. It consists of applying a thermal pulse to one surface of the dielectric by means of a light flash and measuring the electrical response generated by the sample as a function of time while the thermal transient diffuses across the sample (see Fig. 4.14). The form of the time dependence of the electrical response carries the information about the charge or polarization distribution. Such distribution can be obtained only by a deconvolution process. This imposes limitations on the usefulness of the method and necessitates that results should be interpreted appropriately [19].

In Collins' technique [58], a short heat pulse is absorbed on one side of a thermally isolated electret. Heating is achieved by illuminating the sample with a short, intense pulse of light (8-μs duration). The light is absorbed by a thin metal film, 40 nm, on one side of the electret ($x = 0$; see Fig. 4.14). Changes in the voltage across the sample, $\Delta V(t)$, due to the nonuniform thermal expansion and the local change in permittivity, are measured as a function of time t as the heat diffuses through the electret from the target electrode ($x = 0$) toward the opposite electrode ($x = d$) [19].

The voltage response $\Delta V(t)$ is related to the charge and polarization distributions and to the temperature change $\Delta T(x,t)$ by the following relationship [58]:

$$\Delta V(t) = \frac{1}{\varepsilon_r \varepsilon_0} \int_0^d \left\{ \left[A\rho(x) - B\frac{dP}{dx} \right] \int_0^x \Delta T(x')\, dx' \right\} dx \tag{4.19}$$

where $A = \alpha_x - \alpha_\varepsilon$, $B = \alpha_p - \alpha_x - \alpha_\varepsilon$; and $\rho(x)$ and $P(x)$ are, respectively, the internal distribution of charge and polarization. The symbols α_x, α_ε, and α_p are the coefficient of thermal expansion and the temperature coefficients of dielectric constant and permanent polarization, respectively.

For a nonpolar electret, $P = 0$, and Eq. (4.19) can be reduced to a simpler form [19]:

$$\Delta V(t) = \frac{\alpha_x - \alpha_\varepsilon}{\varepsilon_r \varepsilon_0} \int_0^d \left[\rho(x) \int_0^x \Delta T(x', t)\, dx' \right] dx \tag{4.20}$$

The evolution in time of the electrical response $\Delta V(t)$ provides information on the distributions of space charge and polarization in the sample. This information, however, can be obtained only after appropriate deconvolution of the real signal. This requires that the results are appropriately interpreted.

Among the techniques proposed to interpret the data obtained by this method, one of the most interesting is that suggested by DeReggi and Mopsik, which exploits the Fourier series [59, 60]. DeReggi used his version of the thermal pulse technique to measure the space charge induced on XLPE samples cut from AC cables aged under different conditions [61]. In these investigations, the tested sample was 220 μm in thickness and was prepared by peeling the insulation using a lathe. Investigation of space charge in HDPE has been conducted by Suzuoki et al. [62] using a modified technique of the thermal pulse method. The basic principle in the Suzuoki technique was the same as Collins except the current response during a long, steplike heat pulse was utilized. It was reported in DeReggi and co-workers [59, 60] that a resolution improvement can be obtained by collecting data from both sides of two-sided metalized samples. The resolution improvement is seen to come from the large time spread of the even and odd terms that appear separately in the symmetrized responses, compared to their spread in time in the response from each side where odd and even terms appear [19].

4.4.1.2 *Laser Intensity Modulation Method (LIMM)*

The LIMM technique was developed by Lang and Das-Gupta [63]. This method utilizes sinusoidally modulated surface heating of dielectric samples to produce spatially nonuniform temperature distributions through the thickness. In the LIMM technique, each surface of the sample, usually a thin film, is exposed to a laser beam that is intensity modulated in a sinusoidal fashion by an acoustooptic modulator or light chopper. The laser beam is absorbed by the front of the electrode of the sample. The sinusoidal modulation of the laser beam causes a sinusoidal fluctuation in temperature of the front electrode, resulting in propagation of temperature waves into the sample. The temperature waves are attenuated as they progress through the sample and they also are retarded in phase.

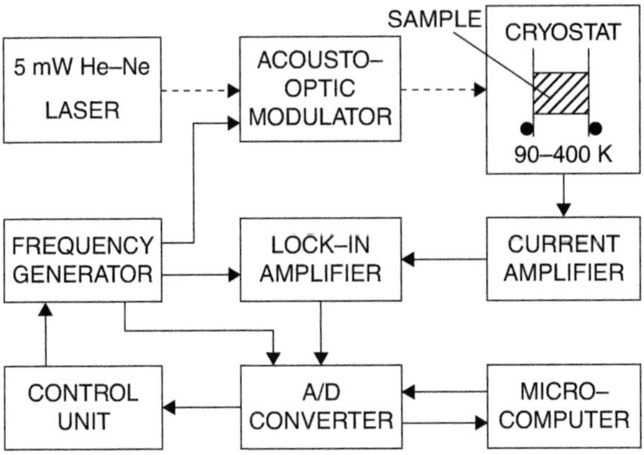

Figure 4.15 Block diagram of setup for a specific implementation of the LIMM (Laser Intensity Modulation Method (after [57]; see also [64]).

Thus, a nonuniformly distributed thermal force acts on the sample. The interaction of the fluctuating temperature and the spatially distributed polarization and space charge produces a sinusoidal pyroelectric current. This current is a unique function of the modulation frequency and the polarization and charge distributions [19].

A specific setup implementation for the LIMM, based on the use of a modulated laser, is illustrated in Figure 4.15 [64]. The sample, coated with opaque electrodes and mounted in a cryostat, is heated with the sinusoidally intensity-modulated laser light. A thermal "wave" penetrates into the sample with the modulation amplitude decreasing strongly with depth and with modulation frequency. The heat absorption causes an electrode current in the short-circuited sample, which is given by [57]

$$I(t) = \frac{S}{d} \int_0^x G(x) \frac{\partial T(x,t)}{\partial t}\, dx \tag{4.21}$$

where S and d are the sample area and thickness, T is the temperature, and $G(x)$ is the distribution function that depends on the pyroelectric characteristics of the material and the total space charge density ρ_T, given by the sum of "real" space-charge density ρ plus the equivalent charge density (or distribution) of permanent polarization, ρ_P:

$$\rho_T(x) = \rho(x) + \rho_P(x) = \rho(x) - \frac{dP(x)}{dx} \tag{4.22}$$

Further details about $G(x)$ can be found in [57].

With this method the measurement current shows a modulation, too. Amplitude and phase of the current are detected with a lock-in amplifier. The experiment is performed for many (~100) modulation frequencies in the range from 100 Hz to 100 kHz and for illumination of both sides of the sample. A typical run takes ~3–5 h [57].

Likewise with the thermal pulse methods, a mathematical deconvolution technique is required to compute the polarization and space-charge distributions from the current–frequency data. The Fourier series denconvolution technique developed by DeReggi and Mopsik was used also for the LIMM to resolve the polarization and space-charge distributions. The resolution of LIMM is limited to $\geq 2\,\mu$m [19].

Lang and Das-Gupta have used LIMM in the determination of the spatial distribution of polarization and space charge in several polymer electrets (see, e.g., [63]). They also used this method in their investigation of the space-charge distributions in thermally poled PE [65].

4.4.1.3 Thermal Step Method (TSM) The thermal step method, developed by Toureille, uses a thermal step that is applied across an insulator to yield a current response to the thermal expansion of the sample (see, e.g., [66, 67]). The thermal step creates a thermal wave that propagates within the insulation thickness, thereby generating a temporary local displacement of the space charge. This is reflected at the electrodes by a variation of the induced charges, which involves a current between the electrodes. The measurement and mathematical processing of such current allows the determination of the distribution of the electric field and charge density across the sample. The TSM is applicable to any sample presenting a geometry for which the space–time distribution of the temperature can be calculated, such as material specimens and industrial components (power cables, capacitors, printed circuit boards, and stator windings) [67].

The thermal step is created generally by heating one sample side relative to the other by application of a constant temperature difference between the two electrodes [57] (e.g., by keeping one electrode at -10°C and then injecting a warm liquid in the other electrode to heat it up to 20°C [19]). The method is nondestructive because the maximum temperature reached is close to room temperature. As pointed out in [57], the TSM was originally used for thick samples (thickness $>$ 1 mm), since heating of one sample surface is performed thermally and was relatively slow in the initial experiments. In [19] it is said to be applicable for thick insulators in the range of 2–20 mm (see Section 4.5.1), but further developments of the method enabled its application to cable insulation of larger thickness. On the other hand, by using faster heating, this method has also been applied to thin (10–100 μm thick) samples [57, 68]. From the fundamental study [67] more details about TSM can be found, which are briefly reported in what follows.

TSM in Short-Circuit Conditions Let us consider an insulating sample of thickness d and of surface S provided with two conducting electrodes of abscissas $x=0$ and $x=d$ (see Fig. 4.16a). The material is considered as homogeneous and

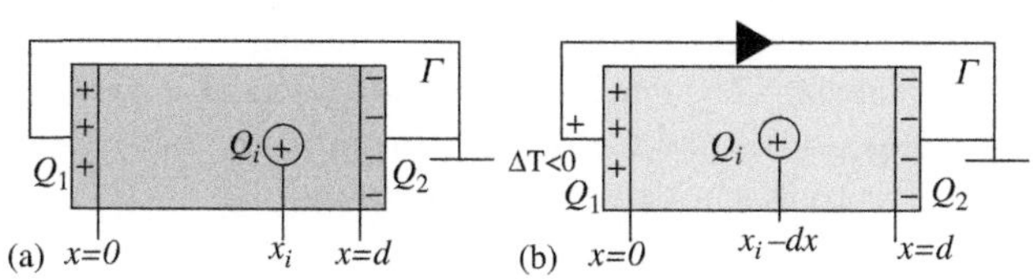

Figure 4.16 Principle of the TSM in short-circuit conditions: sample at equilibrium at (a) T_0 and (b) during application of a thermal step (TS) (after [67]).

infinitely flat ($d \ll \sqrt{S}$); thus, the electric field is assumed as constant in any plane parallel to the electrodes. The sample is placed in short circuit at a temperature T_0. Because the system consisting of sample, electrode, and conducting wire is in electrostatic equilibrium, a charge Q_i situated in the bulk of the insulator at an abscissa x_i induces charges on the electrodes, labeled Q_1 (for the electrode located at $x=0$) and Q_2 (for the electrode located at $x=d$). The dependence of the induced charge on the space charge Q_i and on the geometrical dimensions of the sample can be obtained resorting to:

1. The short circuit conditions $\int_0^d E(x)\, dx = 0$
2. The conservation of electric charge, that is, $Q_1 + Q_2 + Q_i = 0$
3. The conservation of the surface divergence of electric displacement at the interfaces electrode–dielectric $\nabla \cdot s(\varepsilon \mathbf{E}) = Q/S$

thereby achieving [67]:

$$Q_1 = -\frac{d - x_i}{d} Q_i \qquad Q_2 = -\frac{x_i}{d} Q_i \tag{4.23}$$

If a temperature step $\Delta T = T - T_0$ is applied on one side of the specimen (see Fig. 4.16*b*), the propagation $\Delta T(x,t) = T(x,t) - T_0$ of the thermal front across the sample gives rise to local variations of the permittivity and of the abscissas (expansion or contraction), which can be written (as a first-order guess) [67]:

$$\varepsilon = \varepsilon_{T_0}(1 + \alpha_\varepsilon \Delta T) \tag{4.24a}$$

$$dx = dx_0(1 + \alpha_x \Delta T) \tag{4.24b}$$

where α_ε and α_x are the temperature coefficients of dielectric permittivity and linear expansion, respectively. These variations of the abscissas and permittivity cause a modification of the induced charges; with proper rearrangements one gets for $Q_2(t)$ the expression [67]

$$Q_2(t) = -\frac{Q_i x_i}{d}\left[1 + \frac{\alpha}{x_i}\int_0^{x_i} \Delta T(x,t)\, dx - \frac{\alpha}{d}\int_0^{d} \Delta T(x,t)\, dx\right] \tag{4.25}$$

where

$$\alpha = -\frac{1}{C}\frac{dC}{dT} \approx \frac{1}{x}\frac{dx}{dT} - \frac{1}{\varepsilon}\frac{d\varepsilon}{dT} = \alpha_x - \alpha_\varepsilon \tag{4.26}$$

being C the capacitance of the sample before the application of the TS.

The variation of the induced charges is reflected in the external circuit by the appearance of a so-called thermal step (TS) current $I(t) = -dQ_2(t)/dt$. After proper rearrangements according to Eqs. (4.25) and (4.26) and remembering that $\rho(x) = \partial D(x)/\partial x = \varepsilon \partial E(x)/\partial x$, the TS current can be written as follows [67]:

$$I(t) = -\alpha C \int_0^d E(x) \frac{\partial \Delta T(x,t)}{\partial t} dx \tag{4.27}$$

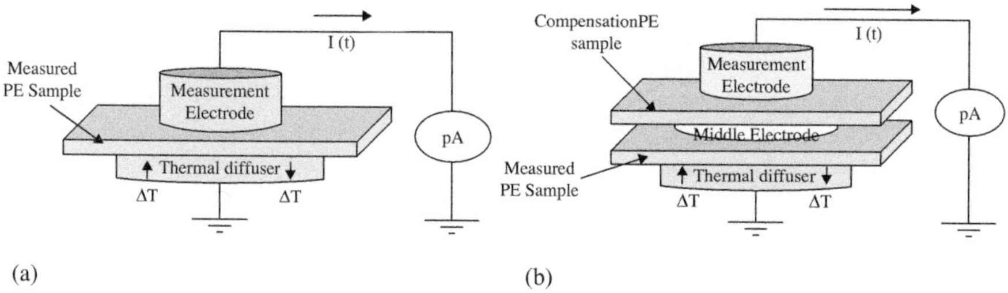

Figure 4.17 Experimental setup of the TSM: (*a*) TSM in short-circuit conditions and (*b*) the principle of the double capacitor (after [67]).

where $E(x)$ is the electric field at the abscissa x. As shown by Eq. (4.27), the TS current amplitude is a function of the amount of charge stored in the sample, of its distribution, and of the material parameters. In practice, the TS current ranges between 1 pA and 1 μA [67].

To determine $E(x)$, the TS current must undergo a lengthy numerical treatment, which requires a perfect knowledge of the temperature derivative $\partial\, \Delta T(x,t)/\partial t$ at every moment and for every abscissa. Numerical processing can be made via either the Fourier series, or successive differentiation, or again the inverse matrix technique (for further details refer to [67]).

In practice (see Fig. 4.17*a*), the TS is created by a cold liquid circulating within a radiator in contact with the sample (or by a transient current that heats a resistive material under the sample). The electrodes are connected through a current amplifier. The latter is connected to a computer, which records the current and does the numerical data processing. After performing a measurement, the sample can be reheated using a warm liquid, and the experiment can be repeated [67].

TSM under Applied DC Field For measurements under DC field, the setup described in Figure 4.17*a* presents two major disadvantages: (1) the current amplifier must not be in contact with the HV, and (2) if a setup with the current amplifier placed between the sample submitted to HV and ground is used (as, for example, conduction current measurements), the conduction and the polarization currents are likely to mask the TS current. A solution to these problems is a "compensation sample" of identical dimensions as the measured specimen and placed opposite to the latter (Fig. 4.17*b*). By connecting one side of the compensation sample to the current amplifier and the other to the measured specimen via an electrode, a "double capacitor" is obtained. Then, a voltage can be applied to the middle electrode, and the TS to the measured sample. The TSM current will be acquired via the compensation sample. Using two identical specimens offers the considerable advantage of compensating the polarization and conduction currents, which can occur under high fields; the measured current is due solely to the internal field of the measured sample.

Actually, the measurement "under applied field" is made in two steps (see Fig. 4.18) [67]:

1. During conditioning, the voltage is applied to the middle electrode and the current amplifier is short circuited. Thus, the two samples constitute two

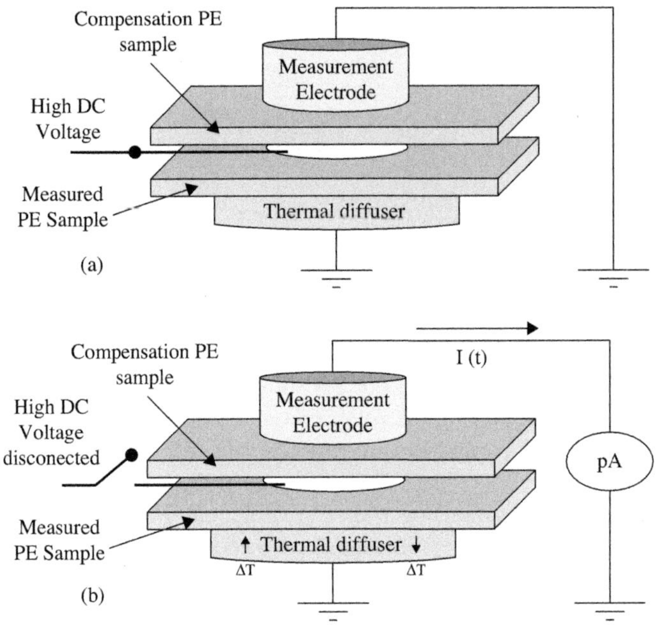

Figure 4.18 TSM under HVDC field: (a) application of HVDC and (b) TSM measurement (after [67]).

identical capacitances placed in parallel with respect to the voltage source (Fig. 4.18*a*).

2. For preventing the transport of induced charges at electrodes via the HV generator rather than via the current amplifier, the power source has to be disconnected during the measurement (Fig. 4.18*b*). The TS current is then measured by exciting thermally the studied specimen (in contact with the thermal diffuser), while the current amplifier is connected to the compensation sample. This time, the two samples are in series with each other and with the current amplifier, and the short-circuit conditions are fulfilled.

If the double-capacitor setup of Figure 4.18 is used, the TS current becomes

$$I(t) = -\alpha C_2 \int_0^d E(x) \frac{\partial \Delta T(x,t)}{\partial t} dx \tag{4.28}$$

where C_2 is the capacitance seen by the current amplifier. During the measurement, the two samples are in series: $C_2 = C/2$. If no voltage is applied to the samples (voltage off), the voltage off TS current should be half that given by a short-circuit measurement [67].

More equations and comments relevant to low fields applied to virgin samples, low fields applied to samples containing space charge, and the calibration with "under field" measurements are important for a thorough application of the method; again, they can be found in [67].

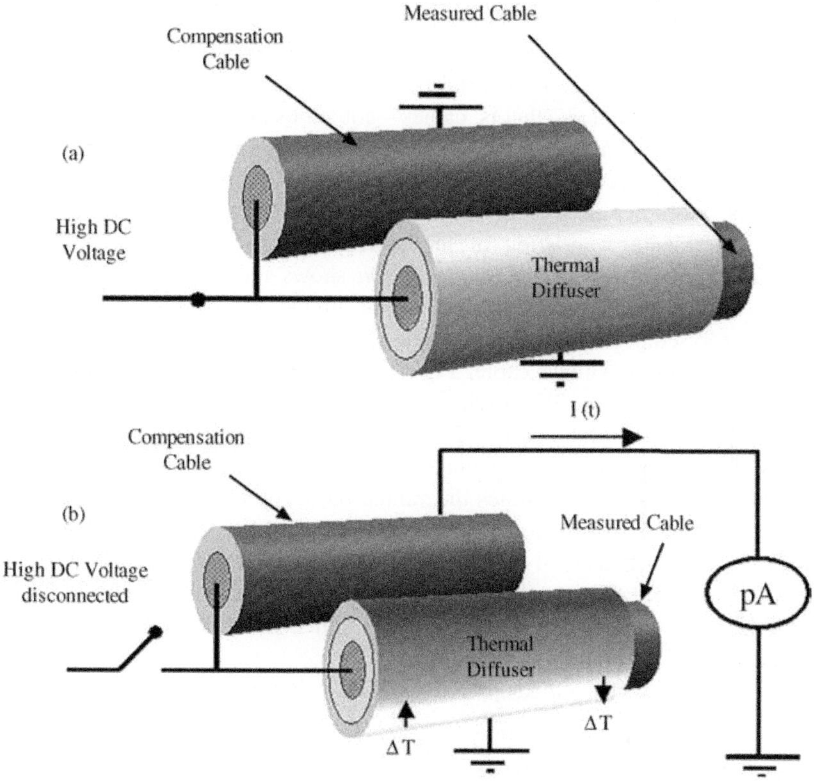

Figure 4.19 "Under field" space-charge measurement by TSM on a power cable: (*a*) application of HVDC and (*b*) TSM measurement (after [67]).

Application to Power Cables The principle of "under field" space-charge measurements on cables is similar to the case of films. The experimental setup for underfield space-charge measurement by TSM on a power cable is illustrated in Figure 4.19 [67].

Equations analogous to those established for flat samples can be written in cylindrical geometry; indeed in the case of a short-circuit cable measurement, the TS current has the expression [67]

$$I(t) = -\alpha C \int_{r_o}^{r_i} E(r) \frac{\partial \Delta T(r,t)}{\partial t} dr \tag{4.29}$$

where r_i and r_o are the inner and outer radii of the cable insulation and r its radial coordinate. If the double-capacitor setup from Figure 4.19 is used, the TS current becomes

$$I(t) = -\alpha C_2 \int_{r_o}^{r_i} E(r) \frac{\partial \Delta T(r,t)}{\partial t} dr \tag{4.30}$$

where C_2 is the capacitance seen by the current amplifier. During the measurement, the two samples are in series: $C_2 = C/2$. As for the case of flat samples, the voltage off TS current should be half that given by a short-circuit measurement [67].

For cylindrical samples—that is, cables—equations and comments analogous to those relevant to flat samples hold for low fields applied to virgin samples, low fields applied to samples containing space charge, and the calibration with underfield measurements. More details are omitted here and reported in [67].

The experimental validation of the underfield TSM on power cables is presented in Figures 4.20*a*–4.20*d*. Figure 4.20*a* shows the division by 2 of the TS current with respect to the classical "short-circuit" measurement when the double-capacitor setup is used. For an under field experiment performed on an unconditioned medium voltage cable, Figure 4.20*b* confirms the proportionality between the amplitude of the TS current and the applied DC voltage holding at low voltages (see [67] for details). Figure 4.20*c* illustrates the same proportionality after application of positive and negative voltages to a sample cut from another medium-voltage cable. Figure 4.20*d* illustrates the validation of the treatment made in [67] about TS current in the case of a low voltage applied to a cable containing charge: the sum of the currents I_f (measured on a cable under $+2$ kV DC before conditioning) and I_0 (due to the space charge accumulated in the cable after conditioning) is equal to the current measured on the cable while submitted to $+2$ kV DC after being conditioned.

More about the TSM method applied to HVDC cables is found in Section 4.5.1.

4.4.2 Pressure Pulse Methods

Some methods utilizing pressure pulses have been developed to study charge and polarization profiles in the thickness of dielectrics. It is an approach proposed by Lewiner and co-workers [20, 69] based on an idea similar to Collins' idea of a nonuniform deformation, but using—instead of a diffusion process—the propagation of a pressure wave. Since all methods depend on the same physical process, namely an ultrasonic pulse propagating through the sample, a general description of such a process will be given first in this Section. Subsequently, these methods will be reviewed individually in Sections 4.4.2.1–4.4.2.3.

Consider a plate of homogeneous dielectric material with zero conductivity, containing space charges and permanent dipoles; the plate has thickness d and is interposed between plane parallel electrodes A and B, orthogonal to an axis z (see Fig. 4.21). Assume all the involved variables are uniform in planes perpendicular to the z axis. The charge distribution is described by a function $\rho(z)$. The permanent polarization can be represented by an equivalent charge distribution $\rho_P(z) = -dP/dz$. At the time scales involved, the induced polarization is assumed to have an instantaneous response to electric field variations and is described by an infinite frequency relative permittivity ε_r [69].

At starting time $t = 0$, a compression is applied uniformly on one face of the sample through electrode A. The wavefront z_f of the pressure wave travels through the sample at the speed of sound u_S, as shown in Figure 4.21. Under this mechanical perturbation, the atomic structure is compressed and results in two effects:

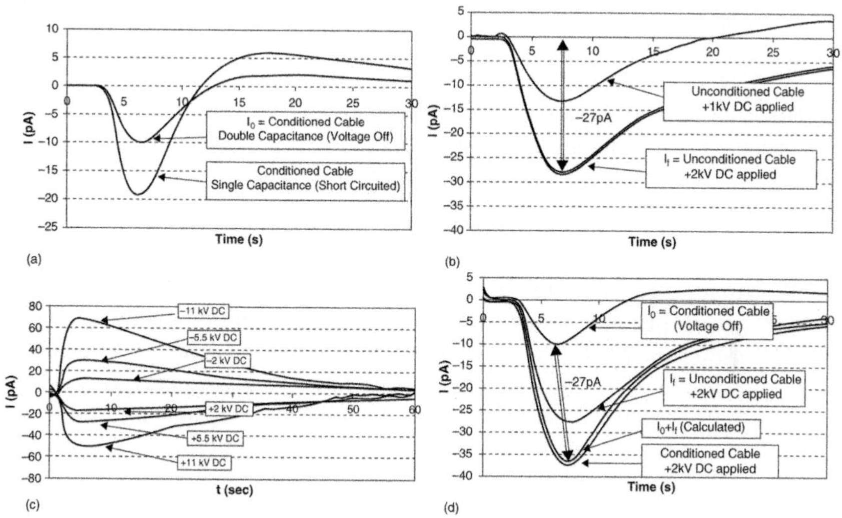

Figure 4.20 "Under field" TS signals obtained on power cables: (*a*) comparison between the measurements in short-circuit and in double capacitor configuration, (*b*) unconditioned XLPE insulated power cable during submission to DC voltages, (*c*) unconditioned power cable during submission to various DC voltages, and (*d*) sum of the currents I_f and I_0 on a DC-conditioned power cable (5 kV/mm max, 60°C, 1 h) (after [67]).

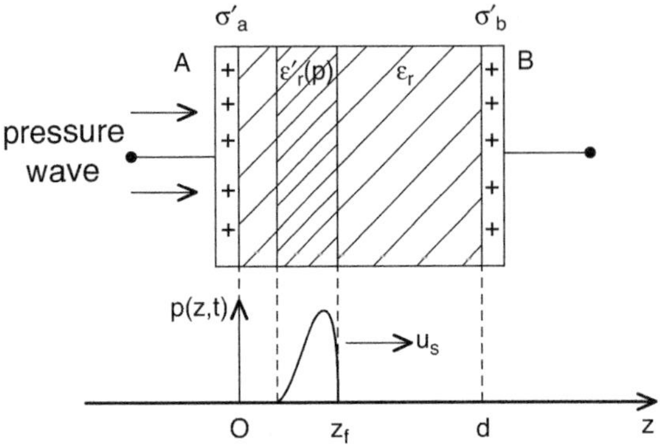

Figure 4.21 Sketch of the pressure pulse propagation methods (after [69]).

1. The charges that follow the atomic lattice are displaced.
2. The relative permittivity ε_r varies because of the variation of the local concentration of dipoles and charges [19].

These effects produce a variation of the charges on electrodes A and B, and this variation depends on the space-charge distribution and on the shape of the pressure wave. Depending on the load impedance, for instance open circuit or short circuit, the variation of the induced charges produces a voltage difference across the electrodes or a current. The evolution in time of this voltage or of this current contains information on the space-charge distribution and piezoelectric activity. The wavefront of the pressure wave acts as a virtual probe sensitive to charge or potential traveling at the speed of sound through the sample. Using the basic laws of electrostatics and considering the boundary conditions and hypotheses already made, it is possible to establish relationships between the measured electrical responses and the space-charge distributions through the samples [19].

Let us consider the case of a sample containing space charges and a permanent polarization P, resulting in a total charge density $\rho_T(z) = \rho(z) - dP/dz$. In open-circuit conditions the voltage $V(t)$, which appears on electrode B when electrode A is grounded, during the propagation of the pressure wave in the sample, can be expressed as

$$V(t) = XG(\varepsilon_r) \int_0^{z_f} E(z,0)p(z,t)\, dz \tag{4.31}$$

where $z_f = u_S t$ is the abscissa of the wavefront, $p(z,t)$ is the pressure profile, $E(z,0)$ is the profile of the electric field within the sample at the instant when compression is applied, and X is the compressibility of the material. $G(\varepsilon_r)$ is a function of the relative permittivity ε_r, containing the dependence of ε_r on pressure [19].

In short-circuit conditions, the current $I(t)$ flowing in the external circuit is related to the electric field distribution by the following relationship:

$$I(t) = XC_0G(\varepsilon_r)\int_0^{z_f} E(z,0)\frac{\partial}{\partial t}p(z,t)\,dz \tag{4.32}$$

where C_0 is the capacitance of the noncompressed sample:

$$C_0 = \frac{\varepsilon_0\varepsilon_r S}{d} \tag{4.33}$$

and S is the sample area [19].

These relationships show that if $p(z,t)$ is known, the electric field distribution $E(z,0)$ can be obtained from the measurement of $V(t)$ or $I(t)$, by solving the integral Equations (4.31) and (4.32). The total charge distribution $\rho_T(z)$ is further obtained through Poisson's equation by derivation. As for the thermal methods dealt with in Paragraph 4.4.1, a separation of $\rho(z)$ and $P(z)$ is not possible unless assumptions about the relevant distributions are made [57].

4.4.2.1 Pressure Wave Propagation Method

The pressure wave propagation (PWP) method [70] is based on the propagation of a pressure wave $p(t)$ inside the sample, mostly by a laser light pulse. In a few cases only is the pressure wave generated in ways other than by a laser, for example, by a piezoelectric transducer to which an electrical pulse $V_P(t)$ of short duration is applied. This is the practical implementation of PWP according to [13, 70] (see Fig. 4.22). In particular in [13] the basics of the PWP are explained as follows: "The acoustic wave acts as a charge probe. The charge moves as the acoustic wave propagates it. This movement causes a change of surface charge on the electrodes. The time signal of displacement current indicates the charge distribution in the sample. By measuring the displacement current between the electrodes, the charge distribution is obtained" ([13], page 523).

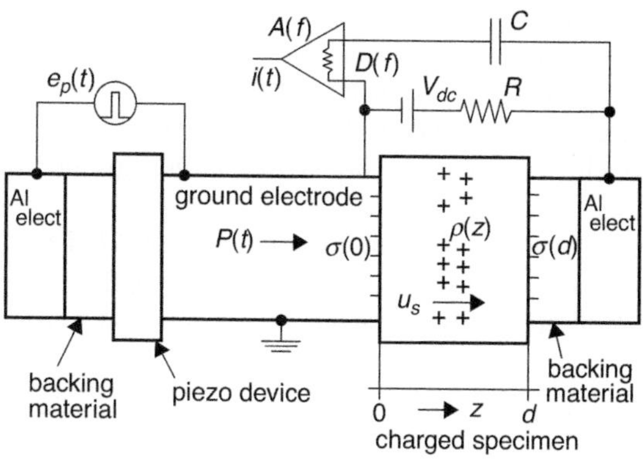

Figure 4.22 Scheme of the PWP method with pressure wave from a piezoelectric transducer (after [13]).

Alternatively—and predominantly—the pressure wave is generated by absorbing a short laser light pulse in a metal target bonded to the dielectric plate under investigation. For a short duration pulse, the electrical response gives directly the spatial distributions of the electric field and charge density. A short-duration (typically 1 ns) pressure wave can be generated by the impact of a 35-ps Nd:YAG laser[9] pulse on a 500-μm-thick aluminum target covering one of the electrodes [19].

In any case, the pressure wave propagates through the sample, thereby perturbing the space charges trapped within the material. The microscopic movement of charges, due to the acoustic impulse, causes an increase of charge $q(t)$ at the electrodes, which can be detected directly via the displacement current $i(t)$. This current is directly proportional to the distribution of space charge within the sample [70].

The PWP method was developed by Alquiè et al. [71–73]. It was used to measure the space-charge and polarization distributions in thick slabs (50–200 μm) of polymer electrets and in PE used in HV cables, as well as to determine surface-charge distributions [19]. It has also been reported that the PWP method allows the study of the evolution of the charge distribution in a sample during the polarization process under HV field and the investigation of the phenomena occurring during a reversal of the polarity of the applied voltage [73].

4.4.2.2 Laser-Induced Pressure Pulse (LIPP) Method In this method [19, 57], one side of a sample containing space charges is irradiated with a very short laser light pulse. The target absorbs the energy of the laser, and the rapid expansion that results produces a brief but intense pressure pulse $p(t)$ in the dielectric. As the very narrow compressed region travels through the sample, a current is induced in the external circuit due to nonuniform changes in dimension and permittivity reflecting the space-charge distribution. Compared to the version of the PWP method where the pressure wave is generated by a piezoelectric transducer, the advantage of the LIPP is to allow pressure pulses with steeper rise fronts, which increases the spatial resolution of the system [16]. When the thickness of the compressed region is small, the profile of the current response is directly proportional to the charge distribution, including the charges on the electrodes [19].

The LIPP (laser-induced pressure pulse) method has been extensively used by Sessler [57] and Gerhard in their investigations of the space-charge and polarization profiles in thin-film polymer materials FEP (Fluorinated Ethylene Propylene), PI (polyimide), PET (polyethylene terephthalate), PVDF (Poli-Vinyli-Dene Fluoride) and silicon oxide, charged with electron beams, corona discharges, or liquid contacts [19, 57].

In the Sessler–Gerhard technique, a laser pulse of 70-ps duration and 1–10 mJ energy over a 0.1-cm^2 area is directed toward an absorbing layer on the front electrode of the sample used for detection of the acoustic pulses. The absorbing layer

[9] Nd:YAG (neodymium-doped yttrium aluminum garnet; Nd:Y3Al5O12) is a crystal that is used as a lasing medium for solid-state lasers. Nd:YAG lasers operate in both pulsed and continuous mode. Pulsed Nd:YAG lasers are typically operated in the so-called Q-switching mode: an optical switch is inserted in the laser cavity waiting for a maximum population inversion in the neodymium ions before it opens. Then the light wave can run through the cavity, depopulating the excited laser medium at maximum population inversion.

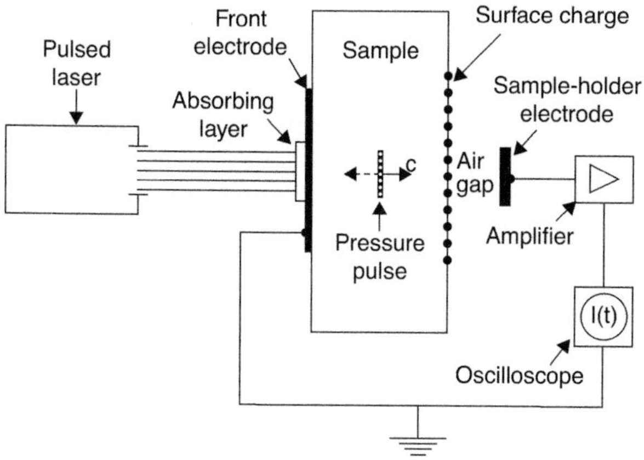

Figure 4.23 Experimental setup for the laser-induced pressure-pulse (LIPP) method on one-side metalized samples (after [19]; see also [57]).

contains either a ~2-μm-thick painted-on or sprayed-on graphite coating or a 100- to 500-nm evaporated Zn, Cd, Pb, or Bi coating. The laser pulses are generated by an actively mode-locked[10] and Q-switched 1.064-pm Nd:YAG laser system. Figure 4.23 shows the experimental setup for the LIPP used by Sessler and co-workers which provides a resolution of ~1 μm [19, 57].

Suzuoki et al. have utilized LIPP in their study of space-charge distribution in 0.1-, 0.6- and 1-mm-thick slabs of low-density PE [74] and a 150-μm-thick film of chemically cross-linked PE [75] under DC stress. The laser used in their technique was a Nd:YAG laser (Quantel 571C, pulse width: 100 ps, 150 mJ, and 12 ns, 1.1 J). The samples had a carbon-loaded semiconducting layer (150 μm in thickness) as an electrode on one side and a vacuum-deposited Au electrode on the other side [19].

It can be argued that the LIPP method is in fact very similar to the PWP method in its most popular setup, that is, when the pressure wave is generated by absorbing a short laser light pulse in a metal target bonded to the dielectric plate under investigation. According to [19] (which, however, dates back to 1997), they still differ strictly speaking, as it can be argued from Table 4.3 in which the comments from Table 4.1 relevant to LIPP and PWP are selected for a more straightforward comparison.

Nevertheless, in [14] (dating back to 2003) it is said: "the pressure-wave-propagation (PWP) or laser-induce-pressure-pulse (LIPP) method ([14], page 18)," thereby acknowledging that PWP and LIPP in practice do coincide. And in the even more recent paper [76] issued in 2009, this argument is further supported, but making

[10] Mode locking is a technique in optics by which a laser can produce pulses of light of extremely short duration, on the order of picoseconds (10^{-12} s) or femtoseconds (10^{-15} s). The basis of the technique is to induce a fixed phase relationship between the modes of the laser's resonant cavity. The laser is then said to be phase locked or mode locked. Interference between these modes causes the laser light to be produced as a train of pulses. Depending on the properties of the laser, these pulses may be of extremely brief duration.

Table 4.3 PWP vs. LIPP

Method	Disturbance	Scan Mechanism	Detection Process	r^a (μm)	t^b (μm)	Comments
Laser-induced pressure pulse method	Absorption of short laser light pulse in front electrode	Propagation with longitudinal sound velocity	Current between sample electrodes	1	100–1000	No deconvolution required
Pressure wave propagation method	Absorption of short laser light pulse in metal target	Propagation with longitudinal sound velocity	Voltage or current between sample electrode	10	5–200	Resolution improved with deconvolution. Also used for surface charge measurements

[a]r is the resolution of the method.
[b]t is the sample thickness.
Source: According to [19].

an observation as to the alternative possible ways to generate the pressure waves other than the laser pulse. Indeed in [76] it is said that:

> Since the introduction of the PWP method in 1976, various techniques have been proposed to generate the pressure wave. When a laser is used to generate the pressure pulse, the method is called the laser-induced-pressure-pulse (LIPP) method. When piezoelectric materials are used, the method is called the piezoelectrically-induced-pressure-pulse (PIPP) method if a pressure pulse is generated, and the piezoelectrically-induced-pressure-step (PIPS) method if a pressure step is generated ([76], page 8).

However, once more in [76], when describing the PWP method explicit reference is made to Figure 4.24, where the wave is shown clearly as generated by a laser and the method is also referenced as LIPP. In conclusion it can be argued that actually the LIPP is the far more common practical implementation of the PWP method at present. More about the recent developments of the PWP–LIPP method will be given in Section 4.5.3.

4.4.2.3 Pulsed Electroacoustic (PEA) Method

Principle of the PEA Method The principle of the PEA method is based on the Coulomb force law: An externally applied pulse field induces a perturbing force density on the material in the presence of resident charges [13, 77]. The perturbation generates an acoustic wave that originates from the charged bulk. Then, the acoustic signal is detected by a piezoelectric transducer mounted on one of the electrodes (see Fig. 4.25). The space-charge profile information contained in the exiting acoustic signal is extracted and calibrated through the use of a digital signal processing. It should be noted here that *the acoustic wave is internally generated by the space charge in contrast to the pressure wave propagation methods where the pressure wave is generated externally.*

The main advantage of PEA, compared to previously described methods, is that the high-voltage circuit and the circuit for detecting the pressure wave are

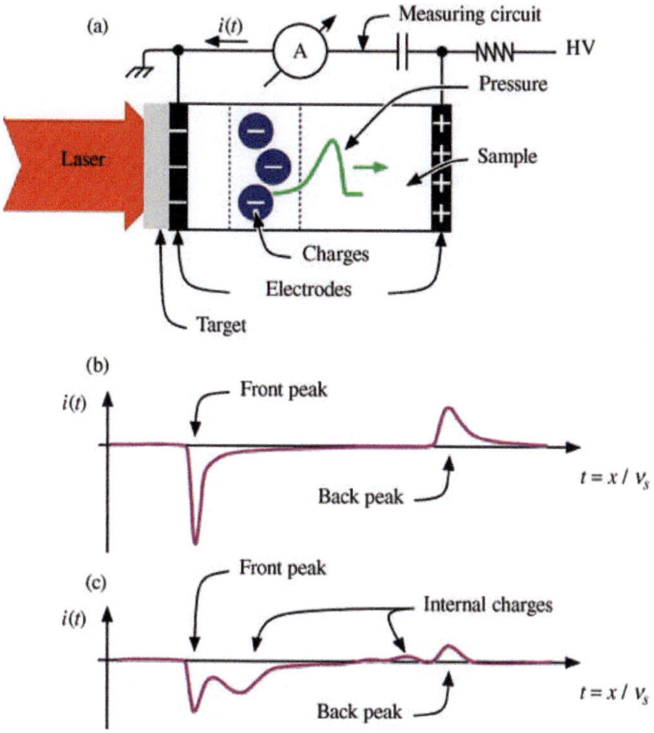

Figure 4.24 Principle of the pressure-wave-propagation (PWP) method when using a laser for generating a pressure pulse (LIPP technique). (*a*) General setup and typical signals under voltage, (*b*) without internal charges and (*c*) with internal charges. The front and back peaks correspond to the charges on the electrodes and can be used to determine the material boundaries in the signal (after [76]).

electrically separated by grounding the electrode placed between the sample and the piezoelectric transducer. Under these conditions, even in case of dielectric discharge, the ground electrode can provide protection to the amplifier and oscilloscope connected to the piezoelectric transducer. This is the reason the PEA is considered the safest method for the study of samples subjected to high voltages.

Initially for the electric field excitation a sinusoidal waveform was used, but such waveform did not make it possible to measure the profile of space charge in the thickness of the dielectric. Subsequently, the sinusoidal waveform was replaced with an electrical pulse of short duration.

Until the late 1980s only samples of laminar form could be tested. In fact, at that time the PEA method was used by Takada and other researchers, for example, in the evaluation of the space-charge profiles in thick slabs of several polymer dielectrics charged by electron beam irradiation, or in scanning charges accumulated on a gas–solid interface (see [19] for a detailed reference list). Subsequently, a new family of piezoelectric transducers based on poly-vinyli-dene fluoride (PVDF) films has made it possible to apply the PEA method also to coaxial cables: Indeed, such

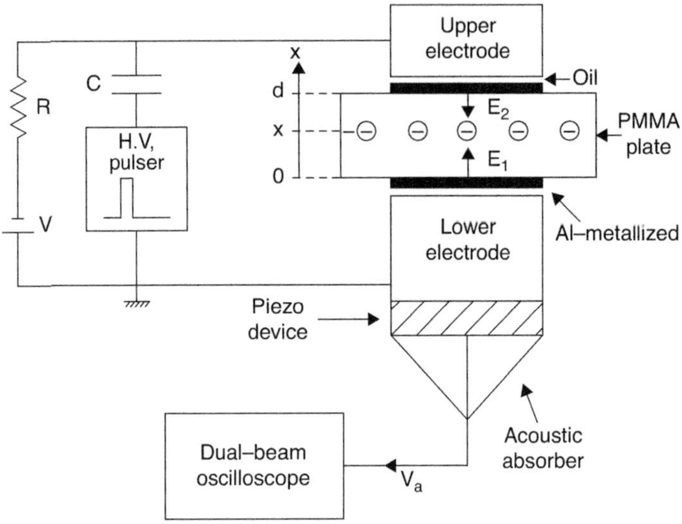

Figure 4.25 Block diagram of the pulsed electroacoustic (PEA) method (after [19]; see also [77]).

transducers have a high level of piezoactivity, a large bandwidth, low acoustic impedance, and are physically flexible, malleable, and relatively cheap.

The resolution of the PEA method depends on the duration of the electric pulse as well as on the thickness of the sample. In [19], a resolution of 100 μm with the application of 90-ns pulses is reported, emphasizing that, in principle, the limitation of this method might come from the difficulty of detecting complex charge and polarization profiles [19].

More details about the PEA method and relevant results obtained can be found in the milestone study [13] by its founder, T. Takada. In [13], dating back to 1999, the principle of the PEA method is described as follows (see Fig. 4.26):

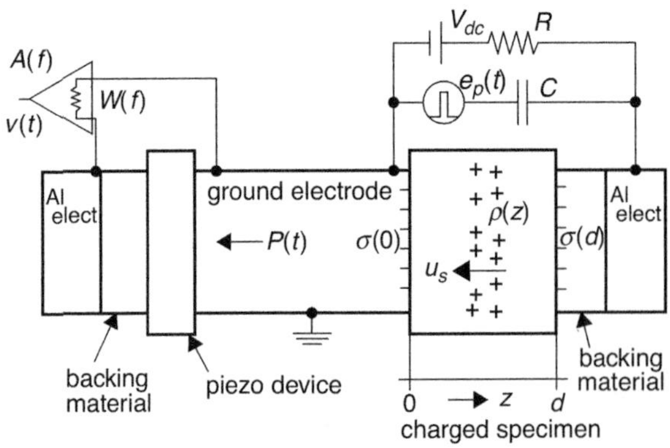

Figure 4.26 Block diagram of the pulsed electroacoustic (PEA) method (after [13]).

Consider a plate sample with thickness d and space charge distribution $\rho(z)$. An externally applied pulsed electric field $e_p(t)$ is applied to the sample and induces a perturbation force on each charge. This force causes the charge to move slightly. This movement launches an acoustic wave that travels at the speed of sound u_S and is proportional to the charge distribution in the sample. A piezoelectric transducer is used to detect the acoustic wave and transform the acoustic wave into an electric signal ([13], page 522).

As to the generation of a pulsed acoustic wave from space charges, more in detail let us suppose a DC voltage is applied to a sample of thickness d. Space charge $\rho(z)$ accumulates in the sample if the amplitude of the applied DC voltage V_{DC} is high enough and the applied period is long enough. At the same time, surface charges $\sigma(0)$ and $\sigma(d)$ (see Fig. 4.26) are induced by both the DC voltage and the space charge $\rho(z)$. When an electric pulse $e_p(t)$ is applied to the sample through a coupling capacitor C, the pulsed acoustic wave related with the charge distribution is generated. The acoustic wave component $p(t)$ propagating through the ground electrode, which is proportional to the space-charge distribution $\rho(z)$, can be expressed in a simplified form if the following conditions are satisfied [13]:

1. Δt_P (width of the HV pulse) $\ll \Delta t_S = d/u_S$ (transit time of acoustic waves across the sample).
2. Z_s (acoustic impedance of sample) $= Z_b$ (acoustic impedance of electrode). To satisfy this condition a piece of conducting "impedance-matching" material is placed between the HVDC electrode and the sample.

Then the relationship between $p(t)$ and $\rho(z)$ becomes

$$p(t) = \frac{Z_{\mathrm{Al}}}{Z_s + Z_{\mathrm{Al}}}\left[\sigma(0)e_p(t) + u_s\int_0^t \rho(\tau)e_p(t-\tau)\,d\tau + \sigma(d)e_p\left(t - \frac{d}{u_s}\right)\right] \quad (4.34)$$

which includes three components: surface charge $\sigma(0)$ on the ground electrode, space charge $\rho(z)$ in the sample, and surface charge $\sigma(d)$ on the HV electrode [13].

Being $Z_s = Z_b$, the total ratio for each acoustic wave component in Eq. (4.34)—both surface charges and space charge—is $Z_{\mathrm{Al}}/(Z_s + Z_{\mathrm{Al}})$, so that by detecting the acoustic wave $p(t)$, charge components $\sigma(0)$, $\rho(z)$, and $\sigma(d)$ are obtained. Attaching a piezoelectric transducer to the back of the ground electrode, the acoustic wave signal $p(t)$ is detected and converted into a voltage signal $v(t)$ (see Fig. 4.26), which is proportional to space-charge distribution. In order to prevent the reflection at the interface between the piezoelectric transducer and the rear electrode, it is necessary to match the acoustic impedance between the piezoelectric transducer and the rear electrode. The voltage signal is then amplified by using a preamplifier with high input impedance. In the frequency domain, the amplified voltage signal is expressed as

$$V(f) = S(f)\left[\frac{\sigma(0)}{u_s\Delta\tau} + R(f) + \frac{\sigma(d)}{u_s\Delta\tau}\exp\left(\frac{-j\cdot 2\pi\cdot f\cdot d}{u_s}\right)\right] \quad (4.35)$$

In (4.35) the first component is the surface charge at the ground electrode ($z=0$), the second component is the accumulated space charge in the sample ($0<z<d$), and the third component is the surface charge at the HV electrode ($z=d$). The time delay of the acoustic wave propagating through the sample is d/u_S. The so-called system response function $S(f)$ depends on the properties of the piezoelectric transducer and of the amplifier. By inverse Fourier transform, the space-charge distribution in the time domain τ or position domain z, $\rho(z=u_S\,\tau)$, is obtained [13].

A good understanding of the frequency response of transducer and amplifier is essential. Indeed, in some cases part of the output signal might be distorted due to the frequency response of the transducer and amplifier (e.g., if the transducer and amplifier do not pass the total frequency content of the original acoustic signal) and might be misinterpreted as space charge. As an example, a negative undershoot of a positive charge pulse in the time domain could not be the result of real measured heterocharge, but of the frequency response of the measurement system; see Figure 4.27 after [17].

Typically, distortion occurs if [17]:

- The amplifier's high cut-off frequency is too low or the frequency response of the amplifier is not flat. Then, a high-quality amplifier with good frequency behavior can be a simple solution.
- Transducer and amplifier in series act as a high-pass filter with a cut-off frequency that may lie in the range of the frequency content of the acoustic signal. This problem may arise because the PVDF transducer has a high capacitance (typically 0.1–5 nF), while low-noise high-frequency amplifiers (500 MHz) often have a low input impedance (50 Ω). It can be solved analytically or via an amplifier with higher input impedance, for example, 50 kΩ (but much higher noise level as well).

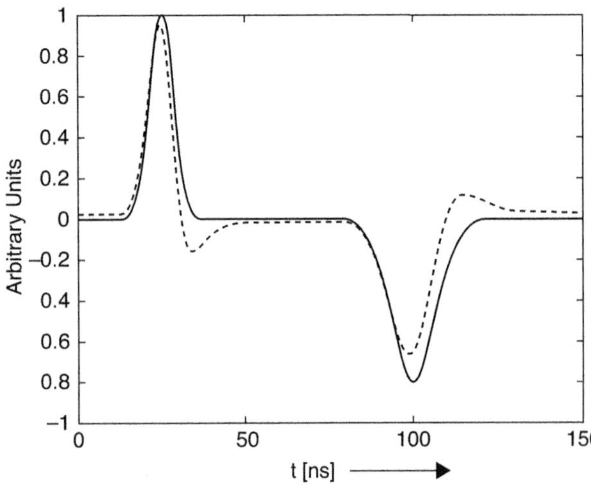

Figure 4.27 Time domain signals: the simulated acoustic signal $p(t)$ (solid line) and the distorted signal $v(t)$ (dash-dot line) (after [17]).

A proper calibration is needed for converting the voltage signal V on the oscilloscope (in volts) into a real charge density estimate (typically in C/m^3 or equivalently in μC/cm^3) [13, 17]. As shown in [17], it holds [see the relationships (4.34) and (4.35):

$$V = K_p p = K_\rho \rho \tag{4.36}$$

where K_p is a proportionality parameter between pressure wave p and voltage output V, while K_ρ is the overall calibration factor.

By comparing (4.36) with (4.35), it can be argued that the calibration factor depends on frequency via the system function, which should account for the full frequency-dependent response of transducer and amplifier. A thorough treatment of calibration, including a convolution-type description of the signals, is reported in [78], but—for a "quick-and-easy" illustration of calibration—a simplified expression of the calibration factor can be derived by assuming that K_ρ is independent of frequency [17]. By considering a sample initially free of space charge and applying across its thickness d a moderate DC calibration voltage $U_{\mathrm{DC,cal}}$ (such that no space charge is injected), only surface charges $\sigma(0)$ and $\sigma(d)$ on the electrodes are generated, but no space charges. In this situation, $|\sigma(0)| = |\sigma(d)| = \sigma_{\mathrm{cal}}$, a quantity that can be easily computed as follows [17]:

$$\sigma_{\mathrm{cal}} = \frac{\varepsilon\, U_{\mathrm{DC,cal}}}{d} \tag{4.37}$$

Thus, given $b = u_S\, \Delta t_P$, from (4.36) and (4.37) with proper rearrangements the following expression can be obtained for the calibration factor K_ρ [17]:

$$K_\rho = \frac{d b\, V_{\mathrm{cal}}}{\varepsilon\, U_{\mathrm{DC,cal}}} \tag{4.38}$$

where V_{cal} is the oscilloscope signal during calibration. Note that whenever the DC voltage U_{DC} or the pulse voltage U_p are changed during a measurement, the calibration factor K_ρ should be recalculated.

Dynamic Electric Charge Behavior As reported in [13] and shown in Figure 4.28, the experimental setup of the PEA method was improved to synchronously measure the space-charge distribution in a plate sample under any shape of applied voltage, such as step voltage, ramp voltage, and AC voltage at any frequency and phase angles. The applied voltage data are set via personal computer (PC) and then converted into the applied voltage waveform through a specific digital–analog converter. The output of the digital–analog converter is transformed to HV of amplitude up to ± 20 kV by an HV amplifier. An electric pulse of 10-ns duration and $+2$-kV amplitude is applied to the sample to stimulate the space charge and to generate the acoustic signal. A piezoelectric PVDF film transducer (9 μm) is used to detect the acoustic signal and then change it into an electric signal. The detected electric signal is then amplified and processed to yield the space-charge profile [13].

Figure 4.29 shows the space-charge distribution in acetophenone-coated low-density polyethylene (LDPE) of 300-μm thickness after various DC voltage

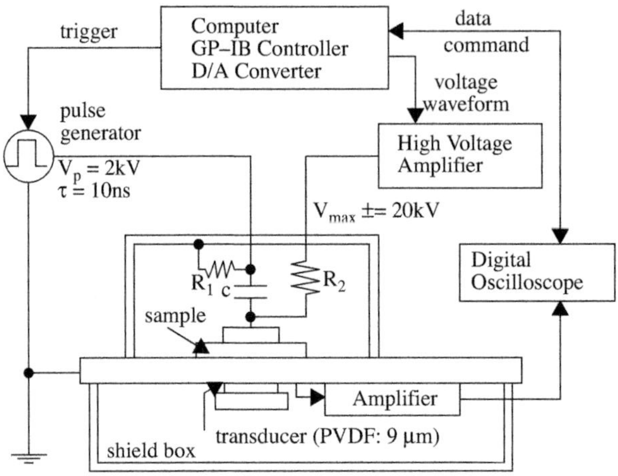

Figure 4.28 Block diagram of the PEA system for space-charge measurements at short intervals and under various shapes of applied voltage (after [13]).

application. The experiment has been conceived because other tests report that cross-linking by-products such as acetophenone help cause hetero- and/or homo-charges in XLPE [13][11] (see also Sections 4.1 and 4.3). The positive signal ① and negative signal ② in Figure 4.29 show the induced charges at anode and cathode, respectively, by the applied DC voltage.[12] It is observed that a large amount of positive charge ③ is injected from the anode at 2 s after voltage application, and the charge drifts toward the cathode side. The top of positive charge reaches the cathode at 2 min; then a large amount of positive charge distributes through the bulk. A small amount of negative charge ④ injection from the cathode is also observed after 30 s [13].

From Figure 4.29, it is argued in [13] that, since the amount of injected positive charge is larger than that observed in LDPE without acetophenone coating and negative charge injection is also observed in acetophenone-coated LDPE, the acetophenone coating reduces the injection barrier and consequently enhances the charge injection. The above results show that the PEA technique provides

[11] The role of cross-linking by-products in causing hetero- and/or homo- charge in XLPE is demonstrated by several more or less recent experiments. For instance, in a thorough experimental investigation "*additive-free non-crosslinked LDPE, additive-free XLPE, and degassed XLPE were soaked in the crosslinking byproducts and the space charge distribution was measured after dc voltage application. . . As a result, the following conclusions were obtained. (1) Cumyl alcohol is responsible for homo-charge layers in front of the electrodes in both low-density and crosslinked polyethylene sheets. (2) Acetophenone is responsible for hetero-charge formation in crosslinked polyethylene, presumably as a synergistic effect with water. (3) α-methylstyrene has no effect on space charge formation in low-density polyethylene, while it assists charge trapping in crosslinked polyethylene. (4) Charge trapping occurs easier in degassed crosslinked polyethylene than in low-density polyethylene, probably because of carbonyl groups induced by crosslinking*" (Y. Maeno, N. Hirai, Y. Ohki, T. Tanaka, M. Okashita, T. Maeno, "Effects of crosslinking byproducts on space charge formation in crosslinked polyethylene", *IEEE Transactions on Dielectrics and Electrical Insulation*, Vol. 12, No. 1, pp. 90–97, February 2005).

[12] Unfortunately, the amplitude of the applied DC voltage is not reported in [13].

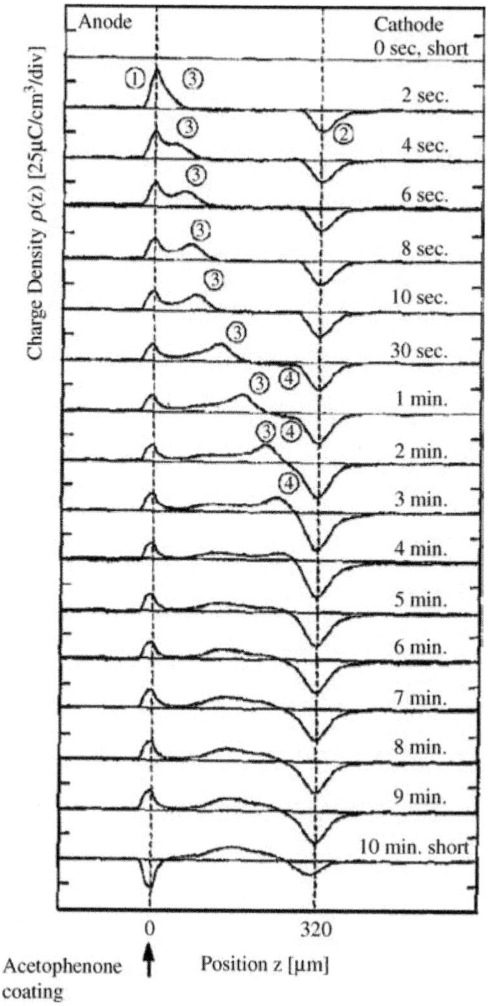

Figure 4.29 Dynamic space-charge measurement in acetophenone-coated LDPE (after [13]).

both short interval (~3 s) and continuous (>4 h) space-charge measurement, thereby monitoring charge dynamics in solid dielectrics under electric stress.

Interfacial Space-Charge Accumulation Space-charge accumulation at the semicon–insulation interface is known to be an issue in HVDC power cables. Figure 4.30 shows the space-charge accumulation at the LDPE–EVA (ethylene vinyl acetate) interface obtained after positive voltage application to the EVA side. Figure 4.30*a* shows the accumulation characteristics of the space-charge distribution during DC voltage application,[13] highlighting that positive charges only accumulate at

[13] Unfortunately the amplitude of the applied DC voltage is not reported in [13].

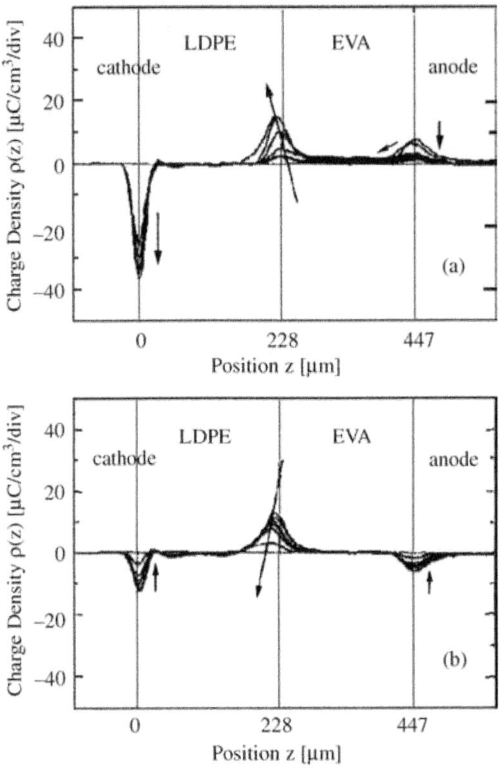

Figure 4.30 Space-charge accumulation at LDPE–EVA interface under DC voltage: (a) during voltage application (volt-on) and (b) under short-circuit (volt-off) (after [13]).

the LDPE–EVA interface. The positive charges induced at the anode decrease with increasing voltage application duration. Furthermore, the positive charge gradually moves into the LDPE layer from EVA. Figure 4.30*b* shows the decay characteristics of space-charge distribution obtained under short circuit, thereby emphasizing that the interfacial charge decreases with time after short circuit. Focusing on the charge decay around the interface, the decay speed in the LDPE layer seems to be slower than that in the EVA layer [13].

Space-Charge Accumulation in Cables As pointed out in previous chapters, the measurement of space charge in the cable insulation walls is fundamental in the development of extruded HVDC cables, since space charges affect the field distribution under DC voltage significantly and space-charge accumulation—resulting from the combination of many factors, such as carrier mobility, carrier density, and trap density—is not easy to estimate from basic relationships and data of the material. According to [13] the first space-charge measurement in a coaxial cable by the PEA method was carried out by Fukunaga et al. [79] through a system similar to that shown in Figure 4.31, where the cable sample electrode is of a coaxial structure, maintaining close contact between the cable and the electrode, and a piezoelectric transducer of

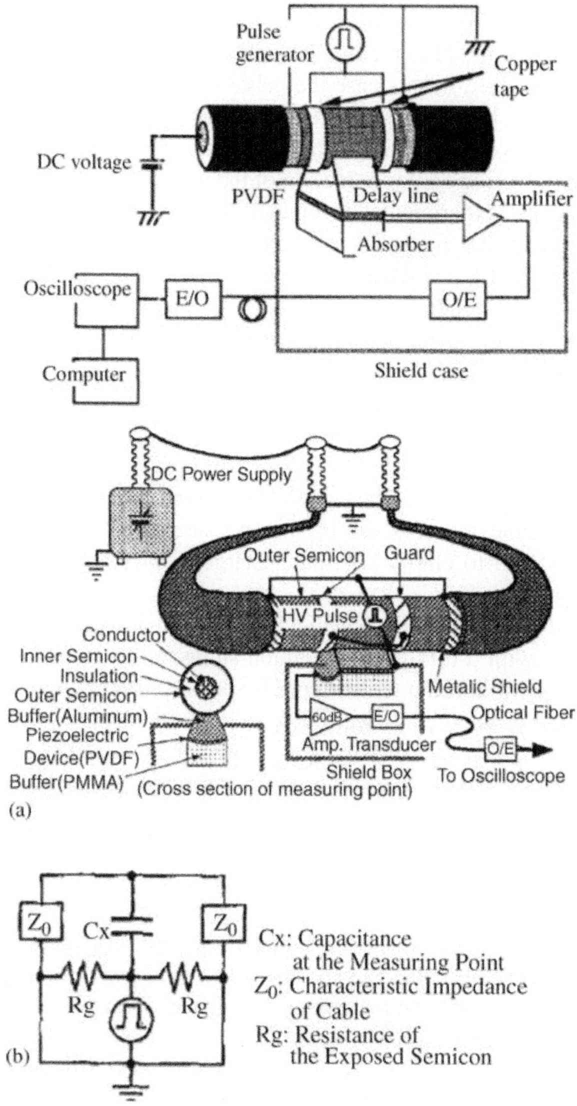

Figure 4.31 Space-charge measurement system for a long cable with the PEA method: (*a*) experimental setup according to two alternative, but equivalent, representations (after [83, 84], respectively) and (*b*) equivalent circuit (after [13]).

PVDF film is tightly wound around the coaxial-shaped electrode in order to prevent acoustic wave reflection at the interface between the electrode and the transducer. Liu et al. [80] analyzed the relationship between the detected signal $v_S(t)$ and the space-charge distribution $\rho_T(r)$. As a result, the space-charge distribution in the coaxial cable was obtained.

Following the successful space-charge measurement in the coaxial cables, it was necessary to carry out the measurement by applying HV to cables. The cable length should be longer than 10 m because large HV terminations have to be attached. If the pulse voltage used in the PEA method is applied at the terminations, it travels through the cable and—since in this case the cable specimen cannot be considered as a lumped capacitance, but rather its characteristic impedance has to be considered—the waveform of the voltage at the measuring point would not be a proper single pulse, due to the distortion and/or reflection of the injected pulse [13].

Hozumi et al. [81] designed an experimental set-up that can solve this problem, as shown in Figure 4.31*a*. The measuring point is at the center of the exposed semiconducting part to apply a pulsed voltage between this point and ground. Typically, the pulse voltage is 2 kV high and 60 ns wide. As observed in [13], page 530, "If the impedance of the capacitance at the measuring point in the frequency range of the pulse voltage is much higher than the characteristic impedance of the cable, most of the pulse voltage is applied at that point (see the equivalent circuit in Fig. 4.31*b*). The acoustic wave is measured using a PVDF film that is typically 110 μm in thickness and is amplified inside the shielded box. The shielded box is electrically isolated because an HV pulse is applied there. The signal is transferred by an optical fiber. This method made it possible to carry out in situ measurements with the application of HV to the cable, and to clarify the space-charge behavior in thick insulation under extremely high fields [13]."

The space-charge evolution—measured with the PEA system illustrated in Figure 4.31—in an XLPE cable with 3-mm-thick insulation subjected to $+70$ kV DC voltage applied to the conductor at room temperature is shown in Figure 4.32*a* [81]. Heterocharges accumulated close to both electrodes, with a trend of the space charge to saturate in $\sim$150 min. The heterocharges are influenced strongly by impurities in the insulation, such as cross-linking by-products, antioxidants, and the like [13], as already mentioned in Section 4.3.2.

It is also observed in [13], page 530, that "when the average field is as high as 1.2 MV/cm (=120 kV/mm), packet charges appear intermittently and move toward the counterelectrode, as shown in Figure 4.32*b*. A similar result for film specimens has been reported by Kon et al., who used the laser-induced pressure propagation (LIPP) method [82]. This phenomenon is believed to be due to the local ionization of impurities under a high field. The relationship between the generation of packet charges and the insulation performance under DC voltages has not yet been clarified. However, *the generation of packet charges leads to instability in the insulation and would not be preferable in designing HVDC cables* ([13], page 530)."

As a concluding remark, in [13] it is observed that:

> A material that does not form these space charges can maintain a uniform electric field, and is necessary in the development of a DC cable. XLPE with fillers is a promising material for DC cable insulation, because it suppresses the movement of carriers. Some materials of this type have been applied to the development of 250 kV DC cables. Their excellent performance has been verified through long-term test and direct space charge measurement in the full-scale cable [83, 84] ([13], page 530).

More details about the space-charge measurements in full-scale cables according to [84] are reported in Section 4.6.

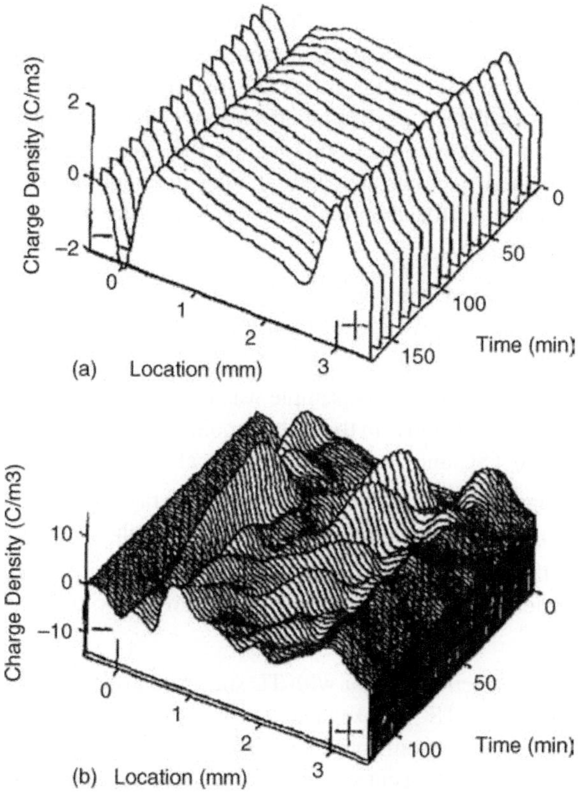

Figure 4.32 Space-charge evolution in XLPE cable with 3-mm-thick insulation under DC voltages. (*a*) +70 kV DC applied to the conductor: heterocharge accumulates in the vicinity of both electrodes; and (*b*) +350 kV DC applied: packet charges appear intermittently and move toward the counterelectrode (after [13]).

4.4.3 Techniques for Estimating the Trap Depth and the Mobility of Space Charges

The methods illustrated in previous Sections yield the space-charge profile in the insulation thickness, but in many cases it could be important to determine the energy levels of the traps occupied by the charge carriers constituting the space charge. The main technique for the measurement of the energy levels occupied by such charge carriers is thermally stimulated currents (TSC) [85], which yield valuable information that supports the designers of insulation systems in selecting the optimum operating conditions [86]. TSC measurements in turn can be classified into thermally stimulated discharge current (TSDC) measurements and thermally stimulated polarization currents (TSPC).

According to the TSDC technique, the short-circuit or open-circuit discharge current is measured while raising the sample temperature at a constant rate. Short-circuit TSDC performs better for dipolar dielectrics than open-circuit TSDC, while the opposite holds for nonpolar dielectrics [85].

As a well-known example of application of the TSC measurements, in [87] carrier trapping sites were analyzed via TSC and thermoluminescence (TL) measurements as a part of an investigation relevant to various DC treeing phenomena in PE. Since such phenomena were explained in [87] based on space-charge storage associated with electron injection or extraction, in this picture the various energy levels of trapping sites where injected electrons are captured play an important role. As reported from [87], the experimental procedure for TSC and TL measurements in PE consists of the following steps:

1. The sample was cooled to a low temperature, T_1 ($-185°C$), in vacuum (10^{-3} Torr) and irradiated with X rays for 20 min, while short circuited in order to fill the traps with carriers.
2. A bias voltage of 180 V was applied and the sample was heated at the rate of 6°C/min from T_1 to a temperature T_2 close to the melting point of the crystalline regions, while measuring the TSC and TL spectra (the so-called first run).
3. The sample was cooled again to T_1 and reheated with no X-ray irradiation under the same experimental conditions (the so-called second run).
4. The difference between the first and second TSC spectra was taken as an estimate of detrapped carriers, since the second TSC spectrum has no trapped carriers.

Typical first and second TSC spectra together with TL spectrum after [87] are plotted in Figure 4.33 [87]. The five TSC peaks denoted as P_1–P_5 with rising temperature belong to the first TSC spectrum (no TSC peak was detected in the second TSC, apart from the steady conduction current at $T>50°C$). These peaks were addressed in [87] to the thermal release of trapped carriers and a comparison with data in the literature led to the following estimates of trap energy levels associated with each TSC peak (see Fig. 4.33 after [87]):

1. P_1 (0.2 eV), shallow traps, detrapping enhanced by the γ-relaxation.
2. P_2 (0.5 eV), traps in the amorphous regions, detrapping enhanced by the β-relaxation.
3. P_3 (1.1 eV), traps in the boundary regions between the crystalline and the amorphouse regions.
4. P_4 (1.2 eV), traps in the crystalline region.
5. P_5 (uncertain), traps in the crystalline region.

It can be argued that these estimates of shallow and deep trap depths are consistent with those reported previously at Section 4.3.2.

A second technique available to evaluate the energy levels of the traps occupied by the charge carriers forming the space charge is the photostimulated discharge (PSD) technique [85]. According to the PSD technique, the trapped charges are detrapped by light impinging on the sample, whose temperature is ideally constant. The PSD can be employed for transparent or translucent, polar or nonpolar dielectrics. The wavelength of the light used to stimulate the discharge ranges commonly from 200 to 800 nm [88].

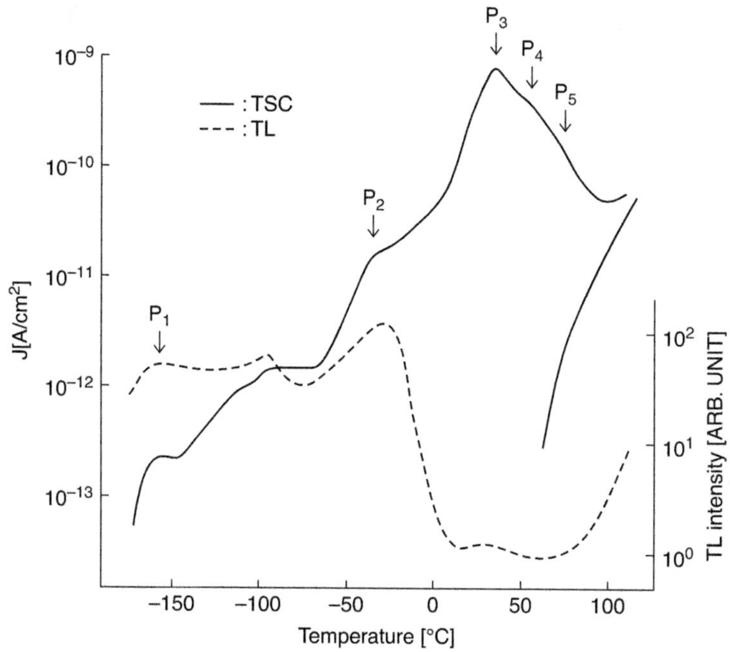

Figure 4.33 TSC and TL spectra of low density polyethylene (after [87]).

Another important parameter as to space charge in dielectrics is charge carrier mobility, which is difficult to measure in polymer insulation. The time-of-flight method, simply applicable to semiconductor samples and to some photosensitive polymers, is often unsuccessful in insulating polymers such as PE [85]; indeed, the time of flight, denoted by a rapid current drop when the carriers begin to reach the counterelectrode, is frequently undetectable because of the very small number of carriers reaching the counterelectrode. The mobility of electrons in a dielectric can sometimes be estimated by studying the transport of an electron packet moving through the sample under an applied electric field, following electron injection from an electron beam, although it has been suggested that the electron mobility may not be constant during transport through the sample [85].

In [89] a theory was developed for the estimation of mobility and trap energy distribution from the results of space-charge measurements. Under strong, but reasonable, approximations the theory arrives at analytical formulas for the so-called apparent trap-controlled mobility and trap energy level (or trap depth). The depolarization characteristic obtained recording space-charge magnitude $q(t)$ at time t elapsed, after poling, when the electric field is removed and electrodes are short circuited and grounded, is the experimental reference for the application of these formulas, whose derivation can be summarized as follows. Let us consider a flat dielectric specimen of thickness d between plane parallel electrodes, short circuited after poling at a given voltage V_0 for a time t_P. At time t of depolarization, it still possesses a mean density of stored charge – averaged over thickness – $\rho_m(t)$.

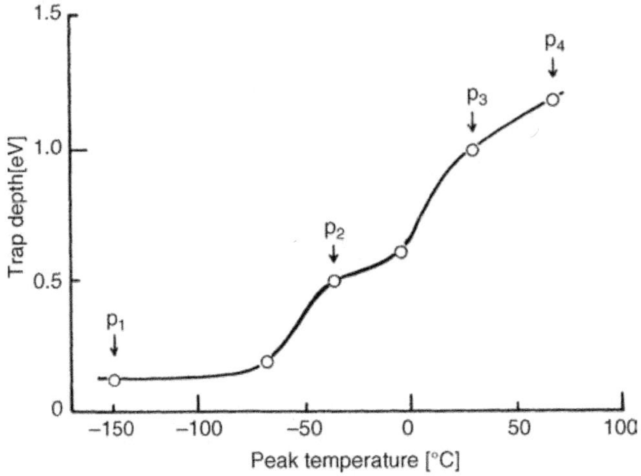

Figure 4.34 Estimated trap depth of trapping centers by applying partial heating method to TSC curve. (after [87]).

Resorting to two main hypotheses [89]—(1) charge relaxation is only due to leakage current while recombination is neglected and (2) charge densities are replaced by their average values (prevailing charge is unipolar and located close to the electrodes)—the following expression for "apparent trap-controlled" mobility $\mu(t)$ is attained:

$$\mu(t) \propto \frac{\varepsilon}{\rho_m^2(t)} \frac{d\rho_m(t)}{dt} \tag{4.39}$$

In [89] mobility according to eq. (4.39) is compared with that estimated by means of:

- Eq. (4.3) based on direct observation of the packet charge dynamics during poling (see Section 4.2.1.1);
- Eq. (4.13) based on the SCLC theory (see Section 4.2.2.2).

These latter are approximated, but yield useful orders of magnitude for a rough evaluation of eq. (4.39).

In [89] mobility is estimated via eq. (4.39) from the high-field depolarization characteristics derived by means of PEA measurements carried out on thin films peeled from HV XLPE cables; Figure 4.35 shows these characteristics for both unaged films (samples labeled as XLPE 1 and 2) and aged films (sample labeled as XLPE 3), poled under DC voltage at 150 kV/mm. The mobility values versus depolarization time calculated through Eq. (4.39) for the "XLPE 1" data set of Figure 4.35 are plotted in Figure 4.36 (referred to as EQUATION 4): They drop from $\sim 10^{-14}$ to $\sim 5\ 10^{-17}$ m^2/Vs as depolarization time goes from 0.1 to 10,000 s, since deeper and deeper traps are involved in the charge-release mechanism. These mobility values fall in a meaningful range compared to the data provided in the literature for

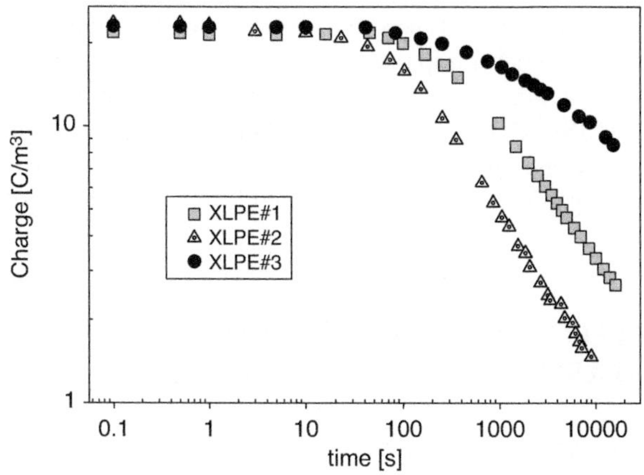

Figure 4.35 Depolarization characteristics for XLPE 1, 2, and 3; poling field = 150 kV/mm (after [89]).

polyethylenes (see, e.g., [12, 34, 90]). Moreover, the other two approaches – i.e. equation (4.3) and equation (4.13), referred to respectively as EQUATION 5 and EQUATION 6 in Figure 4.36) – provide mobility estimates that are in the same range as those derived via Eq. (4.39): This confirms at least the orders of magnitude of mobility determined by the procedure proposed in [89].

The depolarization characteristics were also used in [89] to derive trap depth estimates as a function of depolarization time, obtaining trap depths ranging between ~0.86 and ~1.04 eV for the considered depolarization time range. These estimates are consistent with the deep trap energy level P_3 after [87] quoted previously in this

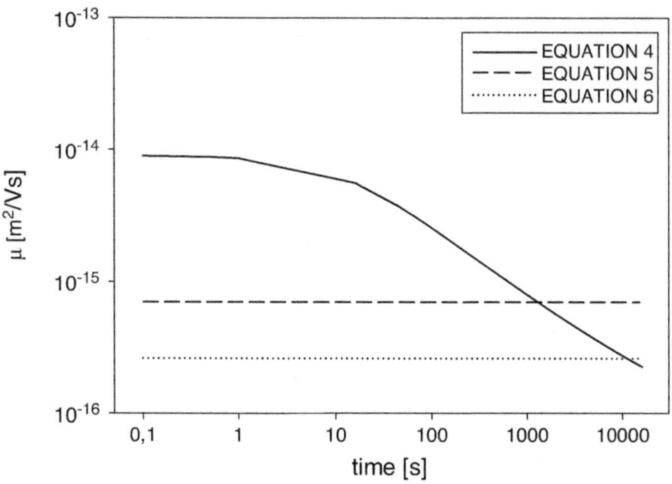

Figure 4.36 Mobility values calculated via Eq. (4.39)—EQUATION 4 in the figure—versus depolarization time, together with the estimates derived from eq. (4.3)—EQUATION 5 in the figure—and eq. (4.13)—EQUATION 6 in the figure—for the data of Figure 4.35 (after [89]).

Section, as well as with deep trap energy levels reported elsewhere in this chapter (e. g., in Section 4.3.1.4 estimates of electron trap energies associated with impurities and structural defects around 1 eV were quoted after [14]). Apparent mobility and trap depth estimates attained in [89] change considerably with aging; thus, on the whole it is concluded in [89] that the proposed approach, even if largely approximated, seems capable of providing quantities sensitive to manufacturing process and degradation.

4.5 UP-TO-DATE DEVELOPMENTS OF THE BEST TECHNIQUES FOR MEASURING SPACE CHARGES

In [19], after describing all the space-charge measurement methods listed at the beginning of Section 4.4, a comparison of these methods was carried out by focusing particularly on the most promising ones, able to give detailed information about the internal charge distribution of a dielectric. These were singled out to be the thermal pulse method (particularly in the further development as the thermal step method, TSM), the pressure wave propagation (PWP), and the pulsed electroacoustic method (PEA). Since 1997—when [19] was issued—further developments and applications have regarded these methods for space-charge measurements, which are at present the most widely used. Such developments are illustrated in this section, with a particular focus on extruded HVDC cable insulation.

4.5.1 Space-Charge Measurements in HVDC Cables with the TSM Technique

In [91], dating back to 2009, a recent approach adopted by Nexans (France and Norway) and the University of Montpellier 2 to study and test newly developed HVDC cable insulation is described. Long-term space-charge dynamics have been investigated through measurements carried out by means of the thermal step method (TSM) on cables with fairly thick insulation ($\sim$5 mm).

A detailed description of the test arrangement and results from [91] is reported below. Indeed, these measurements are an example of an up-to-date approach to space-charge measurements on cables of geometry close to that of full-size power cables; hence, the relevant results can help in understanding conduction and trapping mechanisms in HVDC cable insulation under service conditions.

4.5.1.1 Experimental Facility

Background The space-charge measurement technique is the TSM in a double-capacitor configuration, illustrated in Section 4.4.1.3 and exhaustively described in [67]. Contrasting with the classical setup, when the measurement is made on a single short-circuited cable sample, the double-capacitor technique uses a "compensation cable" connected to the cable under test, as shown in Figure 4.37 (the two cables are assumed to be identical). By connecting the external semicon of the compensation sample to a current amplifier and its core to that of the measured specimen, a double capacitor is created. A voltage can then be applied to both cores,

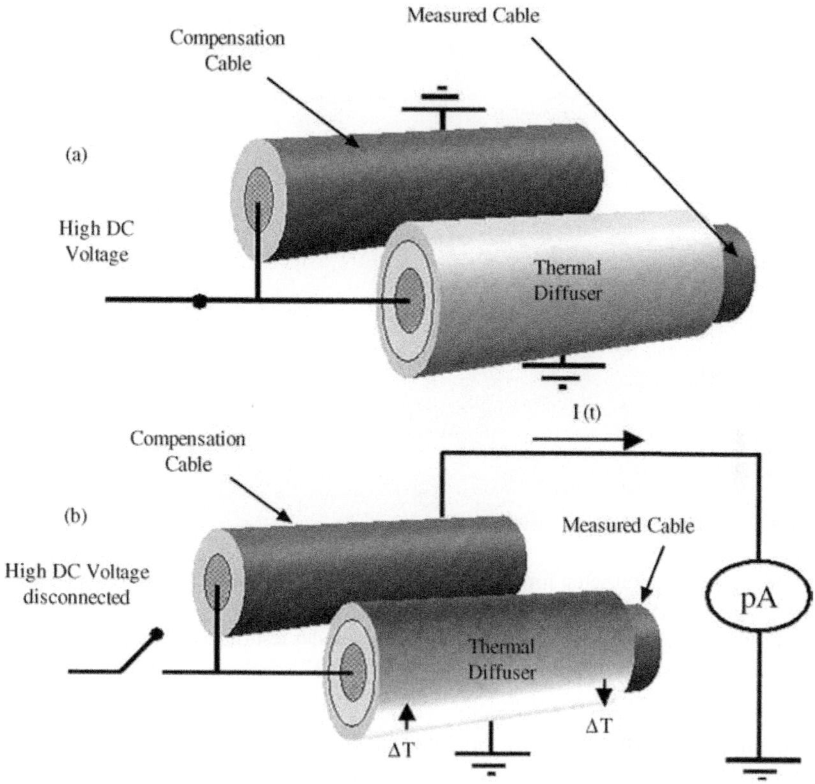

Figure 4.37 "Under field" TSM measurements on cables in double capacitor configuration: (*a*) the DC field is applied prior to the measurement and (*b*) the measurement is performed on the measured cable after disconnecting the DC source (after [91]).

and a thermal step to the cable under test. The current through the compensation sample is recorded (see Fig. 4.37). The current in this double-capacitor configuration, using the outer cooling technique, is given by [67] [see Eq. (4.30) and relevant comments]:

$$I_{\mathrm{TSM}}(t) = -\alpha \frac{C}{2} \int_{r_i}^{r_o} E(r) \frac{\partial \Delta T(r,t)}{\partial t} \, dr \tag{4.40}$$

where $\Delta T(r,t)$ is the variation of the temperature within the cable, C is the capacitance of the thermally excited cable, α is a constant of the material, and r_i and r_o are the inner and outer radii of the insulation. The field is related to the charge via Poisson's equation. The measured current reflects broadly the amount of charge trapped within the insulation. The electric field and space-charge distributions can be obtained by solving Eq. (4.40) [67].

As pointed out in Section 4.4.1.3, using two identical cables offers the considerable advantage of compensating for the polarization and conduction currents, which can flow across the insulation under high DC field. Theoretically, the

Figure 4.38 Experimental setup for TSM measurements of cables under applied field and thermal gradient. (1) HVDC supply, (2) voltage divider and HV switch, (3) cable junction sphere, (4) HV monitoring device (field mill), (5) cables under test (measurement and compensation) fitted with oil-filled lab terminals (after [91]).

measured current is solely due to the internal field of the measured sample. In practice, a slight imbalance between the cable samples always exists and a differential conduction and/or polarization current component is superimposed on the thermal step signal. Obviously, the higher the applied field and temperature, the greater the imbalance. To minimize this problem, the measurement is made in two steps. First, the usual measurement is made while the thermal step is applied. Second, having restored the initial conditions, a second measurement is made without applying a thermal step. In this way, the differential current is obtained and later on subtracted from the current measured in the first step: The result is the signal caused only by the space charge contained in the cable under measurement [67].

Industrial Setup An overview of the high-voltage laboratory dedicated to thermal step measurements is given in Figure 4.38. The thermal step is applied by circulating a cold liquid (the so-called outer cooling technique, OCT[14]) through a cylindrical thermal diffuser in contact with the outer semiconductive layer of the cable. To achieve this, the semiconductive tapes, the copper wire screen, and the outer

[14] Two variants of the TSM can be used: the inner heating technique (IHT) and the outer cooling technique (OCT). The IHT does not require any specific preparation of the cable prior to the measurements and follows the evolution of the mean electrical state of the whole cable insulation (several meters long) versus applied electrothermal stresses. Measurements are performed after AC aging and/or after low DC field poling (sometimes the application of a low DC field is necessary to reveal defects within the insulation). The OCT analyzes the cable locally (20–40 cm lengths) and requires that a cooling exchanger is moved along the cable to detect possible longitudinal differences in space-charge buildup in the insulation [91].

HDPE sheath are removed. The screen of the compensation cable is connected to ground via a current amplifier, which measures the thermal step current [91].

Cables under Test To evaluate the HVDC electrical performance of cables with extruded insulation, small-scale trial cables (but with fairly thick insulation, see Table 4.4) were fabricated using vertical extrusion line and triple crosshead [91].

Experimental Conditions The aim of the study [91] was to determine the field distortion that can be expected in the insulation as a result of space-charge buildup over long-term tests (more than 2500 h). Measurements with applied voltage and thermal gradient were made on cable 1. As to cable 2, given its conductivity, measurements under applied field and temperature were not possible because of strong differential currents and potential decay during measurement. Consequently, cable 2 was measured with voltage off ($V = 0$ V at the high-voltage electrode), the insulation layer being isothermal at room temperature. In this way the deeply trapped space charge and its time evolution can be studied. Aging conditions for tests on both cables are given in Table 4.5. Note that room temperature was 18°C.

All of the measurement conditions are given in Table 4.6. In the case of voltage-off—or volt-off—measurements (cable 2), the samples were grounded and cooled for 3 h prior to each measurement [91].

4.5.1.2 Cable 1: Under field Measurements

Cable 1 was aged for 2600 h under −150 kV and 80°C core temperature. Measurements were performed regularly

Table 4.4 Main Features of the Measured Cables Using TSM

Cable	Insulation	Insulation Thickness	Core
1	Polymer 1	4.6 mm	95 mm^2
2	Polymer 2	4.9 mm	95 mm^2

Source: After [91].

Table 4.5 Aging Conditions for Long-Term Tests on Both Cables of Table 4.4

Cable	Current (A)	Temperature of Conductor (°C)	Temperature Drop over Insulation (°C)	Voltage (kV)	Applied Field near the Conductor (kV/mm)
1	490	80	14	−150	43
2	476	80	14	−230	63

Source: After [91].

Table 4.6 Measurement Conditions for Each Cable of Table 4.4

Cable	Temperature of Conductor (°C)	Temperature of Screen (°C)	Temperature of Cold Liquid (°C)	Thermal Step (°C)	Applied Voltage during Measurement (kV)
1	80	60	30	−30	−150
2	20	20	−10	−30	0

Source: After [91].

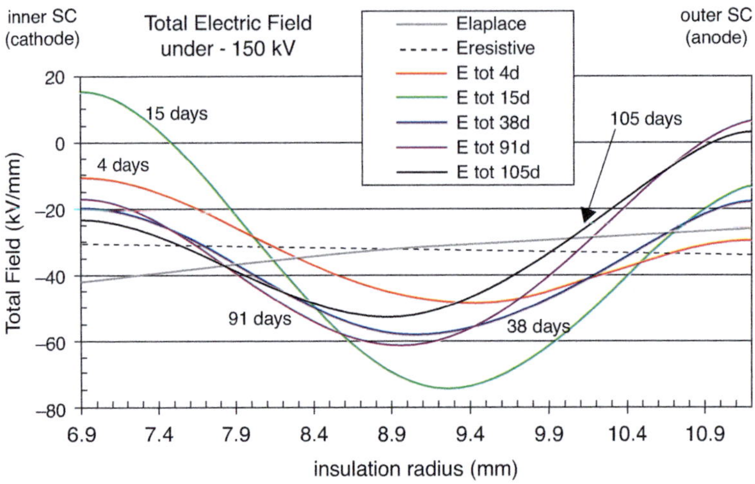

Figure 4.39 Total electric field distribution in cable 1 under applied voltage and thermal gradient. Capacitive (Elaplace) and resistive (steady state) distributions have been included for comparison (after [91]).

in order to study internal field evolution. In each measurement the thermal step signal and the differential current were recorded.

An increase in the thermal step signal was observed up to 15 days, followed by a continuous decrease. This corresponds to maximum field distortion after 15 days, followed by smoothing of the field distribution up to 105 days under stress (Fig. 4.39). The distribution might tend toward the purely resistive distribution, but this is far from certain even after 105 days testing.

Double-charge injection at contacts seems to prevail during aging. However, the density of negative (positive) charge decreases (increases) (Fig. 4.40). Thus the net charge shifts from negative to positive, as shown in Figure 4.41 (charge quantities

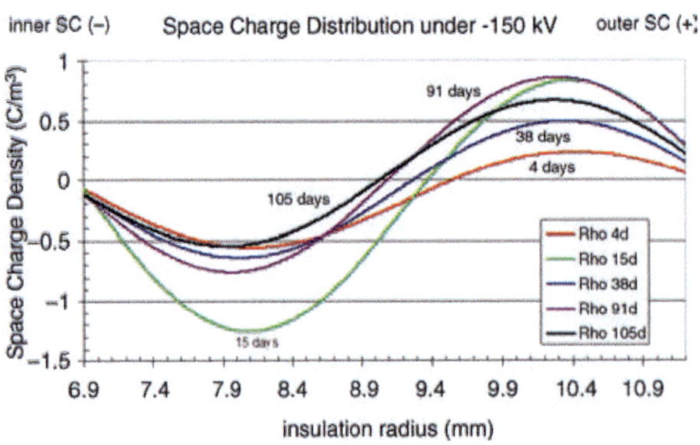

Figure 4.40 Space-charge distributions in cable 1 under applied voltage and thermal gradient (after [91]).

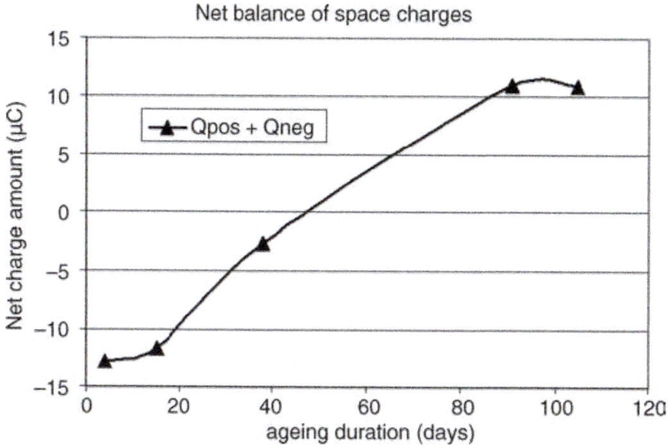

Figure 4.41 Net balance of positive and negative charges in cable 1 (voltage-on measurements under thermal gradient) (after [91]).

have been calculated by integration over the measured volume of the cable). The role of positive carriers seems to be important in space-charge dynamics [91].

4.5.1.3 Study of Cable 2: Volt-Off Measurements Another aging study was carried out on cable 2 under an applied field 50% higher than that applied to cable 1. Since voltage-on measurements could not be performed on cable 2, buildup of residual space charge was studied through volt-off measurements. The negative field close to the anode decreased with increasing poling time and eventually became positive (Fig. 4.42). The density of the injected negative charge close to the cathode decreased, whereas the positive charge expanded from the anode into the bulk (Fig. 4.43). The space-charge densities are lower than those measured in cable 1

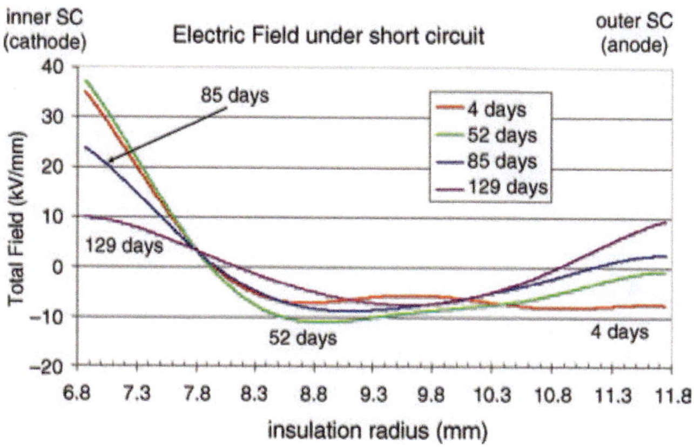

Figure 4.42 Evolution of residual field (volt-off) for cable 2 as a function of aging time (after [91]).

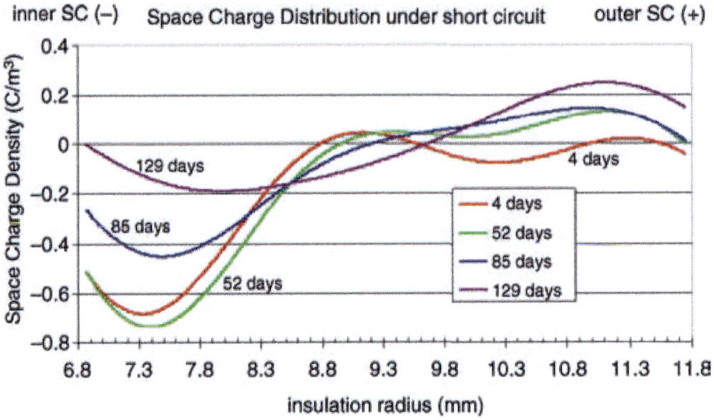

Figure 4.43 Evolution of residual space charge (volt-off) for cable 2 as a function of aging time (after [91]).

under a lower applied field (Fig. 4.40). However, mobile or conduction charge carriers contribute little to the distributions, which relate only to deeply trapped charge (Fig. 4.43, volt-off measurements) [91].

Similar phenomena seem to be occurring in cable 1 and cable 2, despite the differences in applied field and measurement procedures. Negative charge content decays at the later stages of aging, positive charge spreads across the insulation, and the net charge approaches zero (see Fig. 4.44). Since long-term space-charge evolution under stress is important in HVDC cables, these data should be of value.

Because similar trends have been observed in comparable insulating materials, using voltage-on and volt-off measurements (see [91] for detailed list of references) the volt-off mode is considered in [91] to be an attractive alternative for assessment of field distortion under service conditions. The addition of the theoretical field

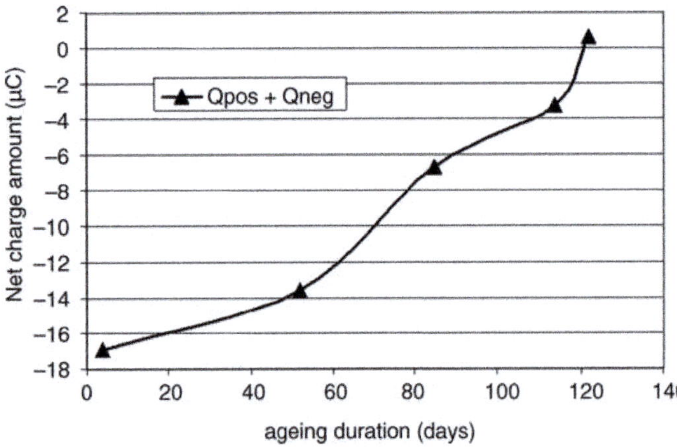

Figure 4.44 Net balance of positive and negative charges in cable 2 (volt-off measurements at room temperature) (after [91]).

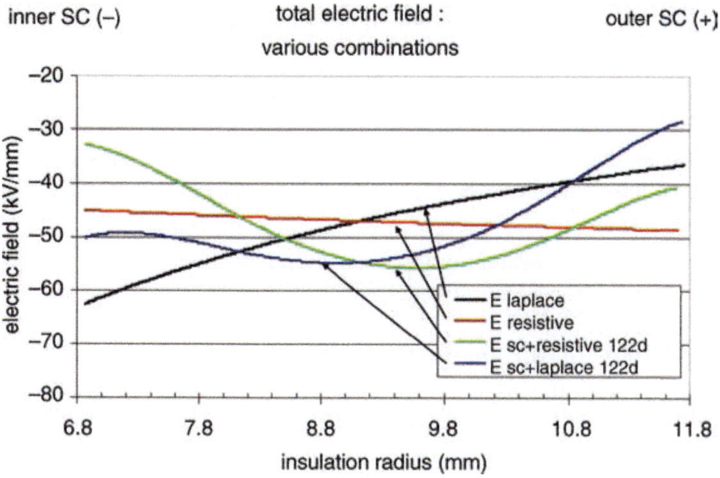

Figure 4.45 Addition of Laplace field and residual field, addition of resistive field and residual field (see legend in the Figure). The residual field was measured in cable 2 after 122 days aging under −230 kV (after [91]).

distribution (Laplace and resistive) to the residual field distribution, caused only by space charge and obtained under short circuit, is shown for cable 2 in Figure 4.45.

Finally, the good performance of the polymer in cable 2 is highlighted in [91]. Space-charge buildup is limited and field distortion stays within acceptable limits, given the high applied field and temperature.

4.5.2 Space-Charge Measurements in HVDC Cables with the PEA Technique

The European HVDC Project (of the fifth framework program), "*Benefits of HVDC Links in the European Power Electrical System and Improved HVDC Technology*", was developed with the goal of showing the potential technical, economic, and environmental benefits of using HVDC polymeric cables and to improve the knowledge of the factors that affect design and reliability of polymeric HVDC cables. The outcomes of the project have provided a better knowledge of the electrical properties at high DC field, through the development of improved test techniques for space-charge and conduction current measurements. Moreover, evaluation of charge accumulation in insulation systems and models for space-charge buildup and life performance useful for reliability assessment have been investigated in detail [92].

This Section illustrates the main results of a series of three articles [92–94] that describe some research activities performed within the HVDC Project about polymeric HVDC cable design and space-charge accumulation, measured with the PEA technique. In particular:

1. In [92], the results of space-charge measurements related to charge injection from the semicon–insulation interface and accumulation in the insulation

bulk, considering different combinations of insulation–semicon, are discussed.

2. In [93], space-charge buildup at the dielectric–dielectric interface is treated, aiming at providing rationale guidelines for an appropriate choice of semicon and insulation among material candidates to be used for HVDC cables, as well as to prompt improvements of the technological processes for materials and cable manufacturing.
3. In [94], the effect of temperature gradient on space-charge accumulation and field profile modification is investigated.

4.5.2.1 Space-Charge Measurements Related to the Semicon–Insulation Interface One of the most important topics dealt with in the HVDC Project was the investigation of the effect of interfaces among electrode–semicon–insulation or different insulating materials on space-charge injection and accumulation in insulation. Interfaces, in fact, can constitute weak points of the insulation in which space charge can be accumulated and/or injected, and electric fields can be enhanced [92].

For the purpose of investigating the space-charge behavior of peroxide cross-linkable insulating/semiconducting materials considered for HVDC cable applications, experimental tests were performed on three kinds of specimens [92]:

- Plaques composed of a sandwich of an insulation layer and a semiconducting layer on one side. The specimens were obtained by press-molding and cross-linked at 180°C, followed by cooling at a rate of 15°C/min.
- Reduced-scale cable models (minicables), having three layers (see Fig. 4.46): inner semicon, insulation layer (thickness = 1.5 mm), and outer semicon. The conductor cross section is 6.1 mm^2.
- Full-scale cables, differing from the second specimen in the insulation thickness (4.5 mm) and conductor cross section (50 mm^2).

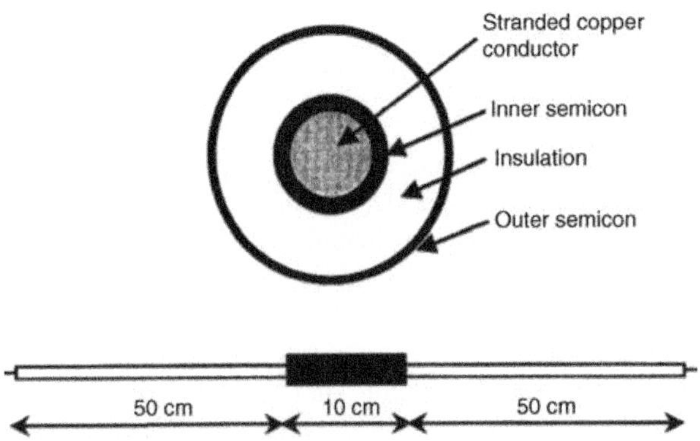

Figure 4.46 Minicable specimen. Conductor diameter = 2.8 mm; inner semicon thickness = 0.7 mm; insulation thickness = 1.5 mm; outer semicon thickness = 0.15 mm (after [92]).

Plaques of three insulation systems (insulation–semicon) and five insulation systems of minicables and full-size cables were tested. Three insulating materials (INS1, INS2, and INS3) and two semiconductive compounds (SC1 and SC2) were tested in this study. INS1 and SC1 represent today's state of the art of material technology used in AC EHV systems. INS3 is an example of a polar insulation with improved wet aging characteristics that has found wide acceptance in the medium voltage (MV) distribution network. INS2 and SC2 are examples of material systems designed to, respectively, minimize the concentration of peroxide reaction products and optimize the control of the semicon–insulation interface. The chosen combinations of insulation and semicon were INS1–SC1, INS2–SC1, and INS2–SC2 for the plaques and INS1–SC1, INS2–SC1, INS1–SC2, INS2–SC2, and INS3–SC1 for the minicables. Specimens of cables and plaques were thermally pretreated for 5 days at 80°C and for 3 days at 70°C, respectively, in order to expel virtually all cross-linking by-products. However, one cable (INS3–SC1) also was evaluated without prior degassing. The different test objects were chosen in order to find the best compromise between the effort involved in manufacture and the capability of experimental tests to be representative of real conditions in power cables. Comparable results were obtained on cables (mini and full-size) and plaques having the same combination of insulation–semicon materials under isothermal conditions. For this reason results obtained mainly on minicables are illustrated in [92].

Space-charge measurements were performed by means of the PEA technique on minicables and plaques at different electric field values up to 60 kV/mm, with the aim of obtaining the space-charge accumulation characteristics for each insulation–semicon combination, thereby obtaining space-charge patterns such as that (enlarged) reported in Figure 4.47: It is the three-dimensional (3D) plot relevant to INS3–SC1 (untreated) at 60 kV/mm and 25°C, where packets of positive charge and negative charge are present. The white is due to saturation in the color scale, thus representing the amount of space charge that exceeded the sensitivity scale (this is the case of electrode charge). The anode and cathode correspond to the inner and outer semicon, respectively.

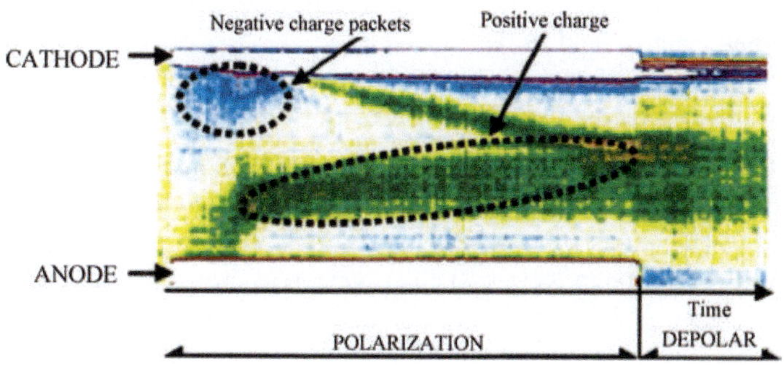

Figure 4.47 Enlarged view of the space-charge pattern relevant to INS3-SC1 (untreated) at 60 kV/mm and 25°C as a function of time. Cathode (outer semicon), anode (inner semicon), and charge packets also are highlighted (after [92]).

Table 4.7 Q_{med} Values Taken from Space-Charge Patterns[a]

	5 kV/mm	10 kV/mm	20 kV/mm	40 kV/mm	60 kV/mm
INS1–SC1	0.015 C/cm^3	0.04 C/cm^3	0.17 C/cm^3	0.3 C/cm^3	0.6 C/cm^3
INS1–SC2	0.013 C/cm^3	0.03 C/cm^3	0.14 C/cm^3	0.29 C/cm^3	0.5 C/cm^3
INS2–SC1	0.015 C/cm^3	0.04 C/cm^3	0.2 C/cm^3	0.4 C/cm^3	0.56 C/cm^3
INS2–SC2	0.014 C/cm^3	0.025 C/cm^3	0.07 C/cm^3	0.2 C/cm^3	0.3 C/cm^3
INS3–SC1	0.03 C/cm^3				0.65 C/cm^3
INS3–SC1 untreated	0.11 C/cm^3				0.85 C/cm^3

[a]Obtained from space-charge measurements performed on minicables at different electric fields (max value at the inner conductor semicon–insulation interface). Test temperature = 70°C.
Source: Reported in [92].

Tests were performed at different temperatures, ranging from 25 to 70°C. They were carried out under isothermal conditions, that is, by heating the specimen uniformly inside an oven, avoiding any temperature gradient across the insulation, which can have an influence on space-charge buildup. Polarization and depolarization lasted 10,000 and 3600 s, respectively. To estimate space-charge accumulation, the medium charge density, Q_{med}, which quantifies the absolute space-charge density accumulated in the insulation bulk, can be trivially calculated from the space-charge profiles. Conduction current measurements also were carried out with the aim of supporting space-charge results and to enable a better understanding of the behavior of the different combinations of insulation–semicon materials to be achieved.

Several measured space-charge patterns are reported in [92], together with the relevant values of Q_{med}. The charge patterns are omitted here for brevity, whereas the Q_{med} values relevant to 70°C are listed in Table 4.7. As a comment to the patterns and Q_{med} values, it is argued in [92] that more space charge accumulates in the insulation bulk on increasing electric field and/or temperature. In particular, charge packets in the insulation bulk, that is, injected charge traveling from one electrode to the other, can be observed, especially at low temperature. Moreover, a significant amount of heterocharge (i.e., charge having opposite polarity with respect to electrode charge) close to both electrodes, increasing with poling time, can be detected, particularly at higher temperature and electric field [92].

After plotting the values of Q_{med} versus maximum applied electric field for tests at 25 and 70°C, the following observations are drawn in [92]:

- The minicable model INS3–SC1, both thermally treated and untreated, shows the largest space charge accumulation at all electric fields and temperatures.
- At low temperature, INS2–SC1 displays the smallest amount of charge, followed by INS1–SC1, INS2–SC2, and INS1–SC2. It can be noted, however, that estimated values of Q_{med} are about the same value for these last three specimens.
- At higher temperatures (e.g., at 70°C), space-charge accumulation increases for all the insulation systems. On the whole INS2–SC2 seems to be the insulation system showing the smallest accumulated charge, followed by INS1–SC2, INS1–SC1, and INS2–SC1 (see Table 4.7).

- The specimens containing semicon SC2 accumulate more charge at room temperature when compared to materials associated with SC1, especially at higher fields. The opposite occurs at 70°C at which the materials associated with SC1 store a larger amount of charge.

Similar results were obtained on plaque specimens [92].

In [92] it is also shown that the accumulated charge amount as a function of the maximum geometric electric field highlights an electric field threshold behavior: The threshold field (below which space-charge accumulation is negligible) decreases as temperature increases, probably because injection from the electrodes is promoted by high temperature. In particular, the threshold almost halves for all the tested insulation systems between 25 and 70°C, switching from 10–12 kV/mm at 25°C to 4–5 kV/mm at 70°C.

After a broad discussion of the results, it is argued in [92] that experimental tests show that the insulating material INS2 accumulates less charge than INS1, especially in combination with semicon SC1. However, at high temperature, the semicon seems to play the major role regarding space-charge accumulation because all the combinations involving SC1 show the larger amount of charge. Thus, it can be speculated that SC1 and SC2 will show a different dependence on temperature as regards charge injection, with SC1 injecting more charge than SC2 into the insulation bulk at high temperatures, while the opposite occurs at room temperature.

To better investigate the dependence of space charge on temperature, the activation energy can be estimated by making Arrhenius plots (see Fig. 4.48) of the

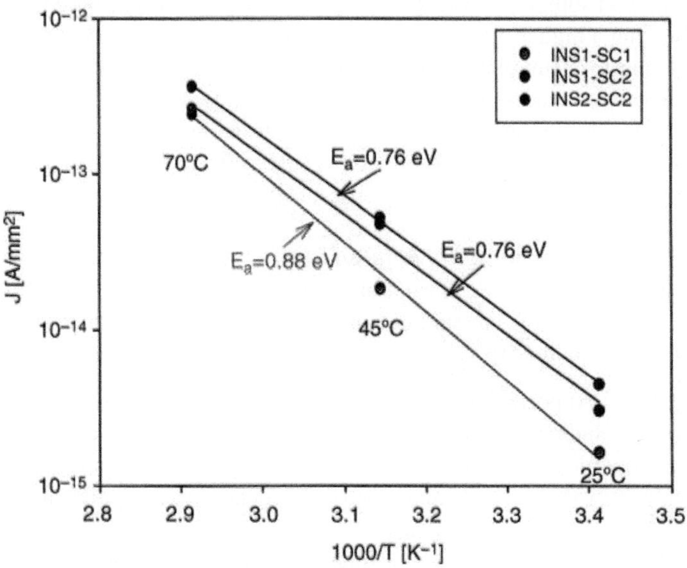

Figure 4.48 Arrhenius plots of the conduction current values measured at three different temperatures (25, 45, and 70°C) for an electric field of 20 kV/mm, with the estimate of activation energy [slope of the lines in log(J) vs. 1/T coordinates]: 0.88 eV for INS1-SC1 and 0.76 eV for INS2-SC2 and INS1-SC2 (after [92]).

conduction current values measured at three different temperatures (25, 45, and 70°C) for a given electric field (20 kV/mm in Fig. 4.48) and for three combinations involving two insulations and two semicons. It can be observed that the larger the activation energy, the greater the influence of temperature on conduction current. The results of Figure 4.48 show that the combinations of semicon–insulating materials having SC1 exhibit a larger activation energy (0.88 eV) than those containing SC2 (0.76 eV), even when the insulation is the same. This demonstrates that injection is an important factor in the steady-state current and may explain why the semicon, particularly SC1, plays an increasing role with temperature and the different dependence of charge injection on temperature for the two semicon materials [92].

It is noteworthy that, because INS2–SC2 shows the lowest activation energy, the electrical conductivity and, thus, the electric field profile of a DC cable manufactured using this material combination, would be less dependent on temperature, which can affect in a positive way the cable reliability. A low activation energy would mitigate the change of the electric field during load changes, namely the well-known—and often mentioned herein—field inversion phenomenon in DC cables, resulting in a more uniform electric field [92].

In [92] it is concluded that the characterization of DC insulation must tackle the evaluation of space-charge accumulation with a "system approach," that is, considering both the semiconductive material and the insulation, in particular, the properties of the semicon–insulation interface. The latter interface, in fact, plays a major role in space-charge injection/accumulation in the insulation bulk. Having analyzed different semiconductive and insulating materials candidates for HVDC cable applications, the best solution to be exploited for HVDC cable design would be the combination showing a high electric field threshold for space-charge accumulation, a small rate of charge accumulation as a function of electric field, and a small activation energy, that is, a space-charge amount less dependent on temperature. Therefore, space-charge measurements provide useful information to cable material manufacturers for tailoring insulation and semicon specifically for HVDC application, thereby improving the reliability of polymeric cables [92].

4.5.2.2 Space-Charge Measurements Related to the Insulation–Insulation Interface Nonhomogeneous electrical insulating materials and interfaces between different materials are frequently encountered in electric insulation systems. In general interfaces can be found [93]:

1. On a microscale in composite insulating materials obtained by a host material containing micro- and/or nanofillers, widely investigated with the aim of improving the properties of the base material. The presence of micro/nanoscale interfaces may affect the physical and electrical properties of the composite material, improving or worsening them with respect to the base material according to the nature and dispersion of micro/nanofiller inside the matrix.
2. On a macroscale in several HV insulating systems, for example, in cables. In particular, *interfaces between cable insulation and semiconductive layers and/or between different insulating materials in cable accessories (e.g., XLPE and EPR used for cable joints and/or terminations)* are very common.

Interfaces are often claimed to be critical within the complete insulation system, affecting electrical, mechanical, and thermal properties. Indeed, interfaces may enhance the effects of electrothermal and mechanical stresses, thereby becoming the weakest points of the insulation system, both in AC and DC [95]. Interfaces can trigger partial discharges (PD) when the contact between surfaces is imperfect; PD should be strictly avoided for cable and accessories because organic insulation (e.g., XLPE and EPR) might not be able to withstand PD activity, even for short times [93], though PD activity is a second-order problem under DC due to the smaller repetition rate with respect to AC. However, under a DC field, interfaces in the insulation bulk may favor space-charge accumulation within the dielectric thickness [96], behaving as favored sites for charge buildup, particularly for low-mobility materials such as extruded polymeric insulation.

Therefore, the investigation of the behavior of interfaces between different insulating polymeric materials in the presence of thermal and electric stresses can help in *improving the design of polymeric cables and accessories*, which takes advantage from a deeper knowledge of the actual field applied to the insulation system under design (see Section 4.3.1.4). For this reason, in [93] space-charge accumulation at the insulation–insulation interface is studied via approximate mathematical models derived from macroscopic physics and used to fit experimental data relevant to model cables having two insulation layers that create cylindrical interfaces (see Fig. 4.49).

The physical explanation for nonhomogeneous dielectrics (e.g., interfaces) as a source of charge accumulation in solid insulating materials under DC or low-frequency fields (besides injection from the electrodes, gradient in the insulation conductivity σ, dissociation of impurities in the insulation bulk) is the so-called Maxwell–Wagner–Sillars (MWS) polarization [34]. According to such a theory, space charge can accumulate in nonhomogeneous dielectrics in the presence of interfaces between insulation layers having different conductivity and permittivity.

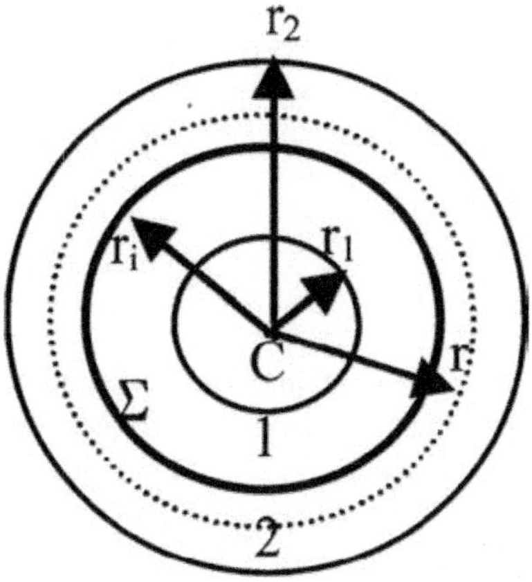

Figure 4.49 Boundary surface, Σ, in a cylindrical geometry, with two insulating layers, 1 and 2. The inner conductor is indicated by C (after [93]).

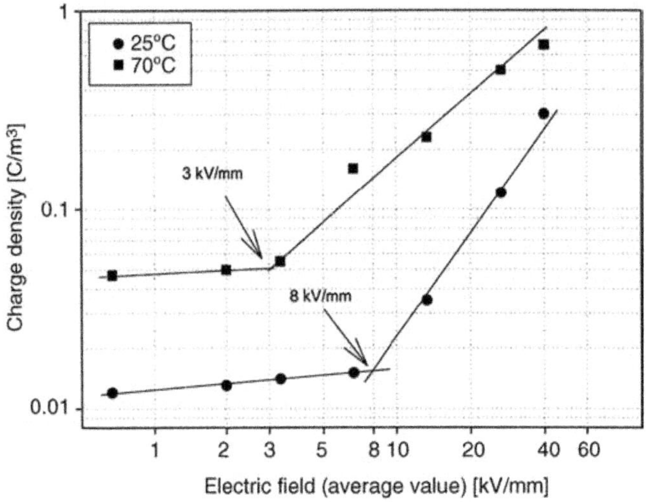

Figure 4.50 Threshold characteristics for XLPE cable at 25 and 70°C (after [93]).

The effect of MWS polarization can be hidden by injected charge at high temperatures (injection is favored by temperature; see Section 4.2.2.1) and/or if the electric field exceeds by a large amount the threshold for space-charge accumulation. Indeed, an electric field threshold for space-charge accumulation exists [97], such that injection from the electrodes is low when the applied field is below the threshold and no net charge accumulates (thus insulation exhibits ohmic behavior), while above the threshold injection from the electrodes rises abruptly and charge accumulates in the insulation bulk according to a power law of the kind $q = A_0E^b$ (see Fig. 4.50).

The extent of space-charge accumulation and electric field modification due to the MWS mechanism, neglecting the contribution of injected charge (i.e., valid only at fields close to the threshold) can be evaluated for the cable cylindrical geometry of Figure 4.49 on the basis of Maxwell's equations. After a cumbersome theoretical treatment (reported extensively in [93]), a first key finding is the following expression for the surface "free" charge density at the interface (i.e., the amount of charge per square meter), ρ_Σ:

$$\rho_\Sigma = \frac{(\varepsilon_2\sigma_1 - \varepsilon_1\sigma_2)V}{\sigma_1 \ln(r_2/r_i) + \sigma_2 \ln(r_i/r_1)} \tag{4.41}$$

where r_1 and r_2 represent the radius of internal conductor and external shield, respectively; r_i is the radius of the boundary surface Σ between two coaxial dielectrics, having different permittivity (ε_1 and ε_2, respectively) and conductivity (σ_1 and σ_2, respectively); and V is the applied voltage. It should be noted that the sign of ρ_Σ depends on the product $\varepsilon_i\sigma_k$ for ith and kth insulation layer. Because conductivity depends on temperature and electric field, and varies from one material to another, it may happen, for example, that $\varepsilon_2\sigma_1 > \varepsilon_1\sigma_2$ at room temperature and $\varepsilon_2\sigma_1 < \varepsilon_1\sigma_2$ at higher temperatures. This can cause the polarity of the interfacial charge, ρ_Σ, to be inverted as a function of field and temperature. More fundamental findings of [93] are the expressions of the electric field profile E_1 in insulation layer 1 (having permittivity ε_1) and the electric field profile E_2 in insulation layer 2 (having permittivity ε_2) of the cable sketched in Figure 4.49, namely;

$$E_1(r) = \frac{\varepsilon_2 V}{[\varepsilon_1 \ln(r_2/r_i) + \varepsilon_2 \ln(r_i/r_1)]r} - \frac{\rho_\Sigma r_i \ln(r_2/r_i)}{[\varepsilon_1 \ln(r_2/r_i) + \varepsilon_2 \ln(r_i/r_1)]r} \tag{4.42}$$

$$E_2(r) = \frac{\varepsilon_1 V}{[\varepsilon_1 \ln(r_2/r_i) + \varepsilon_2 \ln(r_i/r_1)]r} - \frac{\rho_\Sigma r_i \ln(r_i/r_1)}{[\varepsilon_1 \ln(r_2/r_i) + \varepsilon_2 \ln(r_i/r_1)]r} \tag{4.43}$$

The first term of (4.42) and (4.43) is the field appropriate to a permittivity mismatch between the two materials; the second term gives the effect of the free interface charge, ρ_Σ, upon the field discontinuity.

It is noteworthy that, as temperature and/or electric field rise, injected space charge can increase significantly throughout the insulation bulk (see Fig. 4.50) and particularly at the interface, where the trap density is likely higher than in the bulk. As a consequence, electric fields E_1 and E_2 may differ significantly from those calculated via (4.42) and (4.43). However, a general expression for the electric fields in the generic insulation layers 1, 2 ($E_{1,2}$), which also takes into account the space charge accumulated in the bulk, can be derived from [98]

$$E_{1,2}(r) = \frac{\varepsilon_{2,1} V}{[\varepsilon_{1,2} \ln(r_2/r_i) + \varepsilon_{2,1} \ln(r_i/r_1)]r} + \frac{1}{r}\int_{r_i}^{r} \frac{\rho(r)}{\varepsilon(r)} r\, dr \tag{4.44}$$

where $\rho(r)$ is the space-charge density profile obtained experimentally through space-charge measurements. A first estimation of the electric fields $E_{1,2}$ also can be made via (4.42) and (4.43) by substituting for ρ_Σ the free interfacial charge estimated from the part of the measured space-charge profile at the interface.

For verifying the above theoretical expressions, space-charge measurements were performed by means of the PEA technique at field values above and close to the threshold for space-charge accumulation in XLPE. The test specimens, sketched in Figure 4.51, were designed and realized within the European HVDC Project (see beginning of Section 4.5.2) and consisted of triple-extruded, dual-dielectric cables, giving rise to cylindrical chemical interfaces, analogous to those of Figure 4.49. Generally, the poling time was 10,000 s, while the depolarization time was 3600 s. Tests were performed at different temperatures, ranging from 25 to 70°C, mostly under isothermal conditions (more details can be found in [93]).

In [93] the space-charge profiles relevant to specimen 1 of Figure 4.51 (EPR–XLPE minicable) at 25 and 70°C are reported. The poling voltage is 60 kV, which corresponds to a value of the average applied electric field of 29 kV/mm. At 25°C, a positive charge peak corresponding to the position of the interface, increasing with the poling time (up to about 10,000 s), is detected, together with a slight accumulation of homocharge, which is positive in the EPR layer, negative in the XLPE layer. At 70°C, on the contrary, the net interface charge is negative, reaching a steady state

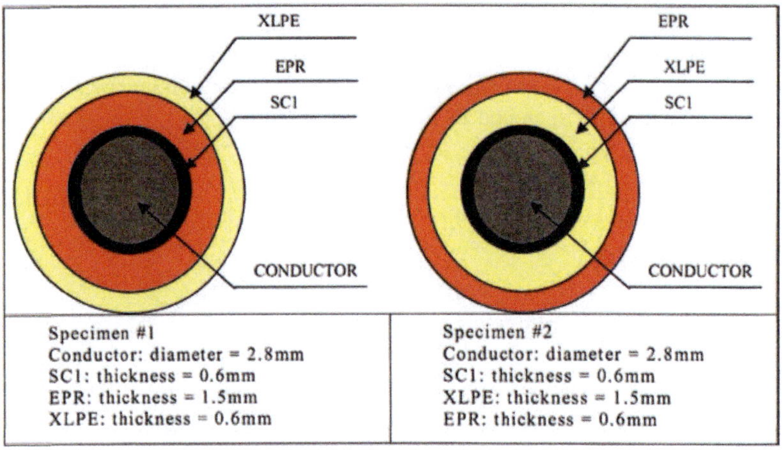

Figure 4.51 Sketch of test cables (after [93]).

after about 5000 s. An amount of homocharge is observed again in the insulation bulk, especially during depolarization (volt off). The interface free charge density, ρ_Σ, estimated from the space-charge profiles at the end of the polarization is $+1.7 \times 10^{-4}\,\mathrm{C/m^2}$ and $-2.0 \times 10^{-4}\,\mathrm{C/m^2}$, for specimen temperatures of 25 and 70 °C, respectively [93].

Space-charge measurements also were performed at 25 and 70°C for a lower poling voltage of 21 kV, which yields an average electric field of 10 kV/mm. The poling time was raised from 10,000 to 250,000 s, as the interface charge needs a much longer time to reach a steady state, due to the smaller conductivity of the insulations at these temperatures and field. This resulted in a positive interface charge density both at 25 and 70°C, with values of about $+1.3 \times 10^{-4}\,\mathrm{C/m^2}$ and $+1.0 \times 10^{-4}\,\mathrm{C/m^2}$, at 25 and 70°C, respectively. Moreover, the amount of interface charge at 70°C decreased with poling time, as debated later on [93].

Furthermore, in [93] space-charge data for specimen 2 are reported, having the opposite chemical interface (XLPE–EPR) at a poling voltage of 60 kV. As expected, the polarity of interface charge is inverted with respect to the charge profiles reported for specimen 1. The interface charge at 25°C is almost the same (in absolute value) as that of specimen 1, that is, $-0.9\times10^{-4}\,\mathrm{C/m^2}$. At 70°C it is significantly larger, that is, $8.0 \times 10^{-4}\,\mathrm{C/m^2}$.

Figure 4.52 shows space-charge profiles for specimen 2 obtained at an average applied field of 15 kV/mm [98], but with a temperature drop over the insulation of 10°C (conductor temperature of 40°C; see Fig. 4.52*a*) and 18°C (conductor temperature of 64°C; see Fig. 4.52*b*). In this case, according to [94] the effect of a temperature gradient on space-charge buildup may be nonnegligible, but mainly at fields below 8 kV/mm, that is, the threshold for space charge-injection (see Fig. 4.50). Results show that, at both temperatures, negative charge accumulates at the dielectric interface. In addition, a significant accumulation of positive charge in the XLPE layer also can be observed at higher temperature [93].

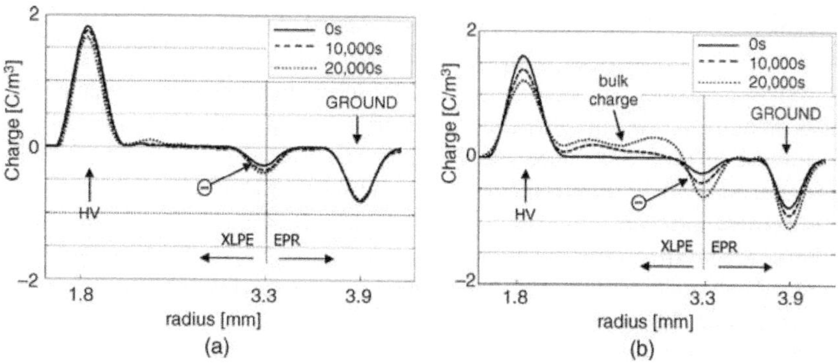

Figure 4.52 Space-charge profiles obtained from specimen 2, poling voltage +30 kV (average applied field = 15 kV/mm): (*a*) inner semicon temperature = 40°C, outer semicon temperature = 30°C and (*b*) inner semicon temperature = 64°C, outer semicon temperature = 42°C (after [93]).

A discussion arises in [93] about the polarity change of net interface charge on raising the temperature from 25 to 70°C that occurs only for specimen 1 and for 2 at high field and isothermal conditions. On the basis of the previous theoretical developments, as well as of the experimental values of conductivity and permittivity derived experimentally for XLPE and EPR, the value of interface charge is calculated via (4.41) and compared with the values obtained by measurements, thereby finding that at low electric fields, close to the threshold for space-charge accumulation, the values of charge derived by the macroscopic MWS model agree quite well with those measured experimentally, confirming that the charge at the interface is mainly due to MWS polarization. On increasing the applied field, a considerable difference between the calculated and measured charge is found; it is speculated in [93] that this is due to injected charge from the electrodes accumulating at the interface.

Another relevant point is the change in the interface charge polarity with temperature. The interfacial charge for specimen 1 and high field becomes negative at 70°C. After discussion, this is explained by hypothesizing that injected negative charge increasing with temperature and applied field is overwhelming the MWS positive charge. A further confirmation of these results is provided by space-charge profiles obtained for tests at lower fields (e.g., average field 10 kV/mm; see above). In this case, space-charge accumulation due to injection is negligible, so the charge at the interface remains positive irrespective of temperature (according to the MWS polarization). Moreover, the amount of interface charge at 70°C decreased with poling time [93], probably due to injection of negative charge from the cathode (the applied field is larger than the threshold at 70°C). As a consequence, one could expect that in service the interface charge could be positive or negative depending on the poling voltage, on the load and on the thermal conditions of the laying environment. This can affect significantly the actual electric field at the insulation–insulation interface [93].

When the order of the insulating layers is changed (specimen 2), the polarity of interface charge is inverted, as would be expected from the macroscopic approach [see (4.41)]. At a higher temperature the polarity of the interface charge changes as occurs in specimen 1. Again, this can be due to injection of positive charge from XLPE, favored by the temperature rise. The interface charge at 70°C is significantly larger for the XLPE–EPR minicable with respect to that of EPR–XLPE. Indeed, the XLPE–electrode interface seems to be more active in charge injection, particularly when the XLPE layer is subjected to the highest field, that is, it is located near to the high-voltage electrode. It should be noted that space-charge tests in nonisothermal conditions showed a different behavior (Fig. 4.52). In this case, a significant amount of injected charge can be observed on increasing the temperature as before. But the polarity of the interface charge is not seen to change, as was the case previously. A tentative explanation is that, because the average applied field is not very high (15 kV/mm) and the temperature at the interface is lower than the maximum value due to the temperature gradient, the injected space charge does not advance sufficiently to hide the MWS interface charge, unlike the situation observed in specimens at a uniform temperature of 70°C [93].

The electric field profiles, obtained in [93] from Eqs. (4.42) and (4.43) for specimen 2 with poling voltage +60 kV (average field = 29 kV/mm) at 25 and 70°C taking the measured space charge at the interface for ρ_Σ, exhibit an inversion at a temperature between 25 and 70°C. At 25°C, the actual field is close to the geometric Laplacian field (i.e., larger in the innermost XLPE layer and smaller in the outermost EPR layer). At 70°C the field is more than doubled in the EPR layer: This can be explained by the polarity inversion of the charge at the interface, due to injected charge from the electrode. It is pointed out in [93] that this can cause reliability loss in polymeric cable systems because the design field may differ significantly from the real applied field if space-charge buildup is not taken into account at the design stage; moreover, the same charge accumulation and inversion effects discussed above may occur also if the conductivity of the two insulation layers shows a different activation energy, that is, a different temperature dependence, so that the difference between $\varepsilon_2\sigma_1$ and $\varepsilon_1\sigma_2$ can change sign with raising (or, perhaps, lowering) the temperature with respect to rated room temperature, and the interface charge can invert its polarity as well, and so may the electric field.

On the basis of the above results, it is concluded in [93] that the electric field distribution of cable insulation systems under HVDC can be affected significantly at interfaces due to space-charge buildup, hence cable and accessory design should take into account a complex scenario where:

1. The nonhomogeneities in conductivity and permittivity can cause interface charge modifying the insulation field distribution that can be evaluated roughly using macroscopic models. Interface charge and field are affected largely by material electrical properties (e.g., conductivity) that are strongly correlated with electrical and thermal stresses and temperature gradient across the insulation. A partial solution could be manufacturing insulating materials with properties showing small temperature dependence.

2. The increase in HV cable design fields requires that microscopic effects, such as injection and extraction processes at the electrodes, be considered when the electric field exceeds the threshold for space-charge accumulation. Significant underestimation in calculations of interfacial charge and field can be avoided resorting to space-charge measurements and models that should also take into account the space-charge accumulation features of the insulating material. In this way models can support efficiently cable and accessory design, thereby improving the reliability of the whole insulation system.

The improvement of cable and accessory design for the whole HVDC extruded insulation cable system according to the guidelines found in this chapter and throughout the whole book is the subject of Chapter 5.

4.5.2.3 Space-Charge Measurements Related to the Effect of Temperature Gradient Sections 4.5.2.1 and 4.5.2.2 have shown that both the inner/outer semicons and the interfaces between adjacent insulating layers (e.g., in cable accessories) have a major effect on space-charge accumulation in extruded HVDC cable systems. Here focus is on a third fundamental factor that affects space-charge storage under DC voltage, namely the temperature gradient within cable insulation that arises when the cable is loaded and heated due to Joule losses in the conductor.

Under load, the steady-state temperature in the insulation wall decreases from the inner to the outer semicon according the following radial temperature profile, which can derived via the thermal Ohm's law [38]:

$$T(r) = T_o + \frac{T_i - T_o}{\ln(r_o/r_i)} \ln\left(\frac{r_o}{r}\right) \tag{4.45}$$

where r_i and r_o are the inner and outer radii of insulation, respectively, T_i and T_o are the corresponding temperatures, and r is the radial coordinate at the generic insulation radius.

Since the electrical conductivity of the insulation depends on temperature and electric field, as broadly discussed in Chapter 3 a conductivity gradient stems from the radial temperature gradient [$\nabla T = dT(r)/dr$ in cylindrical coordinates] that eventually results in a steady-state charge distribution, $\rho_{tg}(r)$, associated with conductivity gradient and modifies the electric field profile with respect to the Laplacian one. Thus it is important to investigate space-charge storage (both of the continuous and of the localized type) due to a temperature gradient in comparison with other charge-supply mechanisms, particularly injection from the electrodes.

For this purpose, in [94]—as well as in other studies, see, for example, [99, 100]—space-charge measurements were carried out on XLPE-based HVDC cable models under application of different temperature gradients across the cable insulation, above and close to the threshold field for space-charge accumulation (see Section 4.5.2.1). Two kinds of specimens were thermally pretreated (in order to expel most of the cross-linking by-products) and subsequently subjected to space-charge measurements via ad hoc PEA test setups [94]:

1. Minicables with $r_i = 1.8$ mm, $r_o = 3.3$ mm (1.5-mm-thick insulation), poled for at least 10,000 s (about 2 and ¾ h) at a voltage V_{DC} ranging from 5 to 60 kV (average fields from 3 to 40 kV/mm), followed by 3600 s depolarization.
2. MV-size cables with $r_i = 4.5$ mm, $r_o = 9$ mm (4.5-mm-thick insulation), poled for 18,000–20,000 s (about 5–5.5 h) at 90 kV (average field 20 kV/mm), followed by 3600 s depolarization.

To distinguish the effect of temperature gradient, space-charge measurements were carried out in an oven at a constant temperature, without and with temperature gradient in the cable insulation. In the former case, the oven was kept at 65°C, while in the latter a temperature drop, ΔT, of 10 or 20°C was realized by keeping the cable inner semicon at 65°C through a current in the conductor and cooling the outer semicon by setting properly the constant temperature of the oven. The measurements reported in [94] showed several puzzling but interesting features.

As far as minicables are concerned, the effect of temperature gradient on space-charge accumulation seems significant in particular at low field, that is, close to the threshold for space-charge accumulation: For example, at the average field of 3 kV/mm (see Fig. 4.50), a small amount of positive charge close to the anode (homocharge) can be observed only when a temperature gradient is applied to the cable (cable conductor with positive voltage—cable screen grounded). A charge increase of 0.02 C/m^3 related to the temperature gradient is observed, in agreement with [38]. In order to support this result, in [94] the charge amount and the electric field profile due to the temperature gradient were estimated in a simplified way with respect to the rigorous treatment reported in Section 3.2.3.2, under the assumption of an ohmic electrode–insulation interface. This assumption is acceptable if the electric field is below or close to the above-mentioned threshold for space-charge accumulation, which corresponds in practice to the transition from ohmic to SCLC conduction. Then, the dependency of electric conductivity σ on E can be neglected and the σ versus T relationship can be expressed as an Arrhenius-type law, as follows [94]:

$$\sigma(T) = \sigma' \exp\left(-\frac{\Delta W}{kT}\right) \tag{4.46}$$

where ΔW is the activation energy for the conduction process in the insulation bulk and σ' is a model constant. Under such an assumption, it can be shown that—once ΔW and the cable temperature profile $T(r)$ are known—the electric field and space-charge profiles associated with the temperature gradient can be calculated through the following relationships [94]:

$$E(r) = E_o \frac{r_o \sigma(r_o)}{r \sigma(r)} \tag{4.47}$$

$$\rho_{tg}(r) = -\varepsilon_0 \varepsilon_r E(r) \frac{\Delta W}{kT^2} \frac{dT}{dr} \tag{4.48}$$

where r is the radial coordinate within the insulation thickness, $E(r)$ is electric field at such a radial coordinate, and E_o is the electric field at the outer insulation radius r_o.

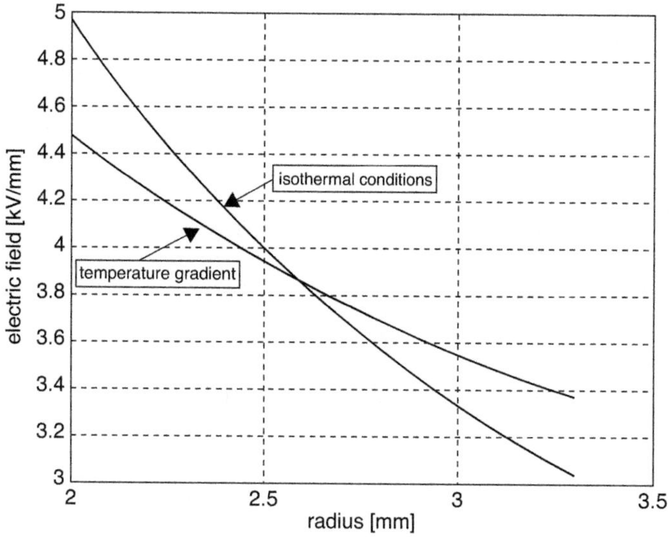

Figure 4.53 Calculated electric field profile under temperature gradient, $\Delta T = 20°C$, in comparison with the geometric field in isothermal conditions. Minicable specimens at poling voltage 5 kV (after [94]).

Considering $\varepsilon_r = 2.3$, $T_i = 65°C$, $T_o = 45°C$, $V_{DC} = 5$ kV, and $\Delta W = 0.1$ eV, Eqs. (4.45)–(4.48) yield a mean charge density value of 0.015 C/m^3, in good agreement with the experimental value as both polarity and order of magnitude. This small charge modifies the electric field with respect to the "Laplacian" field (see Fig. 4.53), but does not invert the field profile: Indeed, the electric field at the outer semicon is 15% larger than the Laplacian field, but the maximum field stays at the inner semicon [94].

When raising the applied field to 6.7 kV/mm, again a small amount of homocharge is observed close to the anode in the presence of a temperature gradient only, but the mean charge density decreases slightly with respect to isothermal conditions. This can be explained with space-charge accumulation due to electrode injection, which cannot be neglected anymore [94].

If the average electric field is increased to 27 kV/mm, that is, well above the space-charge accumulation threshold, the homocharge in the vicinity of the anode disappears, while heterocharge is observed close to both electrodes, which increases further as the field is raised to 40 kV/mm. This can be understood by estimating positive and negative charge contributions (Q^+_{max}, Q^-_{max}, respectively) for a distance of 0.25 mm from the electrodes and comparing them for the temperature gradients applied in the measurements at the same average field of 40 kV/mm, that is, $\Delta T = 0$, 10, and 20°C (see Fig. 4.54) [94].

Figure 4.54 shows that, as ΔT rises (i.e., as the outer semicon cools down; see above), less charge is accumulated in the insulation bulk, namely if the cathode temperature is reduced by 20°C, the negative charge injected by the cathode and accumulated close to the anode is almost halved, whereas the

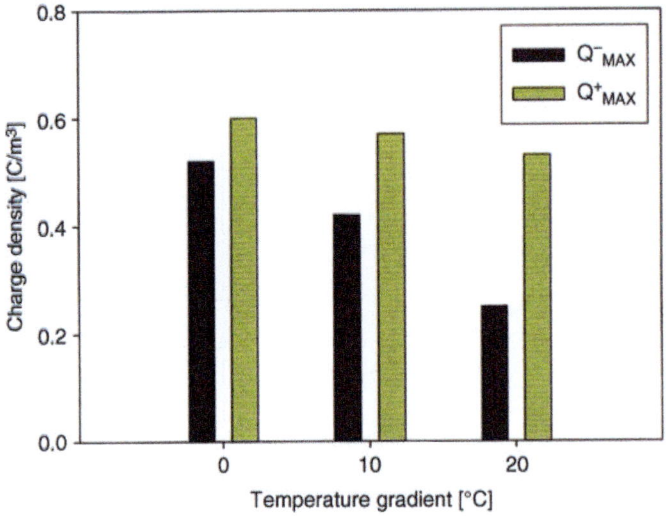

Figure 4.54 Minicable charge density values (separating positive and negative contributions), Q^+_{max} and Q^-_{max}, for a temperature drop $\Delta T = 0°C$, 10°C, 20°C at poling voltage 60 kV (average field = 40 kV/mm) (after [94]).

positive charge remains nearly constant. Similar results are obtained for an electric field of 27 kV/mm.

Therefore, as to minicables it can be concluded that stored charge reduces as cathode cools down because injection and the partially blocking effect of semicon electrodes [21] become important if the applied field exceeds the threshold for space-charge accumulation by a large amount: Schottky injection is assisted by temperature, and lowers if temperature is lowered (see Section 4.2). Hence, at high fields, charge accumulation seems to be more significantly influenced by electrode temperature than by the temperature gradient [94].

As to MV-size cables, the space-charge profiles at an average field of 20 kV/mm (poling voltage 90 kV at the conductor–cable screen grounded), that is, of the order of design field of full-size HVDC cables at present (see Chapter 3), are reported in Figures 4.55*a* and 4.55*b* for $\Delta T = 0$ and 20°C, respectively (inner semicon always at 65°C, see above). The figure shows a larger contribution of temperature gradient to charge accumulation in MV-size cables than in minicables; this might invert the field profile at steady state, depending on field and temperature gradient: For example, the electric field profiles versus poling time for the case of Figure 4.55*b* are plotted in Figure 4.56. This could be explained considering that MV cables have a larger insulation thickness (three times that of minicables). When the applied field does not exceed significantly the threshold for space-charge accumulation, charge injection seems to affect MV-size cables to a lesser extent because injected charge takes a longer time to reach the other electrode. Thus, no heterocharge is observed within poling time close to the cathode, contrary to minicables. Therefore, the field in the neighborhood of the electrodes is not increased by heterocharge (as for minicables), and the injection of opposite polarity charge is not favored [94].

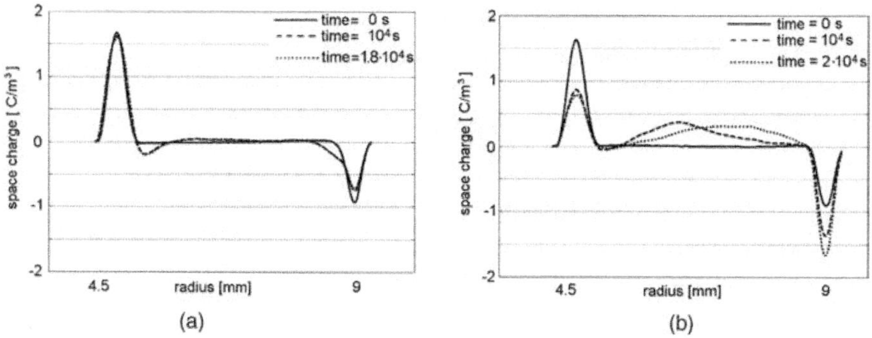

Figure 4.55 Space-charge profiles from tests on MV-size cables with poling voltage 90 kV (average field = 20 kV/mm): (*a*) $\Delta T = 0°C$ and (*b*) $\Delta T = 20°C$, (after [94]).

In summary, the results after [94] prove that temperature gradients affect space-charge accumulation in HVDC cable models at fields not too far from the threshold for space-charge accumulation (among which current design fields of full-size HVDC cables), particularly if cable insulation thickness is large. In fact the effect of temperature gradients is significantly larger in MV-size cables than in minicables, and this calls for a very cautious extrapolation of results coming from mini or MV-size cables to HV, full-size cables.

Similar results were derived in [100] after performing space-charge measurements on commercial 11-kV AC XLPE insulated power cables ($r_i = 5.8$ mm; $r_o = 9.2$ mm; insulation thickness = 3.4 mm; cross section of stranded aluminum conductor = 95 mm^2) by means of a modified PEA system with a flat ground electrode instead of the curved ground electrode of conventional cable PEA systems. Being AC cables, the relevant results are interesting but not so indicative of the space-charge behavior

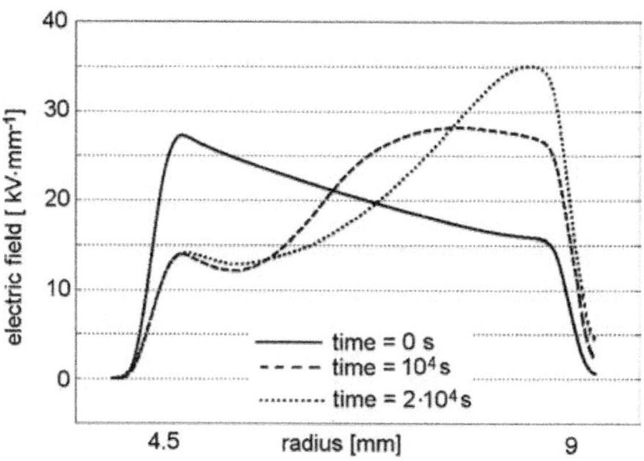

Figure 4.56 Electric field profiles obtained by integrating space-charge profiles reported in Figure 4.55*b*. MV-size cable specimens at poling voltage 90 kV (average field = 20 kV/mm), $\Delta T = 20°C$, (after [94]).

of full-size extruded HVDC cables as those previously illustrated. Moreover, the 11-kV AC XLPE cables did not undergo the degassing process, and therefore contain cross-linking by-products such as methane, cumylalcohol, and acetophenone. These by-products influence the space-charge distributions within the material (see Sections 4.3.2, 4.4.2.3, and 4.5.2.1): Some enhance electron (or hole) injection from the electrode, and, as a result, the formation of homocharge; others support heterocharge formation due to ionic carriers supplied by impurities [100,101].

In the experimental procedure illustrated in [100], three temperature drops, ΔT, were set across the insulation:

1. $\Delta T = 0°C$ (ambient temperature, no temperature gradient)
2. $\Delta T = 10°C$ (~55°C at inner semicon, ~45°C at outer semicon)
3. $\Delta T = 20°C$ (~89°C at inner semicon, ~69°C at outer semicon)

In cases 2 and 3, the cable was first heated up through a current transformer for 2 h until the desired stable temperature gradient was achieved. Thereafter, the current was kept constant throughout the experiment and a DC voltage of +80 kV was applied to the 11-kV XLPE cables for 9 h of poling time, performing space-charge measurements at specific timings of both voltage on and off. Subsequently, charge decay measurements were made after short circuiting the electrodes [100].

The results of volt-on measurements at $\Delta T = 0°C$ and $\Delta T = 20°C$ across the insulation are displayed in Figure 4.57*a* and 4.57*b*, respectively. Focusing first on $\Delta T = 0°C$ (Fig. 4.57*a*), after 1 h of poling a negative space-charge (homocharge) distribution is formed adjacent to the cathode (outer electrode) and a large amount of heterocharge is formed in the vicinity of the anode (inner electrode). After 3 h of poling, positive charge tends to accumulate in the sample bulk toward the cathode. Negative charges seem to accumulate in the sample much faster than positive charges, perhaps because either negative charges get trapped more easily or more negative charge is injected into the sample than positive charge. The buildup of positive charge in the sample bulk toward the cathode can be addressed to the fact

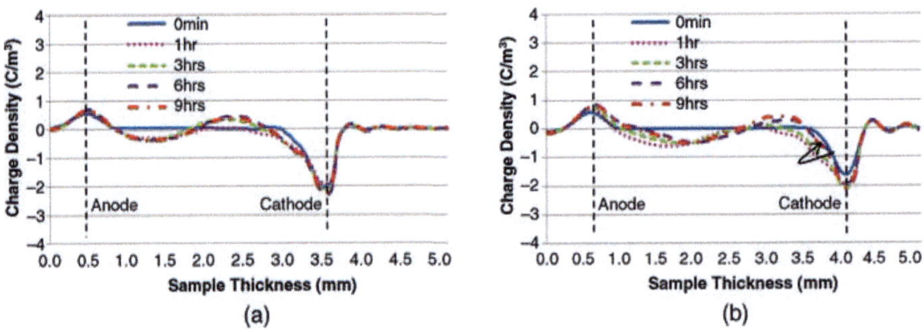

Figure 4.57 Space-charge profiles for (*a*) volt-on space-charge measurements at ambient temperature with T_0~18°C and (*b*) volt-on measurements with $\Delta T = 20°C$ and T_0~69°C, (after [100]).

that positive charge is hardly trapped close to the anode and thus moves toward the cathode, where it neutralizes or recombines with the stored negative charge [100].

Second, when comparing the results at $\Delta T = 0°C$ and $\Delta T = 20°C$, it can be seen that with an increase in temperature gradient, the amount of charge increases across the sample as the relative injection barrier is reduced (see Fig. 4.57*b*). Furthermore, with the increase in temperature, conductivity across the insulation also increases proportionally to $\exp(1/k_B T)$. This causes a drop in the insulation resistivity, a rise in charge mobility, and a faster charge accumulation in the insulation bulk [100].

In [100] the electric field profiles across cable insulation are also calculated, as the sum of:

- The electric field $E_{SC}(r)$ produced by localized space charge, obtained by introducing into the Poisson equation the space-charge profiles measured after 9 h of poling under the different temperature gradients.
- The electric field $E_{Applied}(r)$ arising from the applied voltage, calculated on the basis of the conductivity equation $J = \sigma(E,T)E$, $\sigma(E,T)$ being expressed in the form of Eq. (3.12) with parameters estimated from the results of ad hoc conductivity measurements.

These profiles are shown in Figure 4.58. They are consistent with the fact that, in the experiments, all the three different temperature conditions produced heterocharge formation and heterocharge results in stress enhancement at insulation–semicon interfaces and stress reduction in the insulation bulk. However, the figure demonstrates that the amount of interfacial stress enhancement, as well as the amount and the location of stress reduction in the insulation bulk, is strongly dependent on the temperature drop within the insulation wall [100].

In general, the results after [94] and [100] show an increase of trapped charge in cable insulation under the temperature gradient, particularly in the insulation bulk.

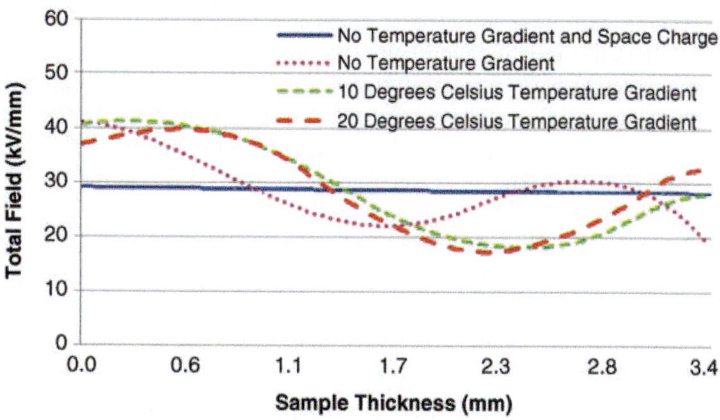

Figure 4.58 Total electric field distribution across cable insulation after 9 h of poling under the different temperature gradients, (after [100]).

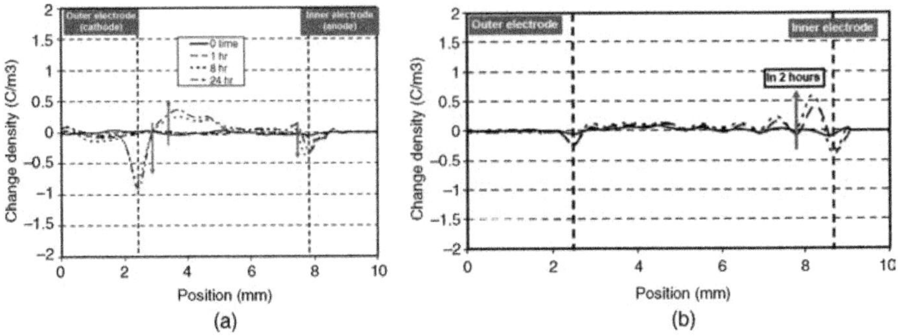

Figure 4.59 Space-charge accumulation under applied voltage (measurement with external voltage temporarily removed) for cable B: (*a*) at room temperature ($T \sim 25°C$) and (*b*) with temperature gradient ($\Delta T = 13°C$ and $T_i \sim 70$ °C), (after [99]).

However, this experimental evidence cannot be regarded as an axiom since in the space-charge measurements after [99] the temperature gradient almost eliminated space charge in insulation thickness. More in detail, in [99] a broad campaign of space-charge measurements was performed on two cable model types, A and B—the former with thinner insulation, the latter with thicker insulation—using the same type of flat ground electrode PEA system employed in [100]. Space-charge measurements under temperature gradients in [99] are carried out on cable type B in un-degassed condition, a prototype HVDC power cable with a 5.7-mm-thick insulation that is of relevance from the viewpoint of the space-charge behavior of full-size extruded HVDC cables. First, however, the space-charge behavior was measured for a uniform temperature (room temperature $T \sim 25°C$) under the application of an external voltage of +80 kV to the inner conductor. The relevant space-charge distribution (see Fig. 4.59*a*) approaches a steady state with a large positive heterocharge at the outer electrode (cathode), while a small amount of homocharge, decreasing with time, is detected at the inner electrode (anode). The time to a stable distribution is much longer than in the thinner insulation of cable A: This might be due, as argued in the case of [94], because the thicker the insulation, the longer is the charge transit time. The longer transit time may also be the reason why a negative heterocharge has not yet appeared near the inner electrode. Under short circuit the positive charge at the outer electrode underwent an almost negligible decay, thereby suggesting that this charge consists of low-mobility ion species deriving from the volatile cross-linking by-products. Under a temperature gradient with the cable conductor heated to 70°C and a temperature drop of 13°C between conductor and outer sheath (see Fig. 4.59*b*), the heterocharge region near to the outer electrode was removed and the homocharge at the inner electrode increased. This behavior is consistent with a situation in which the interface currents have increased more rapidly with the increase in temperature than the current in the body of the insulator, that is, the interface currents are activated with higher activation energy than the currents in the insulation material. The inner electrode is hotter than the bulk insulation and thus has a higher injection current than the bulk current, thereby giving rise to a net homocharge, which reduces the interface field at the inner electrode to bring the two currents into coincidence. The temperature of the outer electrode is cooler than the bulk material, but its temperature rise is sufficient to increase the interface current there and bring it into coincidence with the bulk current without a significant buildup of space charge. From this respect in [99], it is argued that "the injection current involved carrier tunnelling through a potential barrier [and] this result shows that the optimum process involves thermal activation from around the relevant Fermi energy to a state where the barrier width is sufficiently reduced to allow tunnelling to become effective" ([99], page 858) (see Section 4.2).

Hence, it can be concluded that different interfaces and bulk materials may have different conductivity activation energies. This, in combination with different levels of applied voltage and insulation thicknesses, may result in very different effects of the same temperature gradient on the amount, polarity, and location of stored charge close to the electrodes and the insulation bulk. Only a dedicated "real-time" monitoring of space charge under the various load conditions can really give reliable evidence of the space charge and field profiles under temperature gradients within the insulation of a certain HVDC cable design.

4.5.3 Recent Developments in the Pressure Wave Propagation Method

In [76] the latest developments of the PWP–LIPP method are illustrated, which shows very good resolution and has been recently used to study not only uniform but also nonuniform and complex composite materials, for example, epoxy resin samples filled with silica (see Fig. 4.60).

Research is continuing to improve the spatial resolution of the PWP–LIPP method sufficiently to allow its use in the study of integrated dielectrics. In the case of complex materials exhibiting either a very divergent electric field or a heterogeneous structure, the signal is not due only to the presence of charges. Other spurious contributions have to be taken into account. Thus the variation of the electric field caused by its divergence induces a signal contribution originating in the electrostrictive effect, directly proportional to the applied electric field. Such a proportional spurious contribution arises also in the case of heterogeneous materials for which the permittivity changes from one point to another. An additive spurious contribution has also to be taken into account in heterogeneous materials because heterogeneities may include charges or be piezoelectric. Correction procedures are described in [76] to compensate for these spurious contributions, by subtracting measurements, or modeling the sample and performing iterative simulations, or calibrating the spurious contributions to directly eliminate their effects. At the end of the correction procedure, the signal can be treated as obtained from a uniform planar material. As far as the spatial resolution improvement is concerned, a significant change in the measuring setup is required. The necessary signal bandwidth is much broader, and conventional equipment can no longer be employed [76].

From the extensive and very interesting analysis carried out in [76], as well as from the observations made in Section 4.4.2, the PWP-LIPP method is apparently

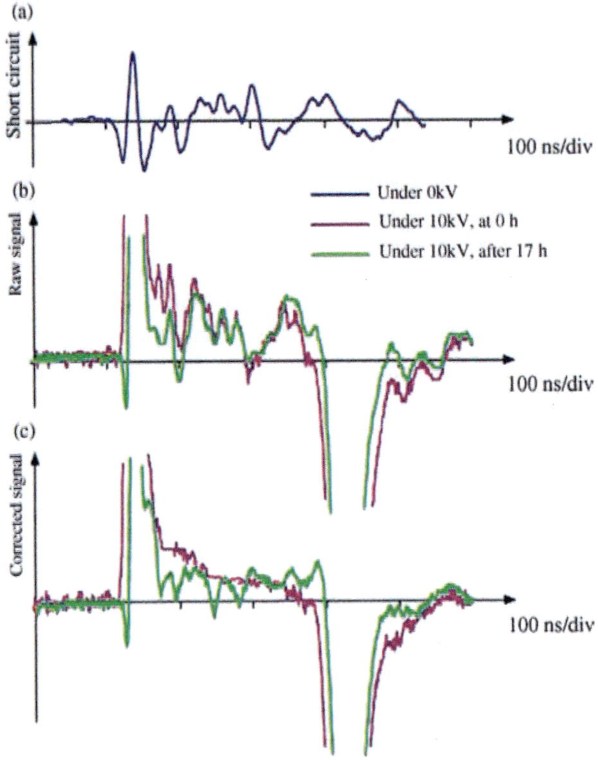

Figure 4.60 Space-charge measurements in a 1-mm-thick epoxy resin sample containing 50-μm mean diameter silica fillers randomly distributed throughout the sample volume. Measurement at 50° C (*a*) under short circuit and (*b*) at the beginning and after 17 h under 10 kV. (*c*) Measurement under short circuit subtracted from the measurement under applied voltage in order to eliminate the additive spurious contributions (after [76]).

applicable mostly to measurement of space-charge profiles in thin or very thin specimens in planar geometrical configuration (though even in strongly nonuniform arrangements with point electrodes; see, e.g., Fig. 4.61) and to cable prototypes (see [102–104]), but not to full-size power cables. Thus, it can be an excellent method at the development stage of new materials for extruded HVDC insulation that are not pure compounds but more or less heterogeneous materials (e.g., with addidives and/or nanocomposites, as will be discussed in Chapter 5).

As to the measurement of space charges and the determination of the relevant electric field in full-size power cables with thick insulation and large conductor cross section, the PEA method and the TSM seem at present the best-suited, as well as the most commonly used, techniques in practice. This observation leads into the summary section.

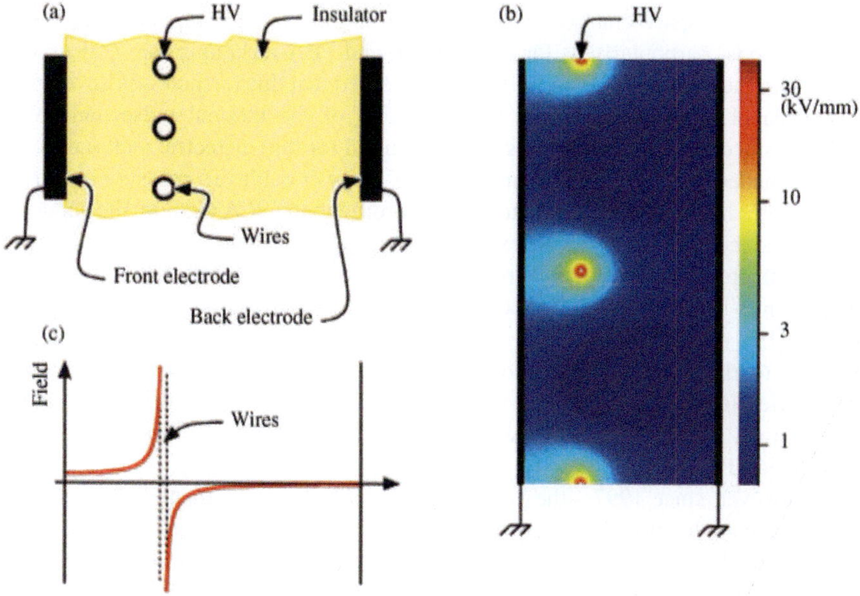

Figure 4.61 Plane-wire–plane-sample geometry generating a divergent electric field: (*a*) sample structure; (*b*) electric field amplitude perpendicular to the wires, which were of 25 μm diameter, evenly spaced at 1.27-mm intervals 365 μm from the front electrode and 835 μm from the back electrode. A voltage of 2 kV was applied to the wires. (*c*) Electric field amplitude in the neighborhood of an embedded wire at high voltage (after [76]).

4.6 FINAL COMPARISON BETWEEN THE BEST SPACE-CHARGE MEASUREMENT METHODS FOR POWER CABLES: PEA VERSUS TSM

In [19], a comparison among the most promising space-charge measurement methods, namely the thermal pulse method, particularly in its development as the TSM, the PEA technique, and the PWP technique was carried out. As to the *thermal pulse method*, in [19] it was argued that, while quick and easy to use, this technique suffers from limited resolution because of its dependence on a numerical deconvolution to interpret the data. Focusing on the *TSM* developed by Toureille [66–68], in [19] it was characterized as applicable for measuring space charges in both thick slabs and thick cylindrical structures such as cables with insulation thickness in the range from 2 to 20 mm. Thus, being also a nondestructive technique, it seemed promising for HVDC insulation already more than one decade ago [19]. On the other hand, it must be emphasized that the power source has to be disconnected while performing space-charge measurement by means of the TSM, as pointed out in Section 4.4.1.3; this involves, for example, that aging tests have to be temporarily interrupted during the measurements and restarted afterward. Hence the measurement of space charges via the TSM is slightly more complex than via the other methods, particularly as far as the PWP and PEA techniques are concerned.

As to the *PEA method*, in [19] it was observed that this method was developed by Takada [13, 77] to overcome the limitation of the thermal pulse and PWP methods, since these latter methods are well-suited for thin dielectric materials, and thus are inadequate for the thick insulation of a power cable. To measure accumulated charges in thick dielectric materials, according to the PEA method, the acoustic pulse is induced internally by an externally applied pulse field. The difficulty here might be in applying the electric stress uniformly. Another limitation to this method might come from the difficulty of detecting complex charge and polarization profiles [19].

As to the *PWP* method and to the several versions developed at several laboratories [69–75], in [19] it was pointed out that this method is not always reproducible, since it is not always certain how well the pressure wave is propagated [19].

However, since 1997—the year when [19] was written—many further developments and applications have regarded the TSM (see, e.g., [91]), the PWP (see, e.g., [76]), and the PEA (see, e.g., [92–94, 99, 100]) techniques, as illustrated in Sections 4.5.1, 4.5.2, and 4.5.3, respectively, with a particular focus on extruded HVDC cable insulation. Anyway, as it can be guessed from what was reported previously in this chapter, the best space-charge measurement methods for power cables, and in particular for HVDC cables, are at present the PEA and the TSM techniques. Indeed, they both enable the measurement to be performed over the whole insulation thickness even of full-size power cables and the subsequent processing of the detected signal in order to derive the distribution of space charge and electric field across the insulation wall. Also the PWP method, especially when implemented according to the LIPP technique, has been considered for space-charge measurements on insulation thicker than that of minicables (say, up to $\sim$3 mm and more) [102–104], but never on insulation thickness tending to that of full-size power cables, at least to the best of our knowledge.

From the viewpoint of spatial resolution, the PEA method seems superior on the whole since it provides very well-resolved space-charge profiles over the whole insulation thickness [105, 106], so that fairly narrow charge storage centers and fast-moving charge packets can be located (see charge profiles and patterns in Sections 4.4.2.3 and 4.5.2, as well as Figure 4.62), and the electric field can be determined with great accuracy (see Fig. 4.63). This is because the whole pressure wave with its time characteristics is acquired as it progresses within the insulation and reaches the transducer. On the contrary, the deconvolution process of the TSM current seems less satisfactory as to the spatial resolution, leading to much more regular and slowly varying field profiles within insulation thickness, which resemble on the whole those computed analytically in Chapter 3 (see, e.g.; charge and field profiles in Sections 4.4.1.3 and 4.5.1).

As to the time resolution, the PEA method exhibits better performances, too. Indeed, the PEA method requires the application of a very short voltage pulse (lasting on the order of tens of nanoseconds), which can be applied to the specimen being tested without interrupting the poling voltage; on the contrary, the TSM method not

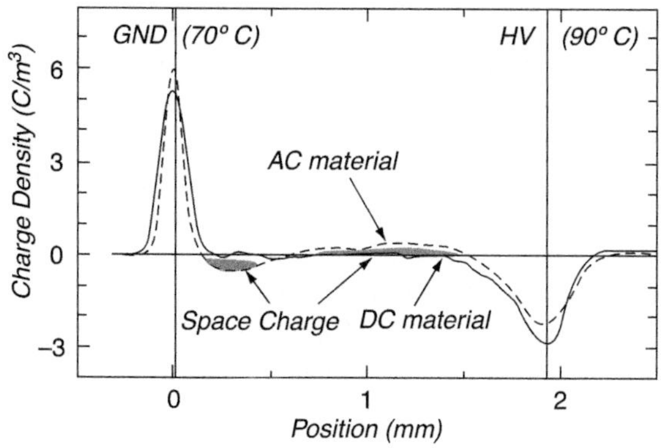

Figure 4.62 Space-charge characteristics of XLPE materials obtained by PEA (after [106]).

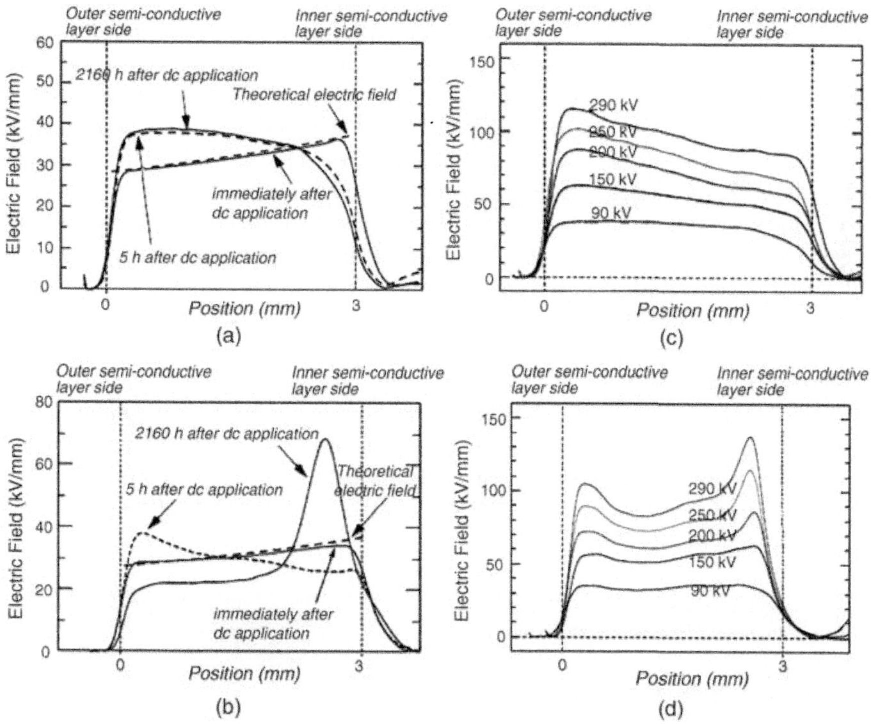

Figure 4.63 Plots of electric field distributions obtained from PEA measurements: (*a*) Electric field distribution in DC XLPE cable under DC voltage; (*b*) electric field distribution in AC XLPE cable under DC voltage; (*c*) electric field distribution in DC XLPE cable versus applied voltage; and (*d*) electric field distribution in AC XLPE cable versus applied voltage (after [106]).

only requires the application of a thermal step to the cable by means of a thermal diffuser—which involves much longer "thermal" time constants for the heat to be delivered from the diffuser to the cable—but also the HVDC source has to be disconnected during the measurement (see Section 4.4.1.3), which means the poling voltage has to be switched off. Hence, one single space-charge measurement via the TSM requires a much longer time than one single space-charge measurement via the PEA technique. Moreover, poling voltage has to be stopped during the measurement via TSM and resumed thereafter: This means that HVDC poling—as well as any other ongoing test that could be run in parallel to DC poling, for example, long-term electrothermal cycles for a cable system prequalification [7]—has to be stopped as well.

However, the PEA technique requires the electrode and underlying piezoelectric PVDF transducer to be in close contact with the insulation semiconducting shield; indeed, the measuring point is at the center of the exposed semicon to apply a pulsed voltage between this point and ground. For this reason the PEA is a destructive technique for a complete cable since it requires that the outer coatings (screen, sheath, outer covering, etc.) be removed along the measuring area (see Fig. 4.31) [13, 83, 84]. As to the TSM method, the thermal step is applied by circulating a cold liquid (the outer cooling technique) through a cylindrical thermal diffuser in contact with the outer semiconductive layer of the cable (Fig. 4.64). To achieve this, the screen and the outer PE sheath are removed. Hence the same damages to the full-size cable are obtained as in the case of the PEA technique. On the other hand, the TSM does not appear to suffer from the problems with acoustic wave propagation and reflection along the cable that affect the PEA for cables, which have been claimed to be solved in [13, 79, 80].

However, due to the differences between the two methods and the complementary performances, when used together they could provide complementary information that could on the whole improve the knowledge of the space-charge characteristics of the cable.

Mostly—in the past and in this chapter—the attention in space-charge measurements has been focused on small-size, small-thickness specimens because of the many difficulties relevant to space-charge measurements in thick insulation [13]. However, since the mid-1990s space-charge measurements on full-sized cables have been satisfactorily performed, especially resorting to the PEA technique. Indeed, PEA measurements on extruded HVDC full-size cables have been performed at

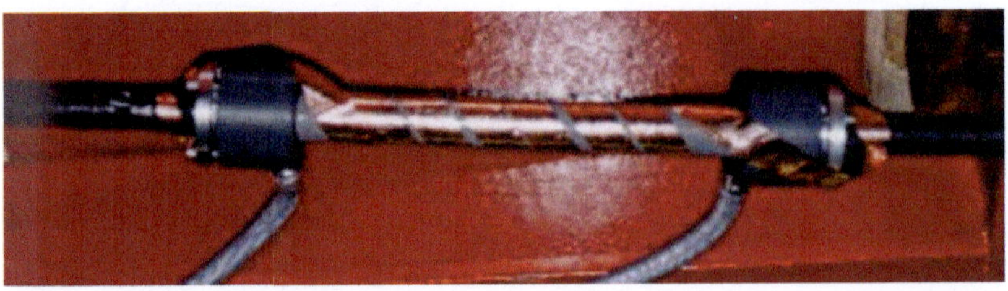

Figure 4.64 Thermal diffuser (400 mm long) used to generate the thermal step from the outer screen (after [91]).

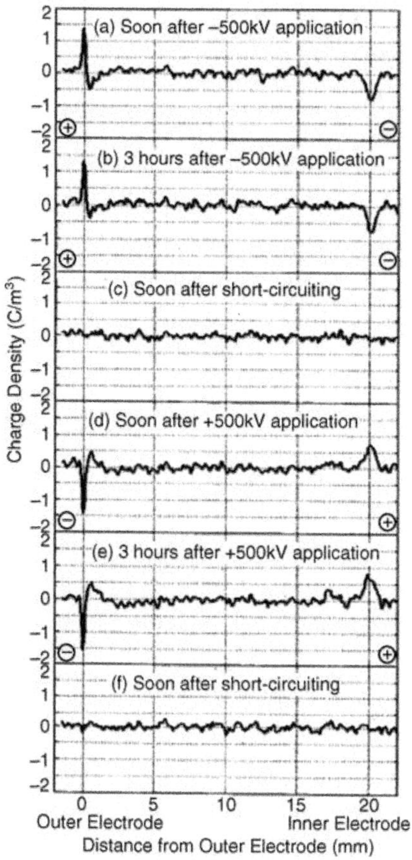

Figure 4.65 Space-charge distributions in sample B at room temperature (after [84]).

high-voltage levels and with large conductor cross sections and insulation wall thicknesses since at least 1998 in Japan, as shown in Figures 4.65 and 4.66 (relevant to two 250-kV cables, both with 800-mm^2 conductor cross sections, 20-mm insulation thickness, either at room temperature – 5–10°C in open air where the cables were laid – or under a severe temperature gradient ranging from 85°C on conductor to 5–7°C as the ambient temperature [84]), and in Figure 4.67 (relevant to a 500-kV cable, with 3000-mm^2 conductor cross section, 23-mm insulation thickness, under temperature gradient—conductor at 90°C, other details are missing [106]).

In particular, in [84] the cables subjected to space-charge measurements are described as follows:

> "Two cable specimens (sample A, B) were prepared for this experiment. Each cable was developed as a DC transmission cable and had a conductor cross section of 800 mm^2, 20 mm-thick insulation, and semiconducting electrodes at the inner side and the outer electrode side. Also, a lead shield was installed as a protective layer . . . Each specimen was 12 m long including terminations ([84], page 28)."

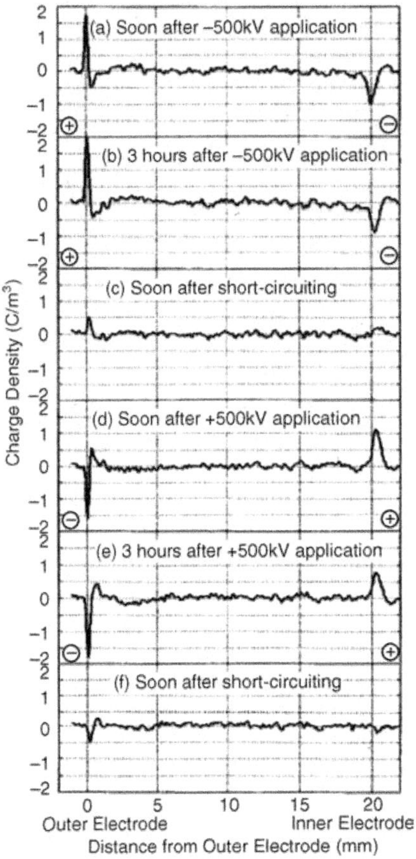

Figure 4.66 Space-charge distributions in sample A at high temperature (after [84]).

As to the insulation, in [83] it is pointed out that the two 250-kV cables used "two kinds of XLPE compound containing inorganic filler to suppress the accumulation of space charge within XLPE insulation ([83], page 7)". Further, in [84] the measurement procedure is described in detail, as follows:

"The authors chose the PEA method for taking space charge measurements. Newly developed equipment was used to take measurements under high voltages of several hundreds of kilovolts . . . The central part of the outer semicon was exposed by removing the lead shield, and the measuring point was fixed on it. After the separate lead shield was connected, a pulse voltage was applied between the measuring point and the ground. Typical pulse voltage is 50 ns wide, and it is as high as 5 kV. Two guard electrodes were set at both sides of the measuring point so that the pulse voltage around this point would be uniform. Measurements were taken at room temperature or maintaining a high conductor temperature via a conductor current . . . A DC voltage of 500 kV, which is twice as high as the running voltage, was applied." ([84], pp. 28–29).

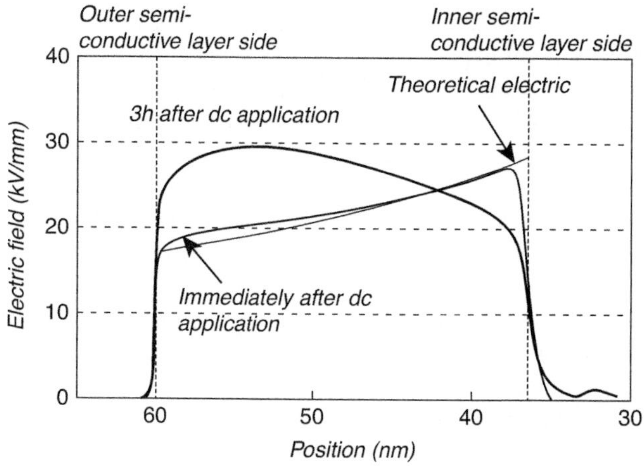

Figure 4.67 Electric field distribution of DC ±500-kV XLPE cable (after [106]).

The space-charge profiles measured in sample B at room temperature and A at high temperature, during a complex experimental sequence (see [84] for details), are shown in Figures 4.65 and 4.66, respectively.

As to Figure 4.67, it "shows the results obtained in the space charge measurements on a DC+/−500 kV XLPE cable under conditions equivalent to actual operation [voltage: 500kV; conductor temperature: 363K (90°C)]" ([106], page 694). As argued in [106], "it can be seen . . . that the field distribution after power application for 3 hours is marked by high field strength on the outer semi-conductive layer side due to homo charge accumulation and a relaxation of the field on the inner semi-conductive layer side."

A new PEA system for full-size cables is introduced in [99], dating back to 2008:

> There have been several reports on the application of PEA for space charge measurement in full size cables in most of which a curve-shaped ground electrode, transducer, and acoustic absorber block were designed to fit around the diameter of the cable sample. For different cable sizes, modification is required to the above components in addition to extra care in the sample assembly in order to ensure a good acoustic contact between the cable and curved electrode. The PEA system employed in this work adopts a flat ground electrode which enables the system to be easily applied to cables with different radii ([99], page 852).

Also following the thermal step method (TSM), space-charge measurements on prototype cables with thick insulation were performed, namely on an XLPE insulated cable with a 95-mm^2 cross-section conductor and ~5-mm-thick insulation, poled at −150 and −230 kV under temperature gradient, as shown in Section 4.5.1 [91], and space-charge measurements via the TSM on full-size cables are being performed at present according to private communications.

Some final remarks about space-charge measurements on full-size power cables are suggested in [99], where it is argued that:

Less attention has been paid to *space charge dynamics in full sized cables* presumably due to a limited range of experimental systems suitable for examining cables. Such investigations *would however be more relevant to practical situations as all the features specific to cable design, such as insulator/semiconducting interfaces, insulating material processing, and divergent electric field would be fully reflected in the space charge behaviour. They would also make it possible to investigate the effect of a temperature gradient in radial insulation on the space charge accumulation, and hence replicate the conditions experienced when the cable is loaded in service* ([99], page 851).

On these grounds, the authors of this book are fostering the measurement of space charges in full-sized HVDC extruded cables, also taking advantage of the many realizations of such cables that are going on at present and in the future all over the world. For this reason, we have proposed a procedure for space-charge measurements in full-size extruded cables, by which is meant a measurement protocol that accounts for the experimental practices and expertises—gained over the years in performing such kind of measurements—in terms of poling time, depolarization time, and heating and cooling of specimens [107].

Such procedure seems particularly useful for assessing the space-charge behavior in terms of electric field of a cable that has undergone a prequalification test. The measurement can also indicate the possibility of avoiding the repetition of prequalification tests and/or type tests in the case of HVDC cables for which the Laplacian field profile cannot be taken—strictly speaking—as a reference for comparing the electrothermal stress level of new realizations with that of previously manufactured cables. In fact—contrary to HVAC cables [108]—standards for extruded HVDC cables do not exist yet, apart [7]. However, the development of many new HVDC projects worldwide requires a deeper understanding about the chance of extending to a new project the prequalification and/or type tests already performed on a similar cable system.

A first step forward in this direction could be the application of the AC case "range of type approval" to the DC case. The four grounding prerequisites (see [108] at Section 12.2) can be deemed as valid for the DC case, but some considerations should be taken into account, in particular, for the fourth prerequisite, which states: "calculated maximum electrical stresses on the conductor and insulation screens, in the main insulation part(s) of the accessory and in boundaries are equal to or lower than for the tested cable and accessory". As mentioned in Chapter 3, the calculation of the DC electric field profile is possible, but the effect of space charge cannot be taken into account accurately. Consequently, a space-charge measurement procedure on full-size cables during prequalification and/or type tests can be a decisive step toward the extension of previous prequalification and/or type tests.

REFERENCES

1. C.K. Eoll. "Theory of Stress Distribution in Insulation of High Voltage D.C. Cables", Part II. *IEEE Trans. Electr. Insul.*, Vol. EI-10 No. 2, pp. 49–54, June 1975.
2. J.O. Boström, A. Campus, U. Nilsson, P. Carstensen, A. Gustafsson, A. Farkas, and K. Johannesson. "Cross-linked Polyethylene Materials for DC Power Cables."

5th International Conference on Insulated Power Cables (JiCable'99), Paper C2.12, June 20–24, 1999.

3. B. Aladenize, R. Coelho, J.C. Assier, H. Janah, and P. Mirebeau. "Field Distribution in HVDC Cables: Dependence on Insulating Materials." 5th International Conference on Insulated Power Cables (JiCable'99), Paper C2.13, June 20–24, 1999.
4. M.J.P. Jeroense and P.H.F Morshuis. "Electric Fields in HVDC Paper-insulated Cables." *IEEE Trans. Dielectr. Electr. Insul.*, Vol. 5, No. 2, pp. 225–236, April 1998.
5. R.N. Hampton. "Some of the Considerations for Materials Operating under High-Voltage, Direct-Current Stresses." *IEEE Electr. Insul. Mag.*, Vol. 24, No. 1, pp. 5–13, January–February 2008.
6. Brochure CIGRÉ 219. Recommendations for Testing DC Extruded Cable Systems for Power Transmission at a Rated Voltage up to 250 kV. Prepared by CIGRÉ Working Group 21-01, February 2003, also published on *Electra*, No. 206, pp. 46–53, February 2003.
7. Brochure CIGRÉ 496. Recommendations for Testing DC Extruded Cable Systems for Power Transmission at a Rated Voltage up to 500 kV. Prepared by CIGRÉ Working Group B1–32, April 2012, also published on *Electra*, No. 261, pp. 88–95, April 2012.
8. M. Salah Khalil. "International Research and Development Trends and Problems of HVDC Cables with Polymeric Insulation." *IEEE Electr. Insul. Mag.*, Vol. 13, No. 6, pp. 35–47, November–December 1997.
9. Z. Croitoru. "Space charges in dielectrics." *Progress in Dielectrics*, Vol. 6, pp. 105–145, 1965.
10. J.J. O'Dwyer. "The effect of collision ionization space charge on high field conduction in solids." *IEEE Trans. Electr. Insul.*, Vol. EI-15, pp. 264–269, 1980.
11. W. McAllister, G.C. Crichton, and A. Pedersen. "Charge accumulation in DC cables: a macroscopic approach." Conference Record of the 2994 IEEE International Symposium on Electrical Insulation, pp. 212–216, 1994.
12. L.A. Dissado, and J.C. Fothergill. *Electrical Degradation and Breakdown in Polymers.* Peregrinus, London, 1992.
13. T. Takada. "Acoustic and Optical Methods for Measuring Electric Charge Distributions in Dielectrics." *IEEE Trans. Dielectr. Electr. Insul.*, Vol. 6, No. 5, pp. 519–547, October 1999.
14. T.L. Hanley, R.P. Burford, R.J. Fleming, and K.W. Barber. "A General Review of Polymeric Insulation for Use in HVDC Cables." *IEEE Electr. Insul. Mag.*, Vol. 19, No. 1, pp. 14–24, January/February 2003.
15. G. Mazzanti, G.C. Montanari, and L. Dissado. "Electrical Aging and Life Models: The Role of Space Charge." *IEEE Trans. Dielectr. Electr. Insul.*, Vol. 12, pp. 876–890, October 2005.
16. G.C Montanari, I Ghinello, F. Peruzzotti, and M. Albertini. "Endurance Characteristics of XLPE Compounds under DC Voltage." ICSD'98, pp. 439–442, Vasteras, Svezia, June 22–25, 1998.
17. P. Morshuis and M. Jeroense. "Space Charge Measurements on Impregnated Paper: A Review of the PEA Method and a Discussion of Results." *IEEE Electr. Insul. Mag.*, Vol. 13, No. 3, pp. 26–35, May–June 1997.
18. Brochure CIGRÉ 86. Overvoltages on HVDC Cables. CIGRE Joint Working Group 33/21/14.16, August 1994.

19. N.H. Ahmed and N.N. Srinivas. "Review of Space Charge Measurements in Dielectrics." *IEEE Trans. Dielectr. Electr. Insul.*, Vol. 4, No. 5, pp. 644–656, October 1997.
20. E. Chapeau, C. Alquiè, and J. Lewiner. "The Pressure Wave Propagation method for the analysis of insulating materials: application to LDPE used in HV Cables." *IEEE Trans. Electr. Insul.*, Vol. 21, No. 3, pp. 405–410, 1986.
21. S. B. Lang and D.K. Das-Gupta. "Polarization and space charge distribution in thermally poled polyethylene and a comparison with polyvinylidene fluoride." *IEEE Trans. Electr. Insul.*, Vol. 21, No. 3, pp. 399–403, 1986.
22. R.W. Greaves, E.P. Fowler, A. Goodings, and D.R. Lamb. "The direct piezoelectric effect in extruded polyethylene." *J. Mat. Sci.*, Vol. 9, pp. 1602–1608, 1974.
23. N. Hozumi, J. Tanaka, A. DeReggi, B. Dickens, and N. Nagusrinivas. "Effect of DC Tests on Induced Space Charge." Conf. Rec. on the 1990 IEEE Intern. Symp. on Electr. Insul., Toronto, Canada, pp. 332–334, 1990.
24. G. Yianakopoulos, J. Vanderschueren, and J. Niezette. "Influence of Physical Aging Processes on Electrical Properties of Polymers." Proc. of the 3rd Intern. Conf. on Conduction and Breakdown in Solid Dielectrics, pp. 197–201, Trondheim, Norway, 1989.
25. L. Ziyu, L. Rongsheng, W. Huiming, and L. Wenbin. "Space charge and initiation of electrical trees." *IEEE Trans. Electr. Insul.*, Vol. 24, No. 1, p. 83–89, 1989.
26. H.R. Zeller and W. Schneider. "Electrofracture Mechanics of Dielectric Aging." *J. Appl. Phys.*, Vol. 56, pp. 455–459, 1984.
27. H.R. Zeller. "Breakdown and Prebreakdown Phenomena in Solid Dielectrics." *IEEE Trans. Electr. Insul.*, Vol. 22, No. 2, pp. 115–122, April 1987.
28. G. Teyssedre and C. Laurent. "Charge Transport Modeling in Insulating Polymers: From Molecular to Macroscopic Scale." *IEEE Trans. Dielectr. Electr. Insul.*, Vol. 12, No. 5, pp. 857–875, October 2005.
29. S. Hwangbo, Y. Kwon, S. Jeon, and M. Han. "Direct Correlation between Space Charge and Conduction Characteristics of Low-Density Polyethylene at Various Temperature." *Jpn. J. Appl. Phys.*, Vol. 37, pp. 4419–4427, 1998.
30. F.N. Lim, R.J. Fleming, and R.D. Naybour. "Space Charge Accumulation in Power Cable XLPE Insulation." *IEEE Trans. Dielectr. Electr. Insul.*, Vol. 6, pp. 273–281, 1999.
31. F. Aida, S. Wang, M. Fujita, G. Tanimoto, and Y. Fujiwara. "Study of the Mechanism of Space Charge Formation in Polyethylene." *J. Electrostat.*, Vol. 42, pp. 3–15, 1997.
32. R.J. Fleming, M. Henriksen, and J.T. Holbøll. "The Influence of Electrodes and Conditioning on Space Charge Accumulation in XLPE." *IEEE Trans. Dielectr. Electr. Insul.*, Vol. 7, pp. 561–571, 2000.
33. I.W. McAllister, G.C. Crichton, and A. Pedersen. "Charge Accumulation in DC Cables: a Macroscopic Approach." In Conf. Rec. 1994 IEEE ISEI, pp. 212–216, Pittsburgh, PA, 1994.
34. D.A. Seanor, Ed. *Electrical Properties of Polymers*. Academic, New York, 1982, Chapter 1.
35. S. Boggs, D.H. Damon, J. Hjerrild, J.T. Holboll, and M. Henriksen. "Effect of Insulation Properties on the Field Grading of Solid Dielectric DC Cable." *IEEE Trans. Power Del.*, Vol. 16, pp. 456–461, 2001.
36. I.W. McAllister, G.C. Crichton, and A. Pedersen. "Space Charge Fields in DC Cables." in Conf. *Rec. 1996 IEEE ISEI*, Vol. 2, pp. 661–665, 1996.

37. R. Coelho, B. Aladenize, and F. Guillaumond. "Charge Buildup in Lossy Dielectrics with Induced Inhomogeneities." *IEEE Trans. Dielectr. Electr. Insul.*, Vol. 4, pp. 477–486, 1997.

38. K.R. Bambery, R.J. Fleming, and J.T. Holbøll. "Space Charge Profiles in Low Density Polyethylene Samples Containing a Permittivity/Conductivity Gradient." *J. Phys. D: Appl. Phys.*, Vol. 34, pp. 3071–3077, 2001.

39. M. Meunier and N. Quirke. "Molecular Modelling of Electron Trapping in Polymer Insulators." *J. Chem. Phys.*, Vol. 113, pp. 369–376, 2000.

40. C. Laurent. "Charge dynamics in polymeric materials and its relation to electrical aging." 2012 IEEE Conference on Electrical Insulation and Dielectric Phenomena (IEEE CEIDP 2012) pp. 1–20, Montreal (Canada), October 13th-17th 2012 (Whitehead Memorial Lecture).

41. M. Meunier, N. Quirke, and A. Aslanides. "Molecular Modelling of Electron Traps in Polymer Insulators: Chemical Defects and Impurities." *J. Chem. Phys.*, Vol. 115, pp. 2876–2881, 2001.

42. R. Hoffmann, C. Janiak, and C. Kollmar. "A Chemical Approach to the Orbitals of Organic Polymers." *Macromolecules*, Vol. 24, pp. 3725–3746, 1991.

43. K.J. Less and E.G. Wilson. "Intrinsic Photoconduction and Photoemission in Polyethylene." *J. Phys. C: Solid State Phys.*, Vol. 6. pp. 3110–3120, 1973.

44. R. Coelho and B. Aladenize. *Les Diélectriques*: Hermès, Paris, 1993.

45. G. Blaise. "Charge Localization and Transport in Disordered Dielectric Materials." *J. Electrostatics*, Vol. 50, pp. 69–89, 2001.

46. T. Hibma and H.R. Zeller. "Direct Measurement of Space Charge Injection from a Needle Electrode into Dielectrics." *J. Appl. Phys.*, Vol. 59, pp. 1614–1620, 1986.

47. S. Boggs. "Very High Field Phenomena in Dielectrics." *IEEE Trans. Dielectr. Electr. Insul.*, Vol. 12, No. 5, pp. 857–875, October 2005.

48. M. Meunier, N. Quirke, D. Binesti, G. Martic, and J. M. Fourmigué. "Identification of Space Charge in Cable Insulation: A New Approach Using Molecular Simulation." Proc. IEEE Conf. Electrical Insulation and Dielectric Phenomena, Minneapolis, pp. 68–71, October 19–22, 1997.

49. M. Meunier, N. Quirke, and D. Binesti. "Electron Trapping in Polymer Insulators: A New Approach Using Molecular Modeling." *Proc. Intern. Conf. on Electric Charge in Solid Insulators, Tours*, pp. 66–72, June 29–July 3 1998.

50. D. Cubero, N. Quirke, and D.F. Coker. "Electronic Transport in Disordered n-Alkanes: From Fluid Methane to Amorphous Polyethylene." *J. Chem. Phys.*, Vol. 119, pp. 2669–2679, 2003.

51. M. Meunier, N. Quirke, and A. Aslanides. "Characterization of Charge Carrier Traps in Polymeric Insulators." Proc. IEEE Conf. Electrical Insulation and Dielectric Phenomena, Victoria, pp. 21–24, October 15–18, 2000.

52. G. Teyssedre, C. Laurent, A. Aslanides, N. Quirke, L.A. Dissado, A. Campus, and L. Martinotto. "Deep Trapping Centers in Cross-linked Polyethylene Investigated by Molecular Modeling and Luminescence Techniques." *IEEE Trans. Dielectr. Electr. Insul.*, Vol. 8, pp. 744–752, 2001.

53. A. Ekstrom and E. Willard. "Effects of Matrix Polarity on the Optical and Electron Spin Resonance Spectra of Trapped Electrons in Organic Glasses." *J. Phys. Chem.*, Vol. 72, pp. 4599–4603, 1968.

54. A. Campus, P. Carstensen, A.A. Farkas, and M. Meunier. "Chemical Defects and Electron Trapping Relevant to Cable Dielectrics." Proc. IEEE Conf. Electrical Insulation and Dielectric Phenomena, Cancun, 20–24, October 2002.

55. M.S. Khalil and B.S. Hansen. "Investigation of Space Charge in Low Density Polyethylene Using a Field Probe Technique." *IEEE Trans. Electr. lnsul.*, Vol. 23, pp. 441–445, 1988.

56. M.S. Khalil and A.A. Zaky. "The Effect of Cable Structure on Space Charge Formation." *IEEE Trans. Electr. Insul.*, Vol. 23, pp. 1043–1046, 1988.

57. G.M. Sessler. "Charge Distribution and Transport in Polymers." *IEEE Trans. Dielectr. Electr. Insul.*, Vol. 4, No. 5, pp. 614–628, October 1997.

58. R.E. Collins. "Practical Application of the Thermal Pulsing Technique to the Study of Electrets." *J. Appl. Phys.*, Vol. 51, pp. 2973–2986, 1980.

59. A.S. DeReggi, C.M. Guttman, E.I. Mopsik, G.T. Davis, and M.G. Broadhurst. "Determination of Charge or Polarization Distribution across Polymer Electrets by the Thermal Pulse Method and Fourier Analysis." *Phys. Rev. Lett.*, Vol. 40, pp. 413–416, 1978.

60. F.I. Mopsik and A.S. DeReggi, "Numerical Evaluation on the Dielectric Polarization Distribution from Thermal Pulse Data," *J. Appl. Phys.*, Vol. 53, No. 6, pp. 4333–4339, 1982.

61. N. Hozumi, J. Tanaka, A.S. DeReggi, and N. Nagusrinivas. "Space charge induced in stressed polyethylene." Proc. 1989 Conference on Electrical Insulation and Dielectric Phenomena (CEIDP 1989), IEEE Cat. No. 89 CH2773-0 pp. 253–258, 1989.

62. T. Suzuoki, S. Katakai, M. Kanaoka, and Y. Sekii. "Dc characteristics of cable insulating materials." Proc. of the 3rd Intern. Conf. on Conduction and Breakdown in Solid Dielectrics, pp. 66–71, Trondheim, Norway, 1989.

63. S.B. Lang and D.K. Das-Gupta. "Laser Intensity Modulation Method: a technique for determination of spacial distributions of polarization and space charge in polymer electrets." Proc. 5th Intern. Symp. on Electrets, Heidelberg, pp. 444–449, 1985.

64. S.B. Lang. "Laser Intensity Modulation Method (LIMM): experimental techniques, theory and solution of the integral equation." *Ferroelectrics*, Vol. 118, pp. 343–361, 1991.

65. S.B. Lang and D.K. Das-Gupta. "Polarization and Space Charge Distribution in Thermally Poled Polyethylene and a Comparison with Polyvinylidene Fluoride." *IEEE Trans. Electr. Insul.*, Vol. 21, No. 3, pp. 399–403, 1986.

66. A. Cherifi, M. Abou Dakka, and A. Toureille. "The Validation of the Thermal Step Method." *IEEE Trans. Electr. Insul.*, Vol. 27, No. 6, pp. 1152–1158, 1992.

67. P. Notingher, Jr. S. Agnel, and A. Toureille. "Thermal Step Method for Space Charge Measurements under Applied DC Field." *IEEE Trans. Dielectr. Electr. Insul.*, Vol. 8, No. 6, pp. 985–994, 2001.

68. S. Agnel, A. Toureille, and C. Le Gressus. "Study of the Aging of Impregnated Paper of High Power Capacitors Using the Thermal Step Method and the Thermally Stimulated Currents." 1996 Annual Report, Conference on Electrical Insulation and Dielectric Phenomena, pp. 190–193, 1996.

69. J. Lewiner. "Evolution of Experimental Techniques for the Study of the Electrical Properties of Insulating Materials." *IEEE Trans. Electr. Insul.*, Vol. 21, No. 3, pp. 351–360, 1986.

70. W. Eisenmenger and M. Haardt. "Observation of Charge Compensated Polarization Zones in PVDF Film by Piezoelectric Acoustic Step Wave Response." *Solid. State Commun.*, Vol. 41, pp. 917–920, 1982.

71. C. Alquie, G. Dreyfus, and J. Lewiner. "Stress Wave Probing of Electric Field Distribution in Dielectrics." *Phys. Rev. Lett.*, Vol. 47, No. 20, pp. 1483–1487, 1981.

72. C. Alquié, J. Lewiner, and G. Dreyfus. "Analysis of Laser Induced Acoustic Pulse Probing of Charge Distributions in Dielectrics." *J. Phys. Lett.*, Vol. 44, pp. 171–178, 1983.

73. E. Chapeau, T. Ditchi, C. Alquié, J, Lewiner, J. Perret, and B. Dalle. "Comparative Study of the Behavior of Two Polyethylene Types under DC Voltage by the Pressure Wave Method." *Conf. JICABLE 1987, Versailles, France*, pp. 81–97, 1987.

74. Y. Suzuoki, T. Furuta, H. Yamada, S.O. Han, T. Mizutani, M. Ieda, and N. Yoshifuji. "Study of Space Charge in Polyethylene by Direct Probing." *IEEE Trans. Electr. Insul.*, Vol. 26, No. 6, pp. 1073–1079, 1991.

75. Y. Suzuoki, T. Mizutani, T. Furuta, H. Yamada, S.O. Han, M. Ieda, and N. Yoshifuji. "Study of Space Charge in Polyethylene by Laser Induced Pressure Pulse Technique." 1990 Ann. Rep. Conf. on Electr. Insul. Dielectr. Phenom., pp. 203–207, 1990.

76. S. Hole. "Recent Developments in the Pressure Wave Propagation Method." *IEEE Electr. Insul. Mag.*, Vol. 25, No. 3, pp. 7–20, May–June 2009.

77. T. Takada, T. Maeno, and Kushibe. "An Electric Stress-Pulse Technique for the Measurement of Charges in a Plastic Plate Irradiation by an Electron Beam." *IEEE. Trans. Electr. Insul.*, Vol. 22, No. 4, pp. 497–501, 1987.

78. Y. Li, M. Yasuda, and T. Takada. "Pulsed Electroacoustic Method for Measurement of Charge Accumulation in Solid Dielectrics." *IEEE Trans. Dielectr. Electr. Insul.*, Vol. 1, No. 2, pp. 188–195, 1994.

79. K. Fukunaga, H. Miyata, M. Sugimori, and T. Takada. "Measurement of Charge Distribution in the Insulation of Cables Using Pulsed Electroacoustic Method." *Trans. IEE Jpn.*, Vol. 110-A, No. 9, pp. 647–648, (1990).

80. R. Liu, T. Takada, and N. Takasu. "Pulsed Electroacoustic Method for Measurement of Charge Distribution in Power Cables under Both DC and AC Electric Fields." *J. Phys. D: Appl. Phys.*, Vol. 26, pp. 986–993, 1993.

81. N. Hozumi, H. Suzuki, T. Okamoto, K. Wakanabe, and A. Watanabe. "Direct Observation of Time Dependent Space Charge Profiles in XLPE Cable under High Electric Field." *IEEE Trans. Dielectr. Electr. Insul.*, Vol. 1, No. 6, pp. 1068–1076, 1994.

82. H. Kon, Y. Suzuoki, T. Mizutani, M. Ieda, and N. Yoshifuji. "Packet-like Space Charges and Conduction Current in Polyethylene Cable Insulation." *IEEE Trans. Dielectr. Electr. Insul.*, Vol. 3, No. 3, pp. 380–385, June 1996.

83. K. Terashima, H. Suzuki, M. Hara, and K. Watanabe. "Research and Development of ±250 kV DC XLPE Cables." *IEEE Trans. Power Del.*, Vol. 13, pp. 7–16, 1998.

84. T. Takeda, N. Hozumi, H. Suzuki, K. Fujii, K. Terashima, M. Hara, Y. Murata, K. Watanabe, and M. Yoshida. "Space Charge Behavior in Full Size 250 kV DC XLPE Cables." *IEEE Trans. Power Del.*, Vol. 13, No. 1, pp. 28–39, January 1998.

85. Y. Zhang, F. Zheng, Z. An, and X. Zhang. "Research on Space Charge in Insulating and Functional Dielectrics at Tongji University." *IEEE Electr. Insul. Mag.*, Vol. 28, No. 3, pp. 14–25, 2012.

86. G.R. Govinda Raju. "Conduction and Thermally Stimulated Discharge Currents in Aramid Paper. " *IEEE Trans. Electr. Insul.*, Vol. 27. pp. 162–173, 1992.

87. M. Ieda and M. Nawata. "DC Treeing Breakdown Associated with Space Charge Formation in Polyethylene." *IEEE Trans. Electr. Insul.*, Vol. EI-12 No. 1, pp. 19–25, February, 1977.

88. Z. Zhu, Y. Zhang, Z. An, and F. Zheng, "Methodological Investigation on Photo-stimulated Discharge to Obtain Accurate Trap Information in Polymer Dielectrics." *Meas. Sci. Technol.*, Vol. 22, No. 8, art. no. 085109, 2011.

89. G. Mazzanti, G.C. Montanari, and J.M. Alison. "A Space-Charge Based Method for the Estimation of Apparent Mobility and Trap Depth as Markers for Insulation Degradation—Theoretical Basis and Experimental Validation." *IEEE Trans. Dielectr. Electr. Insul.*, Vol. 10, No. 2, pp. 187–197, April 2003.

90. A. Many and G. Rakavy, "Theory of Transient Space-Charge Limited-Current in Solids in the Presence of Trapping." *Phys. Rev.*, Vol. 126, pp. 1980–1988, 1962.

91. J. Castellon, P. Notingher, S. Agnel, A. Toureille, F. Brame, P. Mirebeau, and J. Matallana. "Electric Field and Space Charge Measurements in Thick Power Cable Isulation." *IEEE Electr. Insul. Mag.*, Vol. 25, No. 3, pp. 30–42, May–June 2009.

92. D. Fabiani, G.C. Montanari, C. Laurent, G. Teyssedre, P.H.F. Morshuis, R. Bodega, L.A. Dissado, A. Campus, and U.H. Nilsson. "Polymeric HVDC Cable Design and Space Charge Accumulation. Part 1: Insulation/Semicon Interface. " *IEEE Electr. Insul. Mag.*, Vol. 23, No. 6, pp. 11–19, November-December 2007.

93. S. Delpino, D. Fabiani, G.C. Montanari, C. Laurent, G. Teyssedre, P.H.F. Morshuis, R. Bodega, and L.A. Dissado. "Polymeric HVDC Cable Design and Space Charge Accumulation. Part 2: Insulation Interfaces." *IEEE Electr. Insul. Mag.*, Vol. 24, No. 1, pp. 14–24, January–February 2008.

94. D. Fabiani, G.C. Montanari, C. Laurent, G. Teyssedre, P.H.F. Morshuis, R. Bodega, and L.A. Dissado. "HVDC Cable Design and Space Charge Accumulation. Part 3: Effect of Temperature Gradient." *IEEE Electr. Insul. Mag.*, Vol. 24, No. 2, pp. 5–14, March–April 2008.

95. CIGRE Joint Task Force 21/15. "Interfaces in Accessories for Extruded HV and EHV Cables." Electra, No. 203, pp. 53–59, 2002.

96. G. Chen, Y. Tanaka, T. Takada, and L. Zhong. "Effect of Polyethylene Interface on Space Charge Formation." *IEEE Trans. Dielectr. Electr. Insul.*, Vol. 11, pp. 113–121, February 2004.

97. G.C. Montanari. "The Electrical Degradation Threshold of Polyethylene Investigated by Space Charge and Conduction Current Measurements." *IEEE Trans. Dielectr. Electr. Insul.*, Vol. 7, pp. 309–315, June 2000.

98. R. Bodega, P. H. F. Morshuis, U. H. Nilsson, G. Perego, and J. J Smit,"Charging and polarization phenomena in coaxial XLPE-EPR interfaces", in Proc. IEEE ISEIM, Kitakyushu, Japan, pp. 107–110, June 2005.

99. M. Fu, L.A. Dissado, G. Chen, and J.C. Fothergill. "Space Charge Formation and Its Modified Electric Field under Applied Voltage Reversal and Temperature Gradient in XLPE Cable." *IEEE Trans. Dielectr. Electr. Insul.*, Vol. 15, No. 3, pp. 851–860, June 2008.

100. W. Choo, G. Chen, and S.G. Swingler. "Electric Field in Polymeric Cable Due to Space Charge Accumulation under DC and Temperature Gradient." *IEEE Trans. Dielectr. Electr. Insul.*, Vol. 18, No. 2, pp. 596–606, April 2011.

101. T. Andrews, R.N. Hampton, A. Smedberg, D. Wald, V. Waschk, and W. Weissenberg. "The Role of Degassing in XLPE Cable Manufacture." *IEEE Electr. Insul. Mag.*, Vol. 22, No. 6, pp. 5–16, 2006.

102. F.N. Lim, R.J. Fleming, and R.D. Naybour. "Space Charge Accumulation in Power Cable XLPE Insulation." *IEEE Trans. Dielectr. Electr. Insul.*, Vol. 6, pp. 273–281, 1999.

103. F.N. Lim, R.J. Fleming, and R.D. Naybour. "Space Charge Accumulation in Power Cable XLPE. Insulation." Proc. 3rd International Conference on Electric Charge in Solid Insulators, pp. 329–338, Tours (France), June 1998.

104. F.N. Lim and R.J Fleming, "The Influence of O_2, N_2 and H_2O on Space-Charge Accumulation in XLPE under DC Fields." Proc. 1998 International Symposium on Electrical Insulating Materials, pp. 199–202, Toyohashi (Japan), September 27–30, 1998.

105. S. Maruyama, N. Ishii, M. Shimada, S. Kojima, H. Tanaka, M. Asano, T. Yamanaka, and S. Kawakami. "Development of a 500-kV DC XLPE Cable System." *Furukawa Rev.*, No. 25, pp. 47–52, 2004.

106. T. Yamanaka, S. Maruyama, and T. Tanaka. "The Development of DC+/−500 kV XLPE Cable in Consideration of the Space Charge Accumulation." Proc. of the 7th IEEE International Conference on Properties and Applications of Dielectric Materials (IEEE ICPADM), pp. 689–694, June 1–5, 2003, Nagoya (Japan).

107. M. Marzinotto and G. Mazzanti. "A Procedure for Space Charge Measurements in Full-Size HVDC Extruded Cables." 2011 IEEE Conference on Electrical Insulation and Dielectric Phenomena (IEEE CEIDP 2011), pp. 804–807, Cancun (Mexico), October 17–20, 2011.

108. Standard IEC 62067. "Power Cables with Extruded Insulation and Their Accessories for Rated Voltage above 150 kV (Um = 170 kV) up to 500 kV (Um = 550 kV)—Test Methods and Requirements." Ed. 1.1, 2006.

CHAPTER 5

IMPROVED DESIGN OF HVDC EXTRUDED CABLE SYSTEMS

The development activities for an improved design of HVDC extruded cable systems capable of overcoming the technical issues and the performance limits discussed in the previous chapters have followed mainly three routes [1]:

1. From the *converter* viewpoint, developments in the HVDC VSC converter station technology have removed the functional requirement for polarity reversal on power systems rated, say, today up to 1000 MW [1–3] (this power limit is expected to increase in the future).
2. From the *cable insulation* viewpoint, the performance of extruded HVDC cables has been enhanced by improving the behavior of the polymeric dielectrics (e.g., by introducing additives to traditional compounds used for AC applications or by selecting a polymer from a different functional group).
3. From the *cable system* viewpoint, particular care has been devoted to the accessories (joints and terminations), which are often the weakest points of the whole cable line (which consists in fact of cables plus joints and terminations).

The first route has already been treated in Chapter 2, whereas the second and the third routes are explored in the present chapter 5. Indeed, this chapter treats the improvements to the design of extruded HVDC cable systems brought about by the most recent research and development activities relevant to *cable insulation* and *cable system components*.

The main contents of the chapter are as follows:

- An introduction (Section 5.1) about the general R&D trends in improving extruded polymeric insulation for HVDC cable systems followed over the years.
- A subsequent more focused analysis, where some approaches followed historically in improving extruded polymeric insulation—though abandoned at present—are recalled for the sake of completeness, namely:
  - The use of AC LDPE, XLPE or HDPE cables for HVDC applications without any modifications (Section 5.2)
  - The search for a stress inversion-free or -limited DC cable (Section 5.3)

Extruded Cables for High-Voltage Direct-Current Transmission: Advances in Research and Development, First Edition. Giovanni Mazzanti and Massimo Marzinotto.

- A further analysis dedicated to the most common and up-to-date approach in improving extruded polymeric insulation for HVDC cable systems, that is, the suppression of space charge in the insulation (Section 5.4), achieved by modifying the characteristics of the electrode–insulation interfaces or the bulk insulation characteristics via additives or fillers.
- A description of the further requirements for the improvement of HVDC extruded cable design (Section 5.5).
- A detailed illustration of the improved design of practical HVDC extruded cables (Section 5.6) and accessories (joints and terminations, Section 5.7). These sections are based not only on literature but also on valuable material provided by the main manufacturers of HVDC extruded cable systems in the world.
- A conclusive synthesis of the improved design (Section 5.8) and testing (Section 5.9) of overall HVDC extruded cable systems.

5.1 R&D TRENDS IN IMPROVING EXTRUDED POLYMERIC INSULATION FOR HVDC CABLES

5.1.1 Problems to Be Solved for Improving HVDC Extruded Insulation

From the analysis of the peculiar DC features of polymeric insulation—carried out in Chapters 3 and 4—it can be argued that the DC behavior of this insulation is due mainly to the cumbersome dependence of its dielectric properties on temperature, electric field, and accumulation of space charge in the material. In particular, space-charge accumulation is a consequence of the dependence of the electric conductivity on temperature and electric field, as well as of the strongly nonuniform microstructure of polymeric insulation. This nonuniform microstructure gives rise to many types of localized deep traps (especially due to chemical defects, as discussed in Section 4.3.2.2) where charge can stay for a much longer time than in shallow traps, which are associated with the ohmic behavior of conductivity at low fields (see Sections 4.2.1.1 and 4.3.2.1). The "space-charge picture" is made more complex by other processes that are also strongly dependent on temperature and field: (1) the generation of ions from the dissociation of ionic impurities within the insulation bulk (see Sections 4.2.1.2 and 4.3.1.2); (2) the injection phenomena of both electrons and holes from the electrodes (see Sections 4.2.2.1 and 4.3.1.1); (3) the role played by interfaces: both conductor–dielectric interfaces at the semiconducting shields (see Section 4.5.2.1) and dielectric–dielectric interfaces (see Section 4.5.2.2) in cable accessories, that is, joints and terminations. Research and development studies aimed at improving the design and manufacturing of the whole extruded insulation HVDC cable system should consider investigation of all these effects [4–6].

As discussed in Chapter 3 and more extensively in Chapter 4, space-charge storage in polymeric insulation for HVDC cables alters the electric field within the insulation and may lead to local field enhancements, especially under service conditions, that involve load cycling and—when CSC converters are used—also voltage polarity reversals [4]. As a consequence, breakdown strength may be lowered. As reported in [4], dating back to 1997:

> The effect of polarity reversals on the DC breakdown strength of DC cables with polymeric insulation has been reported for a long time by different authors [7–10].[1] During the early developments of ±250 kV DC XLPE cable in Japan, the cause of the failures that occurred in the initial stages of development work was found to be due to the accumulation of space charge in the insulation [7,11]. It was also shown that for LDPE and XLPE experimental DC cables, the ratio "polarity reversal breakdown voltage" over "DC breakdown voltage" ranges from about 60% to 90%, depending on the type and thickness of insulation and temperature [8–10]. The inability to use a submarine HVDC cable with XLPE insulation between Hokkaido and Honshu in Japan was attributed to the reduction of the DC breakdown strength of the cable insulation upon polarity reversal. Such a reduction was explained as a result of space charge accumulation in the insulating material. [11] ([4], pp. 38–39).

Today it must be emphasized that polarity reversal in extruded HVDC cable systems is not only a challenge for cable insulation, as previously stated in [4], but also for cable accessories. Furthermore, as far as undersea extruded cables are concerned, one more difficulty is encountered: namely, the development of undersea joints, especially for water depths higher than some 100 m.

It can be argued that in 1997 a lot of emphasis was still put on the ability of polymeric insulation to withstand polarity reversals since HVDC converters of the VSC type were in the infant stage and the HVDC systems were dominated by the LCC–CSC type (see Section 2.4.1). In fact, the first VSC HVDC cable system with polymeric insulation—the Gotland link—was contracted in late 1997 and passed successfully the after installation tests in the beginning of 1999 [12], and, since then, the need for successful polymeric insulation HVDC cables under polarity reversals was smoothed. Anyway, the success of the polarity reversal can still be of some interest in the overall consideration of polymeric insulation performances since the main goal is to unbind the DC extruded cables to the conversion technology used at the terminal stations. Furthermore, it is important to remember that regardless of the conversion technology used (CSC or VSC type), the cable can be subjected to electromagnetic transients that can give rise to a "temporary polarity reversal": for example, fast transients due to lightning events when the cable is terminated on an overhead line or when faults occur on the DC station bus bars. On the other hand, slow transients occur during switching operations or during faults distant from the converter stations [13]. This kind of momentary polarity reversal cannot be disregarded during the design of an HVDC cable. For this reason, as well as for the commissioning in late 2012 of the Hokkaido-Honshu ±250-kV HVDC extruded insulation cable system with CSC (see Section 7.2.8), the behavior under polarity reversal (permanent or temporary) is of interest in the present chapter [4].

[1] Whenever a citation from another source includes equations, figures, tables footnotes, and references, the numbers of such items are changed to follow the numbering in this text.

5.1.2 Optimum Characteristics of Polymeric Insulating Materials for HVDC Cables

According to all the above arguments, as well as to other investigations about the dielectric properties and the service requirements of polymeric insulation for HVDC cables [14,15], four optimum characteristics of polymeric insulating materials for HVDC cables were singled out in [4], namely:

1. The *insulation resistivity* should be comparatively *insensitive to variation of temperature, electric stress, polarity reversals, and time of electrification*;

2. The *thermal resistivity* should be *minimum*;

3. The material should have the ability of *storing minimum space charge*;

4. The *DC breakdown strength* should be *high and insensitive to temperature rise and polarity reversals* ([4], page 39).

These requirements were essentially confirmed 7 years later in [1], dating back to 2004, where the *need for high DC breakdown strength* was further emphasized, *with particular reference to superimposed impulse conditions.*

The realization of these characteristics is not easy at all, of course: Table 5.1 summarizes the development activities, as well as the relevant tests and experimental techniques, about HVDC cables with polymeric insulation reported in articles published until 1997 [7,15–36]. In Table 5.2 the test results of some of these studies are analyzed, assessing the effects of temperature rise, polarity reversal, insulating material, additive content, and blending on the "vital" DC breakdown strength by means of ad hoc coefficients, proposed for assessing the performance of HVDC cables with polymeric insulation. These coefficients are the temperature rise coefficient K_{TDC} and the polarity reversal coefficient K_{PDC}, defined, respectively, as follows [4]:

$$K_{\mathrm{TDC}} = 100 \times \frac{\text{DC breakdown strength at high temperature}}{\text{DC breakdown strength at low temperature}} \quad (\text{in } \%) \qquad (5.1)$$

$$K_{\mathrm{PDC}} = 100 \times \frac{\text{DC breakdown strength under polarity reversals}}{\text{DC breakdown strength at unipolar voltage}} \quad (\text{in } \%) \qquad (5.2)$$

where K_{TDC} indicates the dependence of the DC breakdown strength on temperature rise (see Fig. 5.1 as an example) and is expected to be affected by polarity reversals. Hence, K_{TDC} under direct polarity, K_{TDC} (D.P), and K_{TDC} under reversed polarity, K_{TDC} (R.P), are said to account for the effects of unipolar and polarity reversal voltage applications, respectively [4]. Analogously, K_{PDC} indicates the dependence of the DC breakdown strength of the insulator on polarity reversals and is expected to be affected by temperature variations. Hence, K_{PDC} at low

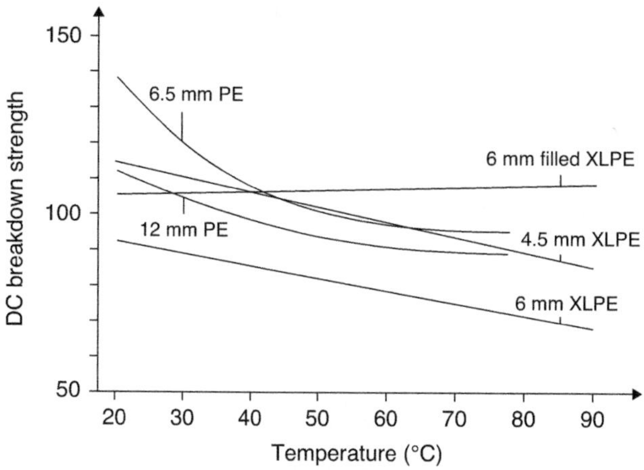

Figure 5.1 Effect of temperature on the DC breakdown strength of some experimental HVDC cables with polymeric insulation (after [4]).

temperature, K_{PDC} (L.T), and K_{PDC} at high temperature, K_{PDC} (H.T), are said to account for the effects of low-temperature and high-temperature performance, respectively [4]. It is observed in [4] that "the higher the coefficients K_{TDC} and K_{PDC}, the better is the performance of the cable. A value of 100% for each coefficient will indicate the independence of the cable performance on both temperature variations and polarity reversals" ([4], page 39).

5.1.3 Historical Development Activities of HVDC Extruded Insulation

The development activities of HVDC cables with polymeric insulation listed in Table 5.1 can be divided into three main groups depending on the target of most research and development work, that is, [4]:

1. *Exploring and establishing the DC insulation characteristics of solid insulation*, where focus was mostly on investigating the dependence of insulation resistivity on temperature and electric field and the field inversion phenomenon. Very little work—if any—was done on space-charge measurements (see, e.g., [15–17,37]), also because techniques for direct observation of space charges in solid insulation were still missing;
2. *Investigating space-charge effects*, while conversely very little work—if any—was done on the dependence of DC insulation resistivity on temperature and electric field, as well as field inversion [31–36].
3. *A more comprehensive approach*: Not only was the dependence of insulation resistivity on temperature and electric field determined but also the space-charge measurements were performed via innovative direct techniques. Manifold analysis methods were employed to measure several significant parameters (see, e.g., [29,30,38])

Table 5.1 Summary of Studies on Developing HVDC Cables with Polymeric Insulation, Together with the Relevant Tests and Experimental Techniques Used.

Research & Development Approach and Authors	Tests and Experimental Techniques Used
1. Basic research and preliminary studies of DC solid insulation characteristics, and stress inversion. (Tellier et al., Oudin et al., France, 1958–1967, Weedy et al., Britain 1978–1984)	DC insulation resistivity and its dependence on temperature and electric field.
2. Development of stress inversion-free DC Cable using fillers to XLPE. (Singh et al., USA, 1973–1980 [15–17])	DC insulation resistivity versus temperature and electric field, DC breakdown strength with direct and polarity reversals at different temperatures, thermal resistivity.
3. Suppression of space-charge effects using additives to XLPE. (Ando et al., Japan, 1976 [7])	Space-charge measurements using thermally stimulated current (TSC) as an indirect method and the cross-cut dust figure (CDF) method. DC breakdown strength with direct and polarity reversals at different temperatures.
4. Suppression of space-charge effects using additives. (Khalil et al., Denmark, 1983, 1987–1992 [18–28])	Space-charge measurements using field probe technique, PWP and thermal step method, DC breakdown strength under direct and polarity reversals, evaluation of DC insulation resistivity, effects of additives on stress inversion, polymer morphology.
5. Suppression of space-charge using an inorganic additive to XLPE. (Ogata et al., Japan, 1990 [29])	DC breakdown tests under direct and polarity reversals, impulse breakdown tests with and without using DC prestress, evaluation of insulation resistance and space-charge evaluation tests using the dust figure method and the pulse electro-acoustic method.
6. Suppression of space-charge using a highly purified special filler to XLPE. (Maekawa et al., Japan, 1992 [30])	Effect of temperature on the DC breakdown strength, Impulse breakdown strength with and without superposing DC voltage, evaluation of the dependence of insulation resistivity on temperature and electric field, and space-charge effects using the indirect method of thermally stimulated currents (TSC).
7. Suppression of space-charge by blending XLPE and HDPE. (Fukagawa et al., Suzuki et al., Japan, 1984, 1989, 1992 [31–33])	Space-charge evaluation using TSC, pulsed electro-acoustic method, Impulse breakdown strength, DC breakdown strength under direct and polarity reversals.
8. The use of AC LDPE, XLPE, or HDPE cables for HVDC applications without any modifications. (M. Pays et al., France, 1988 [34–36])	Space-charge measurements using the pressure wave pulse technique, effects of temperature cycling on the DC breakdown strength with direct and polarity reversals, impulse breakdown tests, impulses superimposed on DC. Effects of hydrostatic pressure as well as sea water in contact with insulation.

Source: after [4]. Reference numbers have been modified to reflect numbering in this chapter.

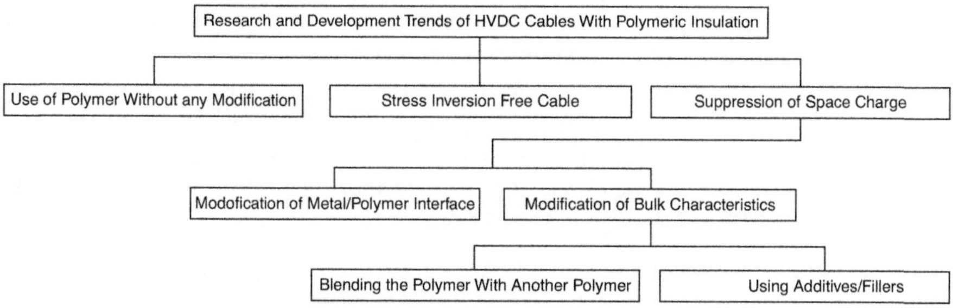

Figure 5.2 Research and development trends of HVDC cables with polymeric insulation (a postprocessing after a figure from [4]).

The scheme reported in [4]—and displayed in Figure 5.2—for illustrating the various methods followed to develop HVDC cables with polymeric insulation still holds, namely:

1. Development of *DC cables using LDPE, XLPE, or HDPE without any modification*
2. Modification of the electrical insulation and thermal resistivities of the polymer aiming at *limiting the stress inversion in the material* (stress-inversion-free cable)
3. *Suppression of space charge* in the polymer and hence reduction of space-charge effects

Details of these approaches are reported in Sections 5.2–5.4. Some of these details, as well as their structure, come from the milestone—although 12 years old—article [4]. However, they have been updated when needed by means of other more recent studies, first of all the fundamental study [5], which dates back to 2003 but which is still mainly accurate.

5.2 USE OF AC LDPE, XLPE, OR HDPE CABLE COMPOUNDS [2] FOR HVDC APPLICATIONS WITHOUT ANY MODIFICATIONS

In order to illustrate the early approach to the development of extruded insulation for HVDC cables, here the results are reported of a few studies on the HVDC

[2] It is important to point out that PE-based insulation materials—either resin or cross-linked—differing because of the amount of other components in the insulation are more correctly called "compounds." The word *compound*, commonly used for materials blended with a nonnegligible amount of fillers and additives like rubbers (e.g., EPR), is also used for PE-based insulation material: for HVAC applications, the small amount of antiaging additives in *n*on-cross-linkable LDPE and HDPE and the small amount of antiaging additives and the cross-linking agent in cross-linkable PE compounds (after cross-linking, XLPE) make such blends compounds.

performance of HVAC cables with LDPE, XLPE, and HDPE insulation compounds without any modification either to the materials or to the cable design [34–36]. Different tests and measurements were carried out on flat samples and cables manufactured with such compounds, namely, space-charge measurements using the PWP method (see Section 4.4.2.1), DC breakdown strength tests at low and high temperature, load cycle tests for assessing the effect of temperature on DC breakdown strength without and with polarity reversals, and impulse breakdown tests with and without superimposed DC voltage. The role played by hydrostatic pressure and seawater in contact with insulation was also investigated. The results of these experimental activities showed that [4]:

1. K_{TDC} is low for LDPE, XLPE, and HDPE HVDC compounds without any modification, about 20% lower than the corresponding values of an XLPE compound (after Ogata et al. [29] (see Table 5.2). HDPE compounds exhibits the lowest K_{TDC}. Thus the DC breakdown strength of these materials decreases greatly with temperature.
2. XLPE and HDPE compounds are affected by thermal instability.
3. XLPE and LDPE compounds are less sensitive to temperature than HDPE compounds.

Table 5.2 Effects of Temperature Rise, Polarity Reversals, Insulating Material, Additive Content, and Blending (+A) on the DC Breakdown Strength of Some Experimental HVDC Cables with Polymeric Insulation

Ref.	Sample Material	Details Thickness (mm)	DC Breakdown Strength of the Material at R.T (kV/mm)	Effect of Temperature		Effect of Polarity Reversals	
				K_{TDC} (D.P) (%)	K_{TDC} (R.P) (%)	K_{PDC} (L.T) (%)	K_{PDC} (H.T) (%)
Ando	XLPE + A	6	100	100	—	—	54
et al. [7]	XLPE	6	87	68	—	—	44
EPRI Project	XLPE + A	5.4	94	85	150	44	79
[17]	XLPE	5.2	69	79	125	31	39
Ogata et al.	XLPE + A	2.5	140	85	114	72	97
[29]		9	122	82	106	64	83
	XLPE	2.5	164	46	102	49	108
		9	105	61	—	—	104
Suzuki et al.	Blended	3.5	—		82	—	
[31–33]	HDPE						
Pays et al.	LDPE	6,5	140	64			
[34–36]	XLPE	6.5	150	65			
	HDPE	6.5	130	38			

Source: After [4]. Reference numbers have been modified to reflect numbering in this chapter.

As argued in [4], the authors of [34–36] drew optimistic conclusions about using PE-based compounds for HVAC applications in manufacturing HVDC cables essentially because they focused on space-charge measurements and neglected the temperature effects on the DC breakdown strength. Indeed, this investigation activity did not mention the evaluation of DC insulation resistivities of the polymers used and their dependence on temperature and electric field, which—as emphasized many times herein—plays a major role on the performance of DC cables. In fact, significant improvements took place in PE-based compound insulated cable production since the 1970s (e.g., extra-clean manufacturing, dry curing, and simultaneous three-layer extrusion; see Chapter 3), but a few investigations show that the ultra-clean PE-based compound grades of newly manufactured cables greatly resemble the older materials as to the inherent dependence of DC insulation resistivity on temperature and electric field [29,39,40]. It seems clear that the insulating material should be modified in order to develop a polymeric HVDC cable with acceptable performance [4].

5.3 STRESS INVERSION-FREE OR -LIMITED DC CABLE

According to a second approach, electrical and thermal conductivities of the insulating compounds for DC cable application were studied from the theoretical viewpoint aimed at stress inversion-free or stress inversion-limited DC cable insulation systems via proper modifications of these conductivities. These studies suggest that this goal could be achieved by resorting to compounds with lower temperature coefficients of electrical conductivity and higher thermal conductivity. Therefore, polymeric compounds with such features were manufactured since the 1970s by including ad hoc mineral fillers into different polymers, among them XLPE, thereby giving a chance for a stress inversion-free cable [4,15–17]. The relevant works do not consider space-charge effects, notwithstanding that an electrical conductivity independent of temperature and electric field involves a reduction of space-charge formation in the material [4].

Table 5.2 synthesizes in the "EPRI Project" row [17] the outcomes of the EPRI project where tests were performed on cable models made of compounds of this type. The values of K_{TDC} and K_{PDC} at high temperature increased with additives, thereby showing that the performance at high temperature was improved. On the other hand, the behavior under polarity reversal was still poor, particularly at low temperatures, as proven by the low values of the coefficient K_{PDC}: This shows that space-charge performances were not satisfactory yet, as expected since space-charge effects were omitted in these developments [4]. It must be emphasized here that such research activity dates back to no later than 1980. It is reported here for the sake of completeness, but it cannot be considered as up-to-date findings.

More recent and up-to-date results appear in Section 5.4, which concerns the current approach to the improvement of the insulation of extruded HVDC cables, namely minimizing space-charge effects on the insulation by suppressing space-charge storage.

5.4 SUPPRESSION OF SPACE CHARGE IN THE POLYMER

As pointed out above and argued in [4], the more recent and widely adopted approach to the improvement of the insulation of extruded HVDC cables consists in minimizing space charge effects on the performance of the cable insulation by suppressing the build-up of space charge. This comment is surely still valid. First of all, space charges can be suppressed in the insulation of extruded HVDC cables by suppressing the injected electrons at the electrodes or neutralizing them, via a modification of the characteristics of the conductor–insulation interfaces. In addition, the characteristics of the bulk materials can be modified so as to achieve space-charge suppression, for example, by altering the trap distribution of PE-based compounds in such a way that the depth of deep traps is reduced and traps are spread as uniformly as possible within the insulation thickness. Following these two paths, various methods have been suggested to suppress space-charge effects on the performance of DC cables with polymeric insulation. Such methods are illustrated in the following Sections.

5.4.1 Modification of the Characteristics of the Electrode–Insulation Interfaces

5.4.1.1 Traditional Approaches It has been demonstrated in the literature (see, e.g., [41]), as well as in Section 4.5.2.1, that the formation of space charges in the insulation is strongly affected by the electrode–insulation interface. It was shown that polyethylene terephthalate (PET) is less prone to undergo charge injection from an electrode than PE, consistent with the contact electrification characteristics of PET and PE. This led to suggest conceptually the adoption of a space-charge suppression layer adjacent to the conductor and between the conductor and the insulation, made of a dielectric that undergoes low charge injection, such as PET [4,42]. Later on, as reported in [4]:

> The influence of electrode material on the build-up of space charge in LDPE was studied using the Thermal Step Method [see Section 4.4.1.3] for space charge measurement. In this work three types of electrodes were used, i.e.: 1) vacuum-evaporated gold and aluminum; 2) plasma-enhanced chemical vapor deposited silicon carbide (PECVD SiC); 3) pressed aluminum. Results of this work indicate that the use of aluminum and (PECVD SiC) electrodes with different combinations considerably reduced the space charge build-up in LDPE [43]. [Anyway, as argued in [4]], the practical application of both of these methods in cable production and the effects of using these techniques on the performance of HVDC cables with polymeric insulation were not investigated ([4], page 40).

According to [5], there is another approach to reducing space charge accumulation via a modification of the characteristics of the electrode—insulation interfaces:

> The modification of the semiconductive layers immediately adjacent to the insulation, [as currently done in practice.] Space charge in the vicinity of the interface, between the semiconductor layer and the insulation layer, may be reduced by the inclusion of carbon

black (10–50 wt%) in the semiconductor layer [44]. An alternative approach is modification of the cable design to include two inner and two outer semiconductor layers [45]. One patent suggests that the addition of zeolites[3] to the semiconductor layers will neutralize space charge sources therein and prevent transfer of space charge to the insulation [46] ([5], page 21).

5.4.1.2 *Effect of Surface Fluorination on the Space-Charge Behavior of PE* As reported in [47], recent results have indicated that direct fluorination[4] is also effective in modifying the electrical properties of PE surface layers, as also in controlling space-charge behavior in PE under high DC voltage.

To illustrate the effects of surface fluorination on the space-charge behavior of PE, let us consider the space-charge profiles of Figure 5.3—obtained by the PWP technique while polarizing a 0.8-mm-thick PE sample at 40 kV DC and 40°C—as well as the relevant comments as reported in [47].

> The space charge profiles in [Fig. 5.3*a*] show homocharge close to both electrodes and charge accumulation (positive and negative) in the bulk of an unfluorinated PE sample. The same sample, after both flat surfaces had been fluorinated at about 55°C for 30 min by a F_2/N_2 mixture containing 20% F_2 by volume (total pressure 225 mbar) and no detectable oxygen, showed positive charge accumulation near the anode within 5 min of the application of high voltage but very little charge accumulation near the cathode [see Fig. 5.3*b*]. However, space charge did not accumulate close to the center of the sample [see Fig. 5.3*b*] at any time during voltage application. On the other hand, a sample in which both flat surfaces had been fluorinated at about 55°C for 30 min by a F_2/N_2 mixture containing 12.5% F_2 by volume and a low level of oxygen impurity (total pressure 1000 mbar) showed considerable accumulation of charge, of both polarities, in the sample bulk [see Fig. 5.3*c*]. Negative charge accumulated close to the anode, but very little charge accumulated close to the cathode ([47], pp. 19–20).

In [47] the comments to Figure 5.3 are summarized by pointing out that if oxygen is present in the fluorination gas mixtures, longer treatment times are required to reduce the space-charge densities by the same amount, as proven by Figure 5.3*d*, which displays space-charge profiles for a PE sample in which both flat surfaces have been fluorinated under the same conditions as those of Figure 5.3*c*, but for 240 min instead of 30 min. The charge densities close to both electrodes are similar to those detected in the sample of Figure 5.3*b*, which has been fluorinated for 30 min at a much lower F_2 pressure but with no detectable oxygen.

Furthermore, the experimental results discussed in [47] suggest that fluorination affects other electrical properties of the electrode–dielectric interface, namely:

[3] Zeolites are microporous, aluminosilicate minerals commonly used as commercial adsorbents. Zeolites have a porous structure that can accommodate a wide variety of cations, such as Na^+, K^+, Ca^{2+}, Mg^{2+}, and others.

[4] Fluorination is one of the most effective approaches to the chemical modification of polymer surfaces: Many polymers can be fluorinated by fluorine gas due to the extreme reactivity and oxidizing power of this gas. Over the last ~40 years direct fluorination has improved from research to industrial applications that take advantage of some properties of polymers, for example, barrier properties, adhesion, wettability, gas separation capability, mechanical strength, and the like [47].

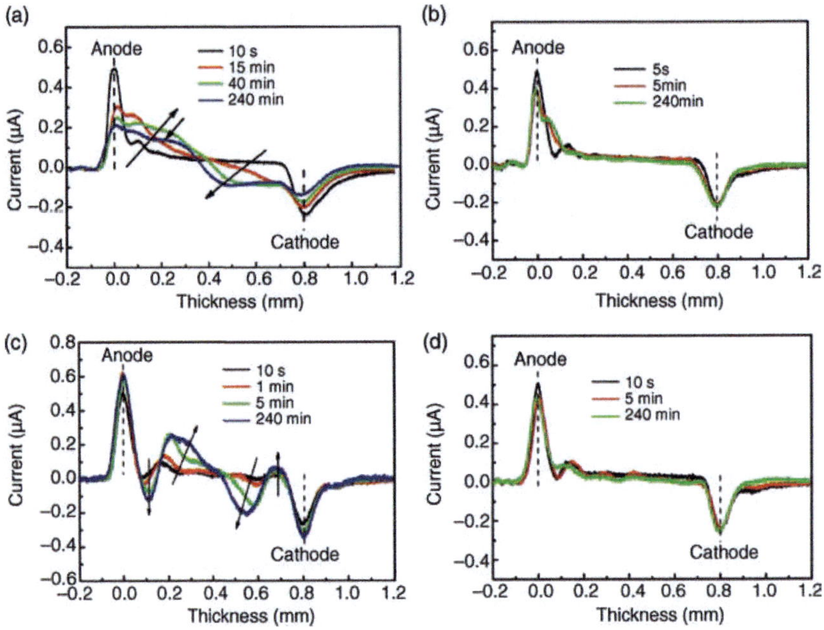

Figure 5.3 Space-charge profiles measured on 0.8-mm-thick PE samples at 40 kV DC and 40°C (poling times in the legends): (*a*) unfluorinated sample; (*b*) surface-fluorinated sample at ~55°C for 30 min by an F_2/N_2 mixture with 20% F_2 by volume (total pressure 225 mbars) and no detectable O_2; (*c*) surface-fluorinated sample at ~55°C for 30 min by a F_2/N_2 mixture with 12.5% F_2 by volume and a low level of O_2 (total pressure 1000 mbars); (*d*) surface-fluorinated sample at ~55°C for 240 min by an F_2/N_2 mixture with 12.5% F_2 by volume (total pressure 1000 mbars) and a low level of O_2 (after [47]).

- It increases the relative permittivity of the PE surface layers because of incorporation of polar groups such as —CHF— [48].
- It increases or decreases the depth of the traps in the surface layer in the absence or presence, respectively, of oxygen [48,49].

5.4.2 Modification of Bulk Insulation Characteristics

As observed in [4]:

> Generally, space charge formation in polymers can be suppressed by modification of the material characteristics, such as reducing the work function and the electron affinity of the polymer, increasing the efficiency of charge transport by annealing charge carrier traps, increasing charge carrier mobility, and/or neutralizing the injected charge carriers ([4], page 40).

See also Sections 4.2 and 4.3. Of course, as argued in [4]: "Modification of the bulk characteristics is by no means an easy task, as it involves a number of considerations in order to keep the basic insulating characteristics of the PE, let alone the other properties of the polymer ([4], page 40)." Essentially, two methods were proposed to suppress space charge by altering the features of the bulk material [4,5]: (a) blending PE with another polymer and (b) using proper additives or fillers in PE compounds.

5.4.2.1 Blending PE with Another Polymer This approach, discussed in [31–33], consists in blending resin HDPE compounds and cross-linkable PE compounds with an ionomer,[5] what is referred to in the following as "blended HDPE" and "blended XLPE," respectively [4].

As to blended XLPE, thermally stimulated current (TSC) measurements indicated a reduced space-charge accumulation and breakdown tests showed an improved impulse breakdown strength with or without superposing DC voltage [4]. As to blended HDPE, space-charge measurements by means of TSC and PEA methods (see Section 4.4.2.3) exhibited a reduced space-charge accumulation and a modification in the charge distribution pattern, while DC breakdown strength with unipolar voltage and polarity reversals at high temperature was improved. However, as observed in [4]:

> The performance of the blended material was not fully evaluated taking into consideration all the factors, such as DC breakdown strength at room temperature, DC breakdown strength with direct voltage and polarity reversals, the effect of blending on the electrical resistivity of the blended material, and its dependence on temperature and electric field ([4], page 40).

On the other hand, as pointed out in [4]:

> In an independent investigation by EPRI, the effect of blending Polyethylene with ionomers on the space charge accumulation was investigated using a Thermal Pulse method [see Section 4.4.1.1]. The results of this investigation indicate that there was no strong evidence of improved space charge dissipation by incorporating ionomers in the PE. Results of this work were supported by measurements of the impulse breakdown strength of the blended polymer prestressed with DC. In this work, it was recommended not to use ionomer-PE blends for extruded DC cable insulation. [50] ([4], page 40).

5.4.2.2 Using Additives or Fillers in PE-Based Compounds Generally, additives or fillers are used in polymeric materials for HVDC cable insulation to improve a certain chemical, mechanical, thermal, or electrical property of the polymer. A broad and sound historical review of these practices dating back to 1997 appears in [4] and is quoted at length in the following because it focuses on this important topic. A further updating of additives used to improve properties of

[5] An ionomer is a polymer comprised of repeat units of both electrically neutral repeating units and a fraction of ionized units (usually no more than 15%). Ionomers have unique physical properties, including electrical conductivity and isoviscosity, that is, an increase in ionomer solution viscosity with increasing temperatures.

polymers for HVDC cables dates back to 2003 [5] and is included here subsequently. Thus, let us start with the broad review after [4]:

> For example, additives were used in cable manufacturing to reduce the possibility of tree generation and polymer oxidation. Lately, the use of additives was suggested by some authors to suppress the build-up of space charge in PE under DC conditions [see, e.g., [18–28] among many studies suggested in [4]]. Ando et al. [7] manufactured experimental DC cables made of XLPE filled with an unspecified mineral additive. Khalil et al. investigated the effect of the additive on space charge accumulation in the polymer using two techniques: TSC as an indirect method, and the Cross-cut Dust Figure (CDF) method, which is a destructive direct method of space charge observation [see also Section 4.4]. TSC results indicated that the additive changed the space charge distribution pattern in the material. According to the authors of that work, CDF results—although not shown in their paper—indicate the presence of a homogeneous space charge distribution in the material. As shown in Table 5.2, DC and polarity reversal breakdown results obtained on cables indicate that the additive improved the DC breakdown strength at high temperature: $K_{TDC} = 100\%$. However, with polarity reversals the breakdown strength was reduced, indicating that the space charge effects were not improved as expected: K_{PDC} for XLPE + Additive at high temperature = 54%.
>
> In another attempt, Khalil et al. investigated the effects of different inorganic additives on the space charge formation in PE. The effects of additives on DC breakdown strength under direct and polarity reversals, DC conductivity, and the morphology of the doped materials were also investigated. The additives used in these investigations were: titanium dioxide (TiO_2), barium titanate ($BaTiO_3$), and iodine" [see, e.g. [18–28] among a broader batch of studies suggested in [4]]. Early space charge measurements were made using a field probe technique and then using the Thermal Step Method ([4], page 41).

The results of these works were summarized in [4] as follows.

> a) The addition of TiO_2, $BaTiO_3$ to LDPE remarkably suppressed the amount of the remnant space charge and consequently the electric field and considerably modified their distribution patterns in the doped material when compared with the corresponding values in the undoped material. Figure 5.4 shows the effect of the addition of $BaTiO_3$ to LDPE on the formation of space charge. Space charge was measured in this investigation using the Thermal Step Method [18–20]. For the $BaTiO_3$ doped LDPE (group A), the space charge distribution pattern is very different from the space charge distribution pattern in the undoped material (group B), let alone the remarkable reduction of the density of space charge in the doped material compared with the undoped material under the same conditions of temperature and electric field. The maximum space charge density in the doped material is about 11% of the corresponding value in the undoped LDPE. As for the effect of iodine on space charge distribution in the iodine-doped material, a dramatic reduction of space charge as well as a change in the space charge distribution pattern was observed in case of iodine-doped LDPE [21].
>
> b) The incorporation of TiO_2, into LDPE did not have a considerable effect on reducing the DC breakdown strength of the doped material. Moreover, TiO_2 seemed to have a beneficial effect on the DC breakdown strength with polarity reversals of the doped material: in this case the DC breakdown strength was found to be almost insensitive to polarity reversals: the coefficient $K_{PDC}(LT) = 98\%$ [22].

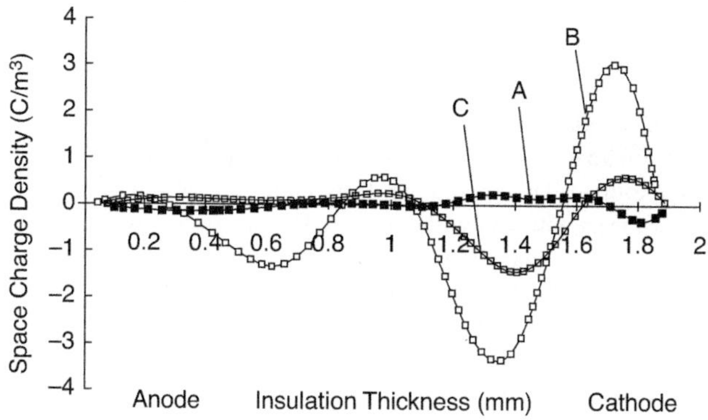

Figure 5.4 Effect of $BaTiO_3$ additive and electrode material on space-charge formation in LDPE. A: $BaTiO_3$ doped LDPE (Au electrodes). B: Undoped LDPE (Au electrodes). C: Undoped LDPE (Al electrodes) (after [4]).

c) The addition of 1 wt% $BaTiO_3$ to LDPE reduced the short-term DC breakdown strength of the doped material by about 16% if compared with the corresponding value for the undoped LDPE [23].

d) The incorporation of iodine to LDPE dramatically reduced the DC breakdown strength of the doped material [24,25].

e) The inclusion of TiO_2 and $BaTiO_3$ into LDPE increased the DC conductivity and the charge carrier mobilities in the doped materials [26,27].

f) The addition of TiO_2 remarkably changed the morphology of the doped material [28].

In 1990, a three-year research project, ADDIS, was started in Denmark aimed at developing new polymeric insulating materials for HVDC cables using additives. The project was sponsored by the Danish Electrical Research Institute (DELRI), the Danish Power Cable company NKT cables A/S, the Danish electrical utilities for generation and distribution of electrical energy (ELSAM and ELKRAFT), and the Electric Power Engineering Department at the Technical University of Denmark. In this project, a certain inorganic additive was used in modifying the characteristics of PE for HVDC cable applications. In addition to space charge measurements, investigations were done to correlate the additives with the space charge accumulation in PE and to explore the various effects of the additive used on the different relevant parameters of the material. Four main kinds of investigations were done in this project: space charge measurements using the Pressure Wave Propagation (PWP) method [see Section 4.4.2.1] and the Thermal Step Method, DC breakdown strength under direct and polarity reversals, DC conductivity and evaluation of insulation resistivity, and the study of the effects of additives on the morphology of the doped materials. Tests were made on both small samples and miniature cables. Although not much could be revealed from the results of this project, preliminary results showed that the incorporation of the inorganic additives into LDPE: i) reduced the amount of the accumulated space charge in the material and modified its distribution pattern; ii) improved the performance of the doped material under polarity reversal DC conditions; iii) improved the stress inversion characteristics of the doped material; iv) changed the morphology of the doped material; v) reduced the insulation resistivity of the doped material, but remarkably improved its dependency on temperature [38].

> In another attempt, a five-year research program was launched in Japan in 1984 aiming at developing HVDC polymeric insulated cable using a conductive inorganic filler in XLPE. The main reason for using the additive was to mitigate space charge effects in the cable under DC conditions. Evaluating tests were made using unfilled and filled XLPE cables of thicknesses 2.5, 9.0 and 13 mm. In this comprehensive work, six kinds of tests were done: DC breakdown tests under direct and polarity reversals, impulse breakdown tests with and without using DC pre-stress, evaluation of insulation resistance, and space charge evaluation tests using the dust figure method as well as the PEA method [29]. Results of this work indicate that the addition of a conductive additive to XLPE insulation lead to: i) modification of space charge distribution in the material, which was proved by measuring space charge distributions in both the filled and unfilled XLPE; ii) improvement of the dependence of DC breakdown strength on temperature rise (according to Table 5.2, higher K_{TDC} and better K_{PDC} at high temperature were achieved for the filled XLPE); iii) improvement of the effects of polarity reversals (as shown in Table 5.2, K_{PDC} is generally improved for the filled XLPE compared with the unfilled material at low temperature, which is not the case at high temperature. However, and especially at low temperature, the effects of polarity reversals are still not satisfactory. From Table 5.2, a relatively low K_{PDC} is achieved in this case, which indicates that the resulting space charge distribution does not lead to the complete independence of the breakdown strength on polarity reversals); iv) lowering the impulse breakdown strength; v) improvement of the dependence of insulation resistivity on DC stress.
>
> In another development, an XLPE compound containing a highly purified special filler was used for manufacturing a prototype cable with 9 mm insulation thickness. The types of tests made on this cable were: effect of temperature on the DC breakdown strength, impulse breakdown strength with and without superposing DC voltage, evaluation of the dependence of insulation resistivity on temperature and electric field, and space charge effects using the indirect TSC method. According to the authors of that work, the performance of the prototype cable was excellent. Furthermore, based on the study of the prototype cable, a 250 kV DC cable with 20 mm insulation thickness was designed, manufactured, and tested ([4], pp. 41–43).

As a summary of this historical review on additives or fillers in PE, in [4] it is finally observed that test results on this cable were reported to be excellent [30]. As a matter of fact these investigations paved the way to the development of the ±250 kV DC cable systems described in Sections 5.6.1, 5.6.5 and at the beginning of Section 5.7.

Later on, in [5]—dating back to 2003—it was observed that:

> Current industrial practice for HVDC cables in the area of polymer insulator additive technology is reflected in the patent literature, the majority of which appears to originate in Japan. Most strategies to date involve an extension of the role of existing polyethylene additives. Insulating grade XLPE used in HVAC cables typically contains additives, such as stabilizers (e.g., antioxidants), cross-linking agents, and other compounds that increase the ability of the material to withstand electrical stresses (e.g., "tree retardant" additives). Lubricating additives, unsaturated compounds to increase cross-linking density, and scorch retarding agents may also be added

[51,52]. With a better understanding of the morphological and electrical processes involved, a more novel approach to the selection of additives may evolve.

Polar compounds, typically aliphatic esters alone or in combination with metals or in admixture with other compounds, may be considered as similar to existing processing aids, the common criteria being compatibility and low migration. Specific examples of these methods are given below ([5], page 20).

Due to their significance, these examples are reported here, too.

Additives for HVAC Polyethylene-Based Compounds Used in HVDC Polyethylene-Based Compounds [5]

Additives traditionally used for HVAC cables can be selectively used in various combinations to produce XLPE insulation for HVDC cables with reduced space charge accumulation. The successful combinations used for insulation thicknesses less than 10 mm were additives A, B, C, and D, see Table 5.3 [51]. To achieve space charge reduction, the insulation produced should not be heated above 230°C for periods less than 10 min; this suppresses the production of polar by-products, such as acetophenone and cumyl alcohol [51]. Only additives A, C, and D were required for insulation of thicker HVDC cables (i.e., greater than 10 mm), the heating regime for cross-linking was 240–350°C for 20–120 min [51].

Another method claimed to reduce space charge in HVDC cables was the *introduction of polar groups bonded to XLPE* [52]. This involves dissolving oxygen gas in the LDPE during the extrusion process. The dissolved oxygen reacted to produce oxygenated free radicals, which were claimed to disproportionate to alcohols, aldehydes, and ketones; these groups were then bonded to the cross-linked structure. This patent seems to contradict the standard approach of minimizing polar entities, thereby minimizing space charge accumulation. The rationale is that an increase in the concentration of polar entities above 1 per 50,000 monomer units facilitates the dissipation of space charge [52] ([5], page 20).

Organic Additives [5]

It has been claimed [53] that the addition of *glycerol fatty acid ester* (0.1–1.0%) to XLPE reduces space charge accumulation. However, the total additive content, up to 10 wt% (organic peroxide ~1.8%, scorch retardant ~0.4% and antioxidants), greatly exceeded the normal levels (1–2 wt%). The glycerol fatty acid esters used have a

Table 5.3 Additives Used for XLPE.

Additive	Type	Chemical Composition
A	Cross-linking agent	Dicumyl peroxide (1.2-1.8 percent)
B	Antioxidant	Di-Stearyl-thio-dipropionate (Irganox PS802®)
C	Antioxidant	Diester of 3-(3,5-di-*tert*-butyl-4-hydroxyphenyl) propionic acid and thiodiglycol (Irganox 1035®)
D	Scorch retardant	2.4-Diphenyl-4-methyl-pentene-1

Source: After [5].

particular structure and are available commercially. A list of examples is provided in the patent [53].

According to a further patent, the addition of *totally or partially sterically hindered phenol type antioxidants* in the polyethylene insulation layer reduces space charge accumulation [54].

Several Japanese patents claim success for various organic additives in polyethylene. One claims that the addition of an organic peroxide, a phenol-based antioxidant, zinc caprylate (0.0005–0.002%) and ethylenic copolymer with dibasic acid anhydride will produce DC cable insulation with excellent dielectric breakdown characteristics [55]. Another claims that the addition of an "organic phosphor" [56] to XLPE is similarly beneficial. It is also claimed that XLPE prepared with 30–40% crystallinity, a conductive additive (0.1–0.8%), an antioxidant and dicumyl peroxide as cross-linking agent will increase the DC breakdown voltage [57].

The addition of 0.3 wt% bis-(*p*-ethylbenzylidene)sorbitol to LDPE decreased the average size of the crystallites as well as the crystalline order within the spherulites. These two effects work to reduce the space charge build-up by allowing a more uniform distribution of impurities throughout the LDPE [58] ([5], page 20).

Grafting of Polyethylene with Maleic Anhydride (after [5])

In the late 1990s, there were attempts to *modify polyethylene by grafting with maleic anhydride*, and thereby reduce space charge accumulation. It was found that space charge accumulation could be minimized in 50–70% cross-linked XLPE grafted with maleic anhydride. Cross-linking was performed using a cross-linking agent with a half-life of 5 hours at 130°C (e.g., 2,5-dimethyl-2,5-di(t-butylperoxy)hexine3) [59]. This procedure reduced the concentration of by-product residues, typically associated with space charge accumulation and, therefore, an overall reduction of space charge was achieved.

Further studies revealed that LDPE grafted with maleic anhydride (0.001–0.01 wt%) had a higher melting point (115°C). The higher melting point avoided the need for cross-linking and the associated adverse effects [60]. An antioxidant additive of tris (2,4-butylphenyl) phosphite increased performance when the graft ratio was high [60,61].

Maleic anhydride is widely used to improve compatibility in thermosets and in polymer blends [62]. Thus when grafted to polypropylene it renders the latter more compatible with polyamides [63]. Reactive extrusion employing this process leads to toughening of commercial grades of nylon by reducing stiffness [64,65]. Increased toughness may also be advantageous for PE, but a limit to the increase in polarity may have to be imposed in order to retain the polymers' excellent insulating properties. Further work in this area appears to be warranted ([5], page 21).

Inorganic Additives (after [5]) As reported above after [4] (see, e.g. [18,20]), it is confirmed in [5] that "*various inorganic additives (TiO_2, $BaTiO_3$, iodine)* reduce space charge accumulation in LDPE and the dependence of its resistivity on temperature and electric field . . . The addition of inorganic additives did not have an appreciable effect on percentage crystallinity, but altered the spherulite size [66,67]"

([5], page 21). Unfortunately, these inorganic additives also reduced the DC breakdown strength of LDPE, as highlighted in [5] which mentions the relevant results of [66,67], more recent than those quoted in [4].

Further remarks in [5] about inorganic additives are as follows:

> It has been found that aggregates of $BaTiO_3$ [66] and carbon black [68] act as breakdown initiation sites. The dispersion of $BaTiO_3$ particles in polyethylene and composite systems is governed by the polarity of the polymer matrix; increasing the polarity of polyethylene ethylene-vinyl acetate increased the dispersion of the $BaTiO_3$ [69].
>
> A further proposal for reducing space charge in thermoplastic resins (including PE) was the addition of *carbon black* at levels of 0.2–5.0 wt%. The carbon black selected had a particle size of 10–100 nm, hydrogen content of less than 0.6 wt%, and specific sorption[6] characteristics [70].
>
> It has been claimed that the addition of carbon black or MgO (0.1–0.8 wt%) with an average grain size of 0.01–0.1 μm improves the insulation properties of LDPE under DC stress [71]. Another patent is based on the addition of "polarized" MgO to PE with "denatured" maleic anhydride and a cross-linking agent [72]. A third patent uses XLPE, cross-linked using dicumyl peroxide, with MgO surface treated with vinyl silane, approximately 1–2 μm in size (max. 15 μm) [73]. Another patent [74] suggests that the addition of filler (e.g., carbon black or MgO) will improve the impulse breakdown strength of XLPE prepared from PE or ethylene-vinylacetate copolymer and a cross-linking agent. It was also claimed that the addition of zinc caprylic acid (0.0001-0.001 wt%) to this type of XLPE would lower the volume resistivity [74]. A recurring theme in several works is an *optimal particle size*, typically below 15 μm. Presumably *larger particles can act as breakdown initiation sites, this being also true of HVAC cables* ([5], page 21).

5.4.2.3 Additives and the Morphology of Polyethylene (after [5])

As reported in [5]:

> The performance of a polymer as an insulating material depends also on its texture, which will be affected by additives. Thus, two or more phases may develop when the additive is incorporated, some richer in additive than others, so that polarity is not uniform. Secondly, inorganics may act as nucleating sites, leading to smaller spherulites and larger interfacial regions. Even though the crystalline regions in XLPE are relatively small spherulites or present as lamellar only, the effects of additives on them must not be overlooked.
>
> In relation to polyethylene it is important to consider the solubility of nonparticulate additives, the chemical nature of the surface (adhesion or wetting of the surface by the polymer), the average particle size and particle size distribution of fillers, the particle shape and porosity, and finally any impurities that may be present. It is also important

[6] Sorption refers to the action of both absorption and adsorption taking place simultaneously. As such it is the effect of gases or liquids being incorporated into a material of a different state and adhering to the surface of another molecule. Absorption is the incorporation of a substance in one state into another of a different state (e.g., liquids being absorbed by a solid or gases being absorbed by a liquid). Adsorption is the physical adherence or bonding of ions and molecules onto the surface of another molecule.

that additives do not catalyze any *side reactions* or oxidation. Thus titanium dioxide, a commonly used white pigment, can also serve as a filler, but in the native uncoated form it is known to catalyze oxidation of polyethylene [75] ([5], page 21).

5.4.2.4 Effect of Nanostructuration One more interesting possibility in order to reduce the storage of space charge in HVDC insulation and the relevant detrimental effects is resorting to nanocomposites with insulating properties. Indeed, high-field conduction and trapping mechanisms in such materials can be favorably modified with respect to conventional polymers [76].

Indeed in [76] it is shown that, in epoxy titania, alumina, and ZnO nanocomposites, PP and EVA-layered silicate nanocomposites, and LDPE titania nanocomposite, space charge (measured by the PEA method) is mitigated by nanostructuration. For instance, charge accumulated in the bulk is considerably large at 60 kV/mm for both EVA and isotactic polypropylene (i-PP), but tends to decrease with increasing addition of nano-fillers from 2 to 6%, as shown in Figure 5.5. Similar results are obtained for i-PP. A common behavior is that the charge is reduced at high field due to the introduction of nanofillers, while it is increased at low field. Consequently, the threshold for space-charge accumulation shifts to lower values when nanofillers are introduced, that is, from 12 to 4 kV/mm and from 14 to 5 kV/mm for EVA and i-PP, respectively. It is argued that such charge at low field is ascribed to ionic impurities included in nanofillers and can be swept away by the pretreatment of nanofiller purification [76].

Charge reduction is explained in [76] by the introduction of a shallow trap band and the resulting increase in apparent carrier mobility. Electronic charge carriers will transport in the shallower band to result in larger mobility in

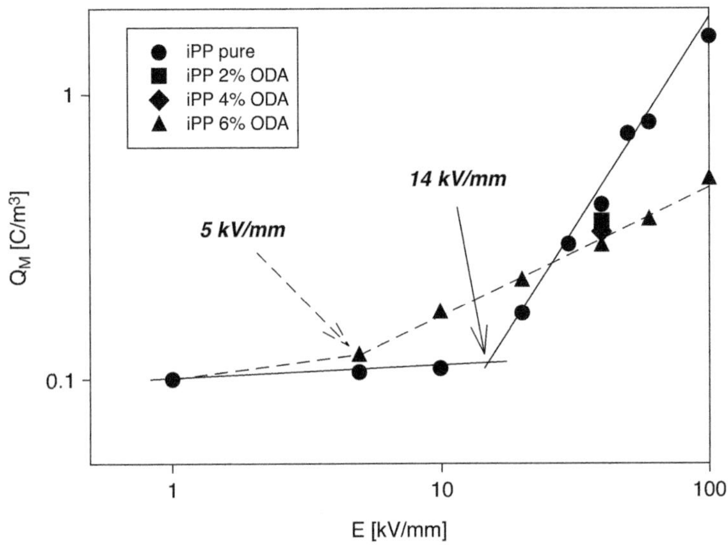

Figure 5.5 Space-charge threshold characteristics for base EVA and 6% nano-filled EVA (ODA: synthetic fluoro-hectorite modified by octadecylamine) (after [76]).

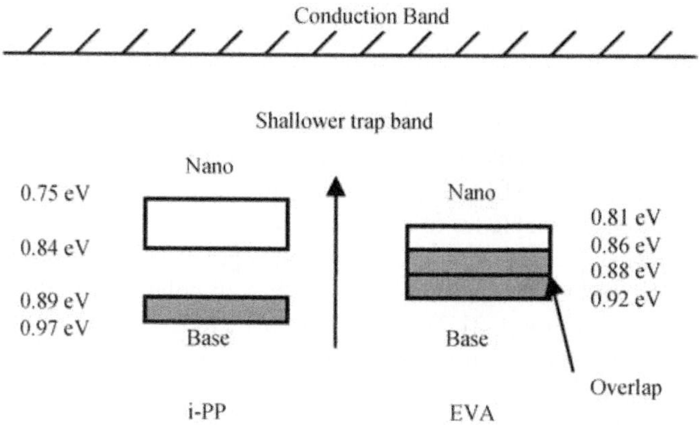

Figure 5.6 Transition of trap band to shallower energy level due to nanostructuration (after [76]).

nanostructured polymers than in base polymers, as shown in Figure 5.6 for EVA and i-PP [76].

5.5 FURTHER REQUIREMENTS FOR THE IMPROVEMENT OF HVDC EXTRUDED CABLE DESIGN

The improvement of the design of HVDC extruded cables is strongly associated with the following main aspects, which are essential for HV cable design in general: (1) a proper selection of the insulation material and (2) a proper selection of the insulation thickness and of conductor cross section.

As to the insulation material, it should be selected among the improved dielectrics developed according to the concepts illustrated in Sections 5.2–5.4. However, all are PE based, as pointed out in Section 3.8. Moreover, the cable insulation selection is influenced by the converter station technology, too, as pointed out in Section 2.4.3. Indeed, in widely established current source converters (CSC) the direction of power flow through the system cannot be changed via a change in current flow direction but rather via a change in the polarity of the two pole conductors in the transmission link, so that polarity reversal becomes a functional requirement of the power cables. Since the earliest applications, extruded insulation for HVDC cables has been known to be strongly affected by such polarity inversions, as pointed out many times herein. On the contrary, in voltage source converters (VSC) the direction of the power flow can be reversed via a change in current flow direction, thus without changing the polarity of the two pole conductors—which is a much more favorable operating regime for HVDC extruded insulation—so that polarity reversal is not a functional requirement.

In the HVDC extruded cable market, a very wide variety of material solutions have been investigated and tested over the years (as broadly discussed in Sections 5.2–5.4), and actually each manufacturer is employing its own polymeric compound as insulation. Therefore, each material requires an ad hoc design approach to grant that the dielectric operates within proper temperature and electrical stress ranges, so that the design system life can be attained as expected [1]. The materials declared by the main cable manufacturers active in the extruded insulation HVDC cable market at the end of 2004 after a questionnaire reported in [1] are: LDPE, proprietary polymer alloy based on PE, XLPE, XLPE from a special functional group, modified XLPE insulation material only suitable for use with VSC converter station technology. Examples of the relevant average dielectric stresses and operating temperatures are as follows: for LDPE, 20 kV/mm and 70°C; for XLPE with polarized filler, as well as with a polar functional group, 21.7 kV/mm and 90°C; for modified XLPE insulation material only suitable for use with VSC station technology, 12.5–14.5 kV/mm and 70°C [1].

Coming now to the rules for the design of cable insulation thickness and of conductor cross section, such rules are quite complex in the case of HVDC cables with extruded dielectric, as already hinted at in Chapter 3. Indeed, insulation thickness is related to rated voltage and rated current since the electric field depends on both voltage and temperature according to the complex insulation design considerations developed in previous chapters. In turn, rated current, also referred to as "ampacity," depends on cable temperature in a rather complex way.

For the sake of illustration, let us explain such concepts starting from the classical (and simplified) approach after International Standard IEC 60287 [77], since in most power engineering books (see, e.g., [78]) the ampacity calculation for HVDC cables up to the highest voltages is based on the prescriptions given in such standards. IEC 60287 states that rated current I_{DC} depends essentially on conductor-rated temperature, cable layer materials and thicknesses, as well as on the burial conditions, according to the following equation [77]:

$$I_{DC} = \sqrt{\frac{\theta_c - \theta_a}{R_{\mathrm{DC}}(T_1 + T_2 + T_3 + T_4)}} \tag{5.3}$$

where θ_c is the conductor temperature, θ_a is the ambient temperature, R_{DC} is the conductor electrical resistance at the relevant temperature (which depends on conductor cross section), and T_i are the thermal resistances of the different thermal insulating layers of the cable and of the laying environment [in Eq. (5.3), $i = 1, \ldots, 4$]. A major role is played by T_1, the thermal resistance of the insulation thickness, and T_4, the thermal resistance of the surrounding medium. As to the latter, referring to an underground bipolar cable, T_4 can be calculated as follows:

$$T_4 = \frac{\rho}{2\pi} \ln\left(\frac{4H^2}{D_e}\right) + \frac{\rho}{2\pi} \ln\left(\frac{\sqrt{4H^2 + d^2}}{d}\right) \tag{5.4}$$

where ρ is the thermal resistance of the soil, H is the burial depth, d is the distance between the two cables of the bipole, and D_e is the outer diameter of the cable.

Referring to the underground bipolar cable, the maximum transmission power P for such system is given by

$$P = 2U_0I_{DC} \tag{5.5}$$

where U_0 is as usual the absolute value of pole-to-ground voltage. Thus, according to the simplified approach of IEC 60287, once U_0 and P are set, the required cable ampacity could be derived straightforwardly via relationship (5.5), from which in turn conductor cross section could be determined via relationship (5.3).

However, things are much more complex than that. First of all, the thermal resistance of the surrounding medium, T_4, depends on the laying conditions. This means that if one wants to express cross section as a function of power and cable voltage, the laying conditions have to be specified, too. This creates one significant difficulty because laying conditions vary more or less along a cable route (see Chapter 7), and it is neither feasible nor economical to change conductor (and cable) cross section whenever the laying conditions change (this would also imply a relevant transition joint, which is a rather critical accessory mentioned in Section 5.6.5).

Second, it must be remembered that—as pointed out above—the thermal resistance of the insulation thickness, T_1, depends on voltage and current via the dependence of electric field on voltage and temperature (see Chapter 3). Therefore, for each assigned laying condition, conductor cross section and insulation thickness should be derived via a non trivial iterative procedure based on Eqs. (5.3) and (5.5) plus Eq. (3.29) of Section 3.1.3.2 under the constraints of the maximum permissible temperature and electric field for the continuous operation of the insulation. Details of such a complex iterative procedure are omitted here for the sake of brevity.

Furthermore, additional considerations relevant to HVDC extruded cable design exist that go beyond the rules illustrated above and in Chapter 3. In particular, some limits such as power losses, drum weight imposed by road transportation constraints, undersea cable weight imposed by cable vessel installation capacity, and the like have a nonnegligible influence on the final cable design and cannot be treated with general rules since they are strongly related to the country and/or site of installation.

In particular, the diameter and weight of cables are important parameters that involve important installation details. For undersea cables, the maximum batch that can be loaded on a vessel depends on these parameters: given the circuit length, a larger cable needs, of course, more batches, vessel loadings, and more joints with respect to a smaller cable. For land cables, which require less outer layers, thus have smaller dimensions than undersea cables of comparable performances, more cable can be loaded on one drum, thereby decreasing the number of joints, cable transport, and right of way [79].

Hence, it can be concluded that the selection of conductor cross section and insulation thickness is the final outcome of a lengthy and cumbersome optimization procedure, which has to be performed not only according to the above considerations but also to economic considerations based on market conditions, prices of raw materials, and the like. However, it can be argued that such a procedure has not become well established for HVDC extruded cables yet.

5.6 IMPROVED DESIGN OF HVDC EXTRUDED CABLES

Practical and up-to-date information about the improved design of HVDC extruded cables can be derived from published studies, some of which deal with HVDC extruded cable systems already in service [78–88]. Some others describe cable systems that either have been qualified and are ready to be installed or are being qualified at the time of this writing [89–96]. Several details about cable design taken from such studies are illustrated in this section. More about the overall projects implemented can be found in Chapter 7. All the details reported in this section either come from published studies or appear by courtesy of cable manufacturers.

5.6.1 First Example of Improved Design of HVDC Extruded Cables

As a first example of improved design of HVDC extruded cables, let us consider the cable design after [89,90], which is relevant to a ±250-kV XLPE insulated undersea cable already treated herein in Section 4.6 as to space-charge measurements and Section 6.1.1 as to accelerated life testing and life modeling. The cable has the following main characteristics: copper conductor with a 800-mm^2 cross section, insulation thickness of 20 mm made of XLPE compound containing inorganic filler to suppress the accumulation of space charge, one single steel-wire armor ($\phi = 8$ mm $\times$ 39 wires), outer diameter 124 mm, weight 42.5 kg/m (see Figure 5.7).

Also cables and factory joints were designed and manufactured. To check the DC electrical performance and reliability of this 250-kV DC undersea cable system, electrical, mechanical, and long-term aging tests were performed (details can be found in [89,90]).

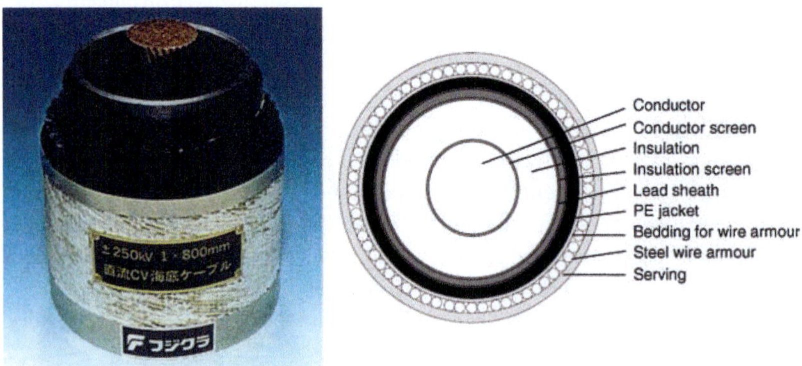

(a) (b)

Figure 5.7 (*a*) View and (*b*) cross section of ±250-kV DC 800-mm^2 XLPE insulated undersea cable (courtesy of VISCAS)

5.6.2 Cable Design Relevant to the Gotland Project

The Gotland HVDC link is a ±80-kV, 50-MW, ~2 × 70-km-long underground extruded HVDC cable system that was commissioned in 1999 [80]. Even though its voltage rating is not very high, this cable is historically important because it seems to be the first commercial HVDC cable system with extruded insulation commissioned anywhere in the world [81] (for details about this link see also Section 7.2.1).

As to the cable development for this link, it began with the selection of the optimal insulating and semiconductive compound, as illustrated extensively in [82]. Various polyolefin-based and EPR-based material compounds were tested among several other compounds: At a first stage, their space-charge behavior was analyzed (the PEA technique was employed for investigating space-charge accumulation); at a second stage, the volume resistivity and electric breakdown strength (see Figure 5.8 after [80]) of the best-behaving compounds were evaluated [80]. Later on, full-size prototype cables made out of the insulations chosen via this preliminary screening were tested, selecting as prototype a classic cable design featuring 24-kV rated voltage, 95-mm^2 conductor cross section, and 5.5-mm insulation thickness (see Figure 5.9 after [80]). Through these tests, the number of candidate optimal insulation compounds was restricted to a few [80].

The subsequent development stage consisted in manufacturing longer cable batches and subjecting them to high-voltage breakdown tests consisting in a step voltage test starting at 600 kV DC and a polarity reversal test also starting at 600 kV, until breakdown. The insulation ultimately selected on the basis of these breakdown tests and employed in the Gotland project is a peroxide cross-linkable polyethylene compound, called XDC, that can be manufactured in conventional extrusion lines [80].

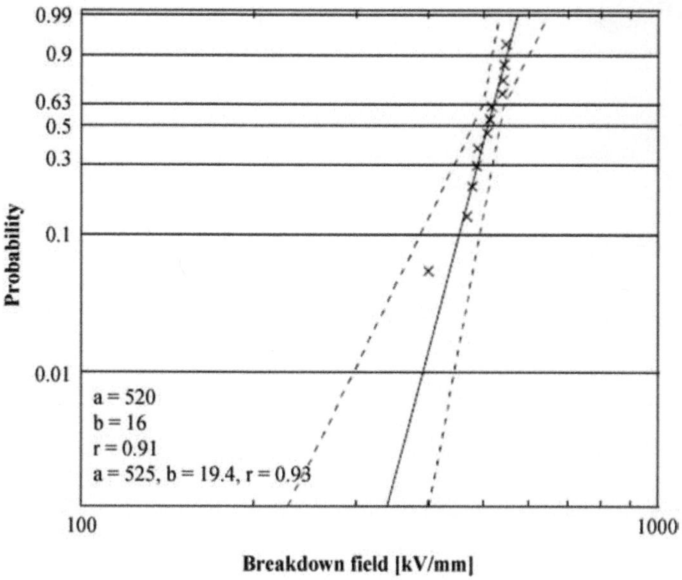

Figure 5.8 DC breakdown field strength data shown in a Weibull plot (140 ± 20 μm press-molded films, ramp test at 500 V/s) (courtesy of ABB).

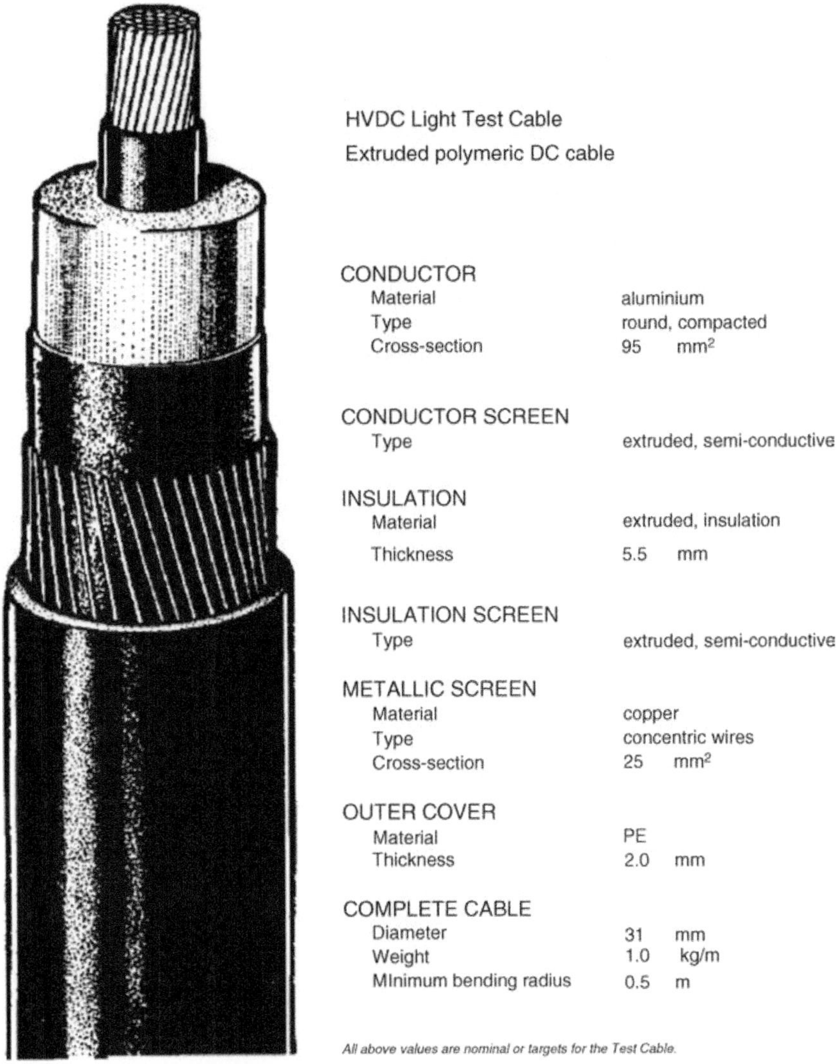

Figure 5.9 95-mm^2 cable prototype used for selecting the optimal material for the extruded HVDC cables (courtesy of ABB).

5.6.3 Cable Design Relevant to the Murraylink Project

The Murraylink HVDC Project is a $\pm$150-kV, 200-MW, $\sim$2 $\times$ 180-km-long underground extruded HVDC cable system that was commissioned in 2002 [78]. More about the Murraylink link can be found in Section 7.2.2 at Chapter 7.

The cable structure is as follows (see Figure 5.10*a* after [78]):

Compressed aluminum conductor

12-mm-thick solid insulation, made in a triple-extrusion process with inner semicon, insulation wall, and outer semicon from materials specifically designed for DC applications

30 mm^2 cross-section copper screen

Radial water blocking system, made of a layer of swelling tape plus a watertight aluminum/polyethylene laminate

Extruded HDPE jacket as thermoplastic outersheath

As far as the conductor cross section is concerned, two cross sections were installed, that is, 1200 and 1400 mm^2. A comprehensive cost optimization avoided a wider variety of cross sections—although the broad changes of thermal resistance along the cable route might have suggested it—since [78]:

1. The easier logistics and spare part inventory overcomes the moderate savings of sub-optimized conductor sizes.
2. Using only two conductor cross sections allowed employing only one joint design for the whole transmission line.

As to the insulation, the design temperature is 70°C; the electric stress distribution at 150 kV DC service voltage within the insulation wall is displayed in Figure 5.10*b* (after [78], where the absolute field level is omitted) and differs when passing from no-load to full-load conditions, although no field inversion is observed.

As to the thermoplastic outersheath, the customer asked for "termite-proof" cables because of the laying environment. Cable samples with different pesticide-free[7] outersheath materials were exposed to real termite attacks: both polyamide and HDPE jackets were found to be termite proof. Finally an HDPE jacket was selected [78].

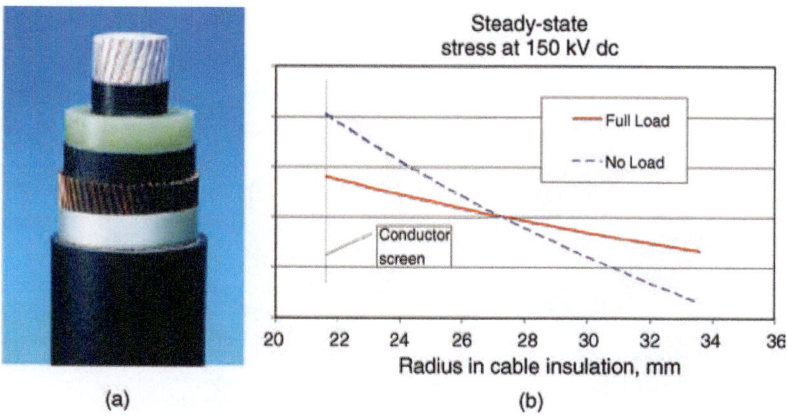

Figure 5.10 Design of the Murraylink cable: (*a*) view of the cable (courtesy of ABB) and (*b*) electrical stress in cable insulation (courtesy of ABB).

[7] Pesticides were prohibited in this case, which were also considered ineffective after a short time.

5.6.4 Cable Design Relevant to the Trans Bay Project

The Trans Bay Project is a ±200-kV, 400-MW, ~2 × 83-km-long undersea extruded HVDC cable system realized in the San Francisco Bay, California. It was the first extruded HVDC cable system project at the 200-kV level implemented in the world for commercial power transmission [85–88]. More about the Trans Bay Project can be found in Section 7.2.7 at Chapter 7.

As to the cable design for this link, shown in Figure 5.11, the cable structure of the undersea cable is as follows [86]:

1. 1100-mm^2 cross section of compressed copper conductor
2. Inner semicon
3. Extruded insulation
4. Outer semicon plus a layer of swelling tape
5. Lead sheath
6. Extruded PE thermoplastic sheath
7. Bedding of polypropylene yarns
8. Armor steel wires
9. Outer serving of polypropylene yarns

The 200-kV XLPE cables supplied for this project were of the highest voltage produced at that time. As to the cable insulation, it is an XLPE compound with high electrical resistivity and low dependence of volume resistivity on electric field and temperature. Model cable testing demonstrated the high performance of the selected insulation system. Thereafter, a prototype cable was produced for the project, whose testing was interrupted at 70 kV/mm due to limits of the laboratory terminations. As to

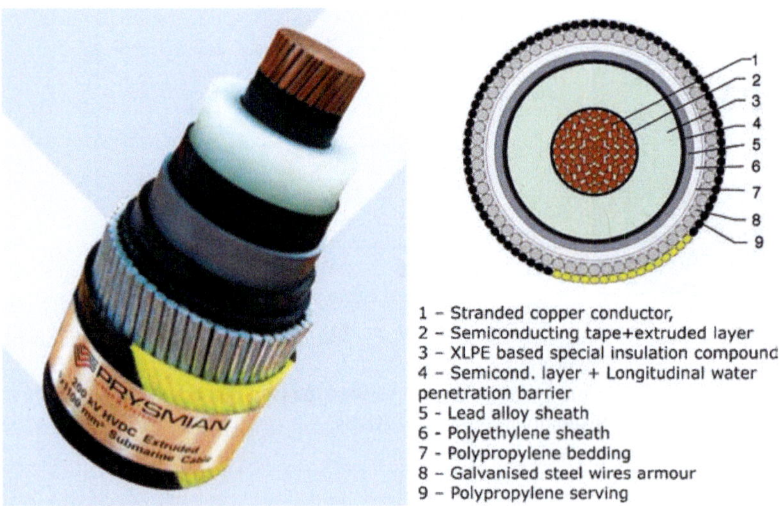

Figure 5.11 Trans Bay cable design (courtesy of Prysmian).

accessory design and development, accessories and relevant installation procedures were optimized to provide the required performance. Both premolded and flexible joints were developed.

5.6.5 Other Improved Cable Designs

Other improved designs of HVDC extruded cables were tested and manufactured [91–96]. As a first example, let us consider a ±250-kV HVDC extruded insulation cable for monopolar transmission with coaxially integrated return conductor (IRC), referred to as coaxial extruded cable from now on [91]. This cable has the structure of a single HVDC cable, plus return conductor and the relevant insulation arranged coaxially around the main cable core. The characteristics of this cable allow other environmental and economic advantages in addition to the well-known advantages of extruded cables versus oil–paper insulated cables (no risk of oil leakage, higher rated temperature, cheaper manufacturing and installation, etc.; see Chapters 1 and 2), namely [91]:

1. The outer magnetic field is practically zero, since the magnetic fields generated by the main and the return cable cancel each other, thus avoiding the magnetic deviation (or "compass error") that could affect ships sailing close to the HVDC cable [97].
2. In the coaxial extruded cable the return current flows in the return conductor, not through sea and/or earth, thereby preventing electrolytic corrosion of neighboring objects and hazards to life at sea and/or earth.
3. The construction cost is less than that of the "stand-alone" main and return cables, especially for small–medium power transmission capacity.

According to [91], for the same rated voltage (±250 kV) and conductor temperature (90 and 75°C for the main and return conductor, respectively) two medium-power transmission cable designs were developed having the same insulation, but differing as to the power transmission capacity, thus as to the conductors cross sections: ±250 kV/180 MW and ±250 kV/300 MW [91].

The first one, that is, ±250 kV/180 MW, is conceived for power transmission to isolated islands. In this cable design (see Figure 5.12, after [91]), the cross sections of the main and return conductors are, respectively, 400^2 and 215 mm^2 for the undersea cable and 500 and 463 mm^2 for the land cable. The return conductor consists of copper wires laid around the main insulation core (diameter and number of the copper wires are, respectively, 2.6 mm and 55 wires for the undersea cable and 5.0 mm and 30 wires for the land cable).

The second one, that is, ±250 kV/300 MW, is conceived for the interconnection of the main power grid: the increase of power capacity from 180 to 300 MW requires the enlargement of the main and return conductor cross sections, which implies a special care in the design and manufacturing of the return conductor due to the increased stiffness of the copper wires. Hence, now the cross sections of the main and return conductors are, respectively, 1000 and 660 mm^2 for the undersea cable and 1600 and 865 mm^2 for the land cable. The diameter and number of the copper wires of the return conductor are, respectively, 5.0 mm and 38 wires for the undersea cable and 5.0 mm and 47 wires for the land cable.

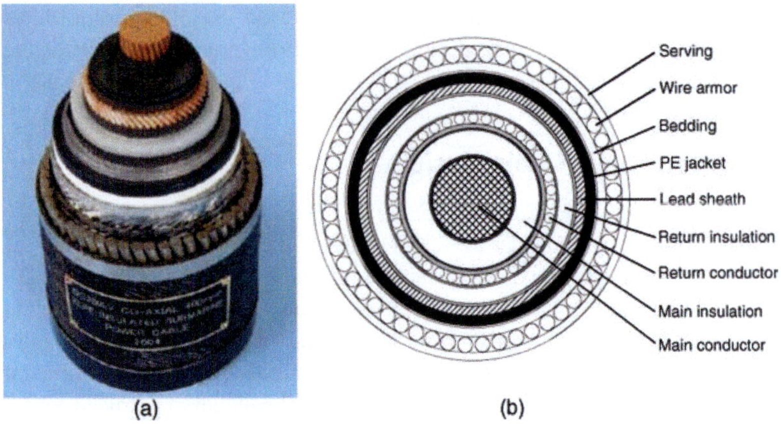

Figure 5.12 (*a*) View and (*b*) cross section of the ±250-kV/180 MW DC 400-mm^2 XLPE insulated coaxial extruded undersea cable (courtesy of VISCAS).

In both designs, the insulation of the main and of the return cable is extruded XLPE with inorganic filler, and the insulation wall thickness of the main and return cable is 12 and 6 mm, respectively.

More details about the cable design parameters (e.g., impulse voltage withstand levels), the system accessories [which include gas-insulated substation (GIS) terminations, land–sea transition joint between the land cable and the submarine cable and the repair joints], as well as the results of manufacture and electric tests are reported in [91].

Finally, for the sake of completeness let us hint at one more cable design, namely the 500-kV DC class land cable after [94–96,98,99]. This cable, which features a 3000-mm^2 copper conductor with Milliken formation and 23-mm-thick XLPE insulation with conductive inorganic filler (see Figure 5.13), is at present

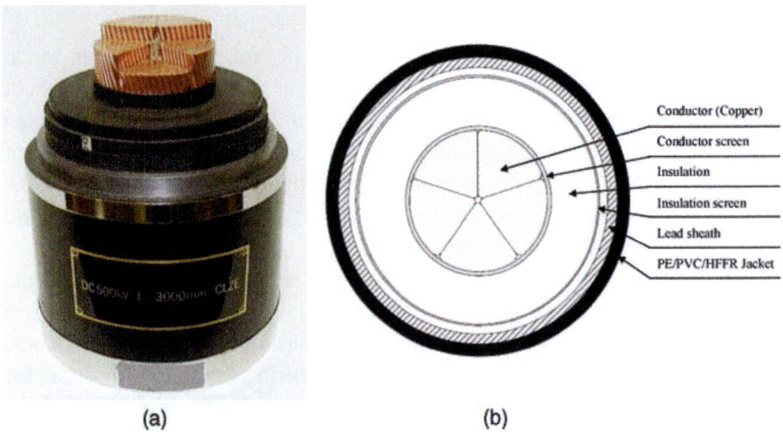

Figure 5.13 (*a*) View and (*b*) cross section of ±500-kV DC 3000-mm^2 XLPE insulated land cable (courtesy of VISCAS).

perhaps the highest voltage laboratory-qualified HVDC extruded cable system in the world as pointed out in [83], since according to [94,99] it passed a long-term load cycle test at the 700-kV DC voltage level for 101 days with polarity reversals.

5.7 IMPROVED DESIGN OF ACCESSORIES FOR HVDC EXTRUDED CABLE SYSTEMS

Cable accessories are sometimes the most critical part of a cable system since they are made of many subcomponents that will be assembled in field, thereby giving rise to the risk of contamination due to pollutants from the environment that might reduce the accessory lifetime. Furthermore, the reliability of accessories is severely affected by the ability of the installer that exposes the accessories of a cable system to human error vulnerability, an aspect that needs to be strictly controlled. For this reason, installers must be suitably trained by means of adequate factory trials before starting in-field assembling activities, in order to gain the proper experience so as to minimize the risk of error during field installation.

HVDC interties are usually very long links—typically hundreds of kilometers—and when only cable lines are used to interconnect the terminal converter stations (without overhead line sections) a high number of joints is required. In case of land links, only field joints are installed, while for undersea links the number of field joints is small, but a high number of factory joints is needed.

Factory joints (see Fig. 5.14) are made at the factory to interconnect cable sections of different production lengths—usually ~20 km long [1]—in order to replace extrusion filtration packs. Their installation is performed in a strictly controlled environment, and this condition strongly reduces the risk of pollutants in the joint assembly; hence, factory joints can be deemed as less vulnerable than field joints. It can be argued that land cable links have field joints only, since the transportation of cable drums with cable length in excess of 1–1.5 km is not an easy task due to both the weight and size of the drums for both roads and railroads.

On the contrary, undersea cable links consist of very long cable lengths (typically longer than 100 km, directly loaded onto the cable laying vessel) and consequently the number of field joints (often referred to as repair joints, too, since

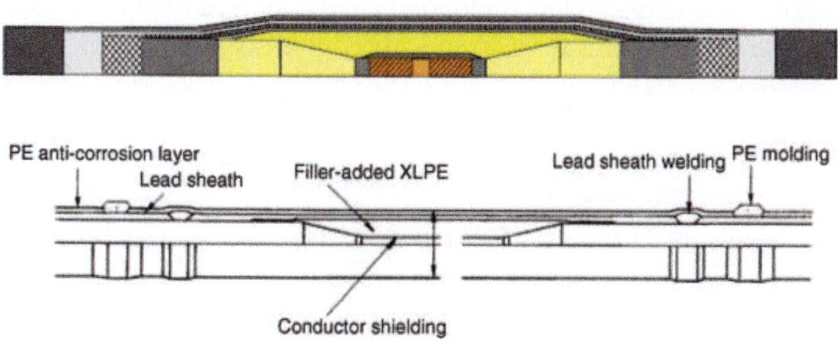

Figure 5.14 Factory joint of molded type for HVDC XLPE undersea cable (±250-kV, 800-mm^2 conductor cross section) (courtesy of VISCAS).

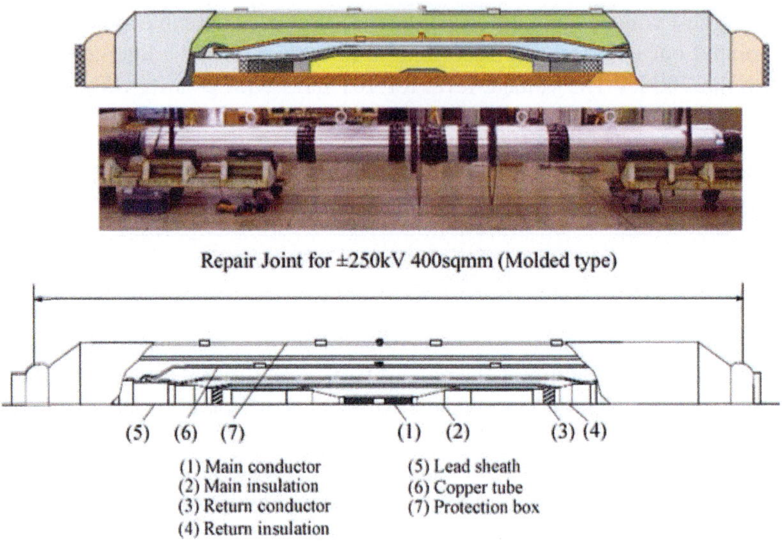

Figure 5.15 Repair joint for ±250-kV/180-MW HVDC coaxial extruded cable (courtesy of VISCAS).

the same type of joint and installation procedure is commonly set up both for field and repair joints) is very small. On the other hand, undersea field/repair joints (see Fig. 5.15) are much more critical than land joints (see Fig. 5.16), since their manufacturing on board the vessel installing the cable is a quite critical operation that might require several days even in good weather conditions [97].

For the sake of illustration and comparison, let us consider a 200-km-long HVDC intertie: In the case of a land cable link, the joints are all field joints and their number can be around 200; in the case of an undersea cable of the same length, the number of joints can be around 20, with 1 field joint in the middle and about 9 factory joints on either side. From these values, it seems that the ratio "number of joints in a land link" over "number of (factory and field) joints in an undersea link of the same length" is ~10. Nevertheless, this value has to be regarded with great care and could be even misleading, since as to field joints different considerations hold between land field joints and undersea field joints, the latter being subjected to mechanical stresses and water tightness requirements that increase with the depth of installation.

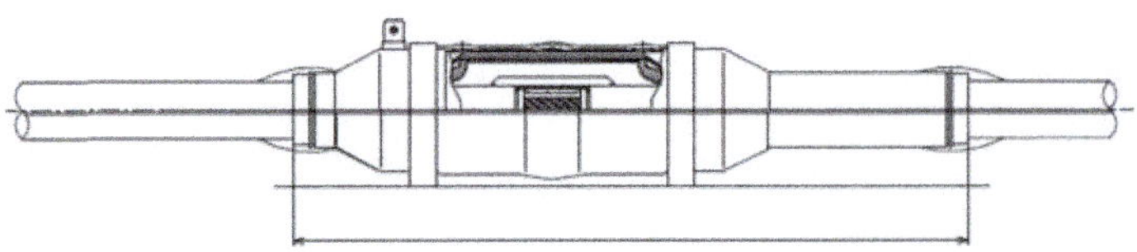

Figure 5.16 Land joint of premolded type for ±320-kV, 2500-mm^2-HVDC extruded cable (courtesy of VISCAS).

The type of joints used for both land and sea links, as well as for both factory and field installation, differ from manufacturer to manufacturer since each one has its own experience. Molded joints are usually preferred as factory joints, while premolded and tape joints have been used for both land and sea applications.

Terminations can be either of the indoor or outdoor type (see Fig. 5.17) and can have either a ceramic (porcelain) hollow insulator [100] or a composite hollow insulator consisting of [101]:

- A load-bearing insulating tube made of resin-impregnated fibres
- A housing outside the insulating tube, made of elastomeric material (e.g., silicone or EPR)
- Metal fixing devices at the ends of the insulating tube

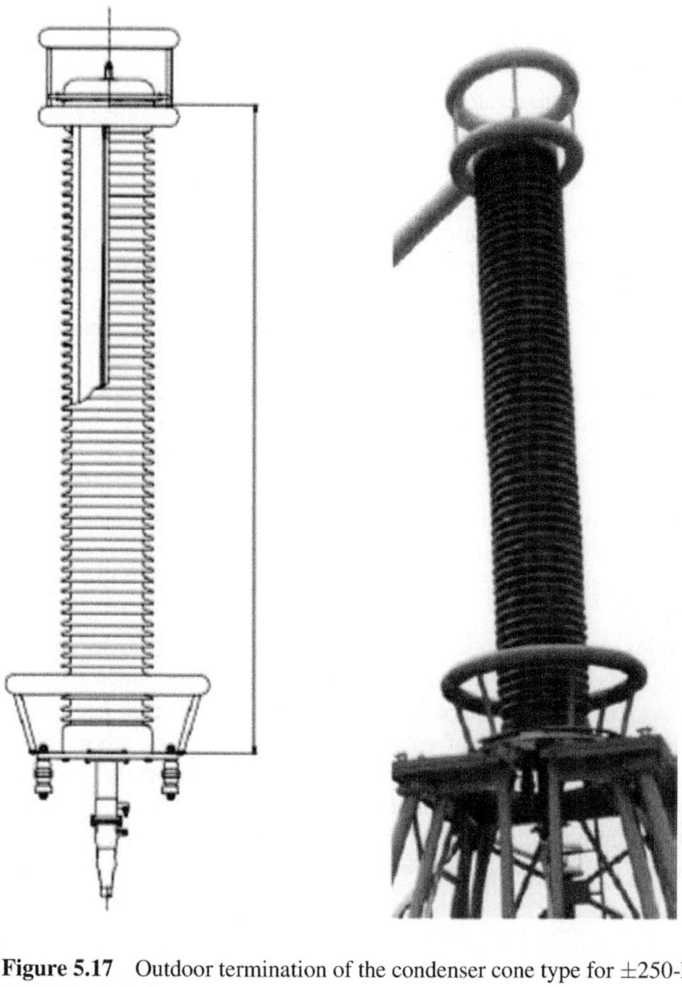

Figure 5.17 Outdoor termination of the condenser cone type for ±250-kV, 500-mm^2 HVDC extruded cable (courtesy of VISCAS).

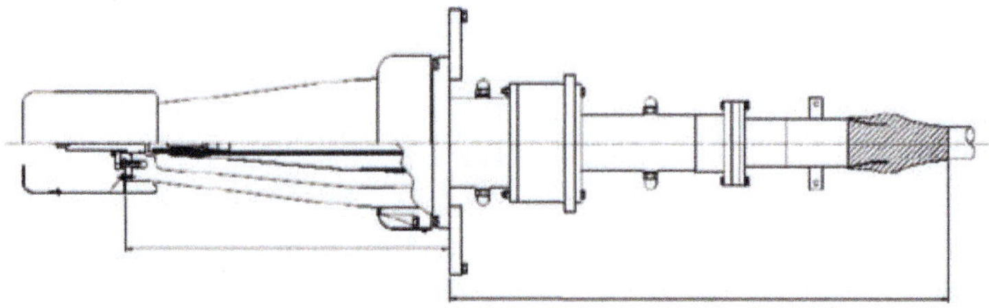

Figure 5.18 GIS termination of the condenser cone type for ±320-kV, 500-mm^2 HVDC extruded cable, intended for use in both land and undersea cables (courtesy of VISCAS).

In both cases, the hollow insulators of terminations are provided with sheds, that is, insulating parts projecting from the insulator trunk intended to increase the creepage distance.[8]

Terminations for HV cables are generally intended for use as pressurized terminations; "pressurized" means a permanent gas or liquid pressure >0.05 MPa (0.5 bar) gauge. The gas can be dry air or inert gases, for example, sulfur hexafluoride (SF_6), nitrogen (N_2), or a mixture of such gases.

Terminations for both land and undersea HVDC extruded cables can be also designed to be connected to a GIS. Such a termination design is displayed in Figure 5.18. According to this design, the main insulation of GIS termination consists of the condenser type insulation made of layers of paper impregnated with silicone oil, epoxy bushing, and SF6 gas immersed in the GIS case. More details can be found in [91].

Publications relevant to accessories for HVDC extruded cable systems are less common than those relevant to HVDC extruded cables. However, since the first commissioning of an extruded HVDC cable in 1999, several articles have been published about the design solutions of cables, accessories, and the whole cable system [78–96,98,99]. In particular, [102] (already dealt with in Section 4.5.2.2) investigates the effects of unbonded insulation–insulation interfaces—as those occurring, for example, in premolded joints or in terminations between the molded stress cone and the bare insulation—on field enhancement due to space charge or the

[8] The creepage distance is the shortest distance or the sum of the shortest distances along the surface on an insulator between two conductive parts, which normally have the operating voltage between them (after [101]).

thermoelectrical properties; however, it must be pointed out that the unbonded interface analyzed in [102] (namely an XLPE–EPR interface) was arranged in a coaxial cable structure not specifically conceived for representing either a joint or a termination, although some useful information for the behavior of accessory interfaces can be derived from that work.

After the details just reported here, the following sections illustrate other useful information about accessories for HVDC extruded cable systems taken from [84], as well as from published works that report on extruded HVDC cable system projects realized since 1999 [78,80–82].

5.7.1 Accessories Design Relevant to the Gotland Project

Since the beginning of the Gotland (HVDC extruded cable system) Project, the development of accessories was a priority because accessories are a critical part of a cable system that affect the behavior of the whole system. Indeed, when testing the cable system, no breakdown of the cable insulation was achieved since all breakdowns occurred in the terminations, although special terminations withstood a DC voltage in excess of 900 kV [80].

5.7.1.1 Joints for the Gotland Project As far as joints are concerned, to test different designs and materials, step voltage tests were performed similar to those carried out when screening the cable insulation and semicons. These tests proved the key role played by the interfaces, where electric field distributions with high stress levels gave rise to previously unknown processes, for example, related to space-charge effects [80]. Three types of joints were tested [80]:

1. *Taped self-amalgamating joints*, made with standard self-amalgamating EPDM tapes. These joints behaved in consistently during the various tests, thus they were discarded.
2. *Tape molded joints*, made with extruded tapes of the same materials as the cable itself. These joints (see Fig. 5.19 after [80]), after several tests and developments of the molding process, were successfully type-tested and approved for installation in the Gotland project.
3. EPDM-based *premolded* (or *prefabricated*) *joints*. These joints have a double advantage: Each single joint is an identical replica of the base design and can be electrically tested in the factory; their installation is faster.

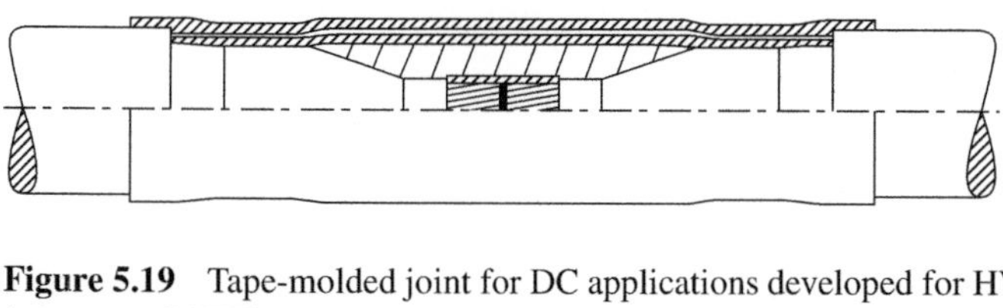

Figure 5.19 Tape-molded joint for DC applications developed for HVDC extruded cables (courtesy of ABB).

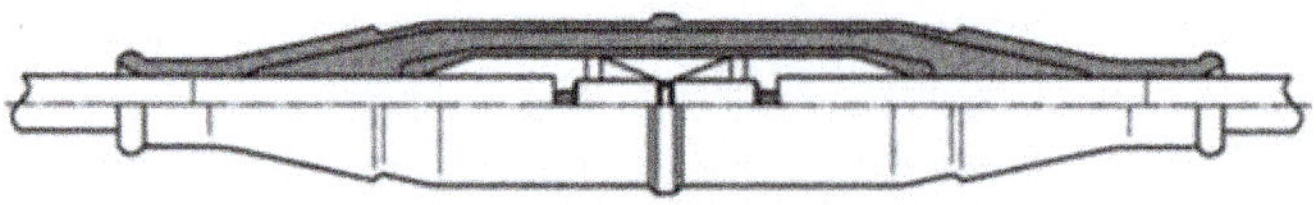

Figure 5.20 Prefabricated joint for DC applications developed for HVDC extruded cables (courtesy of ABB).

Since the realization of tape-molded joints is lengthy (each single layer has to be taped and cured separately) and a quicker jointing method was needed for field jointing of land cables, the development was focused on *premolded joints*, although flexible tape-molded joints are required for undersea cables. The design of the premolded joint is based on the "slip-over concept" (see Fig. 5.20 after [80]), a simple but robust joint type whose installation requires a low number of manual operations, which is realized with separately molded layers of EPDM for obtaining a smooth field distribution (see Fig. 5.21). The joint development was successful.

5.7.1.2 Terminations for the Gotland Project As to the cable terminations for the Gotland project, a new termination for the extruded cable was developed (see Fig. 5.22) that employs both resistive and geometric field control. The optimal design relied on detailed finite-element method (FEM) analysis of the electric field distribution and intensive experimental investigations that showed—similarly to the joints—the crucial role played by the interfaces between the termination and the cable [80].

As emphasized in [80], due to the high intrinsic breakdown strength of the extruded insulation of HVDC cables, such cables have relatively thin insulation walls, which implies that special attention must be paid to the very high interfacial electrical field in joints and terminations.

5.7.2 Accessories Design Relevant to the Murraylink Project

5.7.2.1 Joints for the Murraylink Project In the Murraylink HVDC Project, simple design and assembling procedures were fundamental requisites for the joints since this link features almost 400 joints. The joint design features a screwed connector for the conductor and a prefabricated single-piece EPDM rubber mold

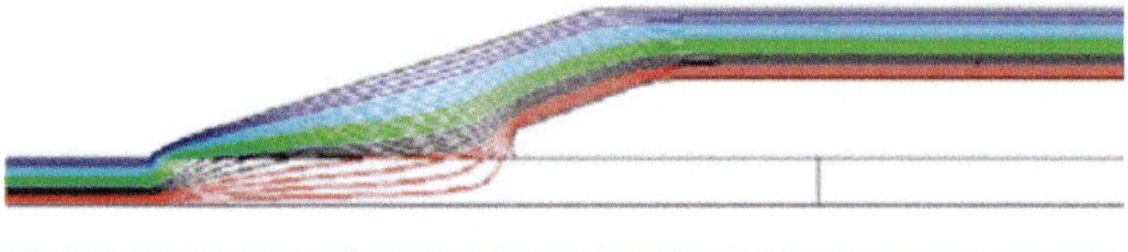

Figure 5.21 FEM calculation showing the equipotential line distribution in a prefabricated joint (courtesy of ABB).

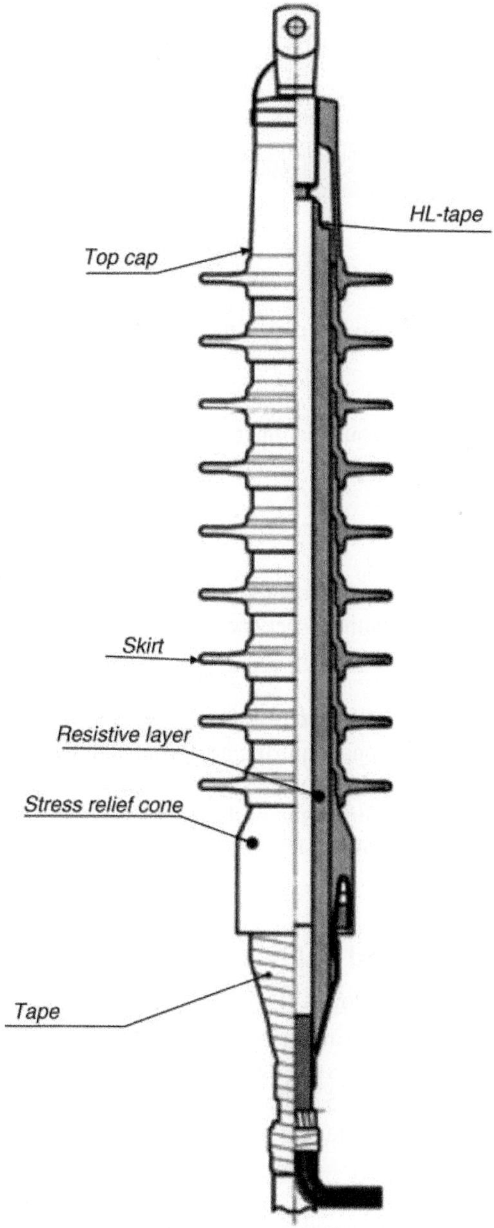

Figure 5.22 Prefabricated EPDM termination developed for the Gotland project (courtesy of ABB).

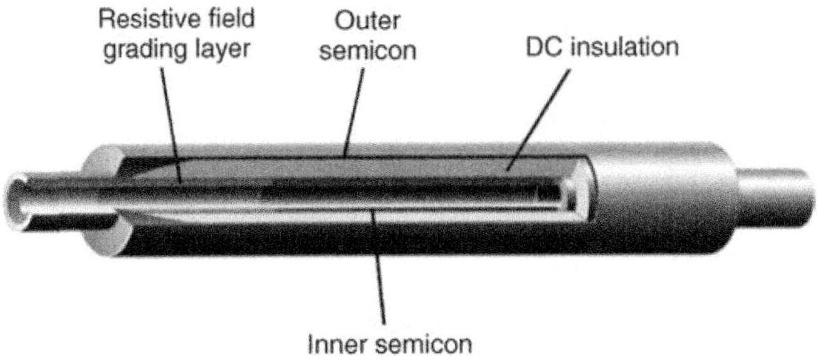

Figure 5.23 Prefabricated DC joint used in the Murraylink project (courtesy of ABB).

with a nonlinear resistive field-grading layer and a built-in inner semicon deflector (see Fig. 5.23, after [78]). The single-piece design of the joint mold is characterized by:

- A lower number of critical interfaces
- Easier pretesting procedures in the factory
- The capability of fitting both the 1200-mm^2 and the 1400-mm^2 conductor cross-section cable, as well as the transitions between these cables, with remarkable installation advantages [78]

5.7.2.2 Terminations for the Murraylink Project As to the terminations for the Murraylink HVDC Project, all-dry terminations requiring no special assembling tools were selected for all HVDC cables (both transmission cables and cables within the converter stations) [78]. The main components of these terminations are: (a) weather sheds and a stress relief cone made of EPDM rubber and (b) a patented nonlinear resistive field control [78,103].

5.7.3 State of the Art of HVDC Extruded Cable Accessories

Some recently published studies (see, e.g., [84,91,92]) describe in detail the state of the art of HVDC cable accessories. Although referring to different manufacturers, it can be argued from [84,91,92] that these accessories are designed consistently with the concepts introduced at the beginning of Section 5.7. For the sake of explanation, broad citations and some figures from [84] are included in this section[9] in order to illustrate as clearly as possible many issues and possible engineering solutions:

> The accessories for the HVDC [extruded cables illustrated in [84] utilize a combination of geometric and refractive field grading for capacitive field distributions and nonlinear resistive field grading for DC field distributions. The geometric shapes of the different materials layers in the accessories ensure a proper level of electric field at time varying voltages such as lightning and switching impulses. DC field control is achieved by a continuous layer of Field Grading Material (FGM), which is placed between the live and

[9] By courtesy of ABB.

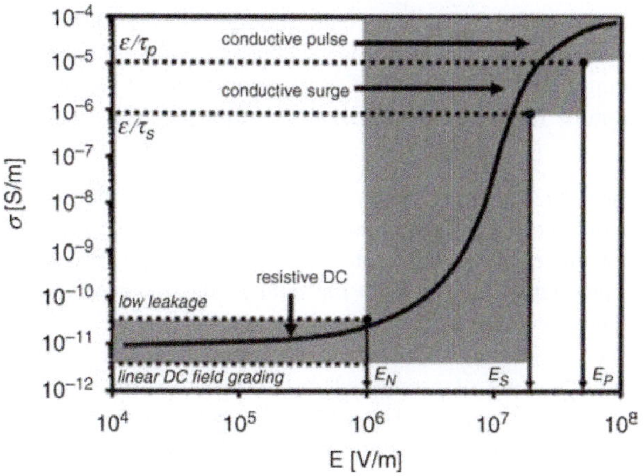

Figure 5.24 Typical FGM resistivity characteristic (courtesy of ABB).

earth electrodes. The nonlinear FGM has a suitable field dependent conductivity. The characteristics are restricted to regions (see Fig. 5.24) that are determined by the electric design fields at nominal (E_N), surge (E_S), and pulse (E_P) conditions on the one hand, and on leakage current and heating conditions on the other hand. The FGM acts as a nonlinear resistive voltage divider and controls the potential drop from the live electrode to ground. One important property is that the FGM becomes active after a time ε/σ ($\varepsilon = \varepsilon_0\varepsilon_r$ and σ are the FGM permittivity and conductivity, respectively; note that σ is field dependent), which has to be compared to the time of the applied voltage (τ_S and τ_P for surge and pulse, respectively). Moreover, the FGM has a high ε leading to additional refractive stress control. With today's FGM technology based on SiC and, in particular, ZnO microvaristor filled polymers, the FGM characteristics can be fine tuned by appropriate choice of the process parameters. Know-how in the area of FGM is a prerequisite for the proper design and development of HVDC extruded cable accessories ([84], page 1–2).

5.7.3.1 Prefabricated Joints

The prefabricated joint body developed for HVDC [extruded cables described in [84] consists of four different material layers (Fig. 5.25). An inner deflector made of a

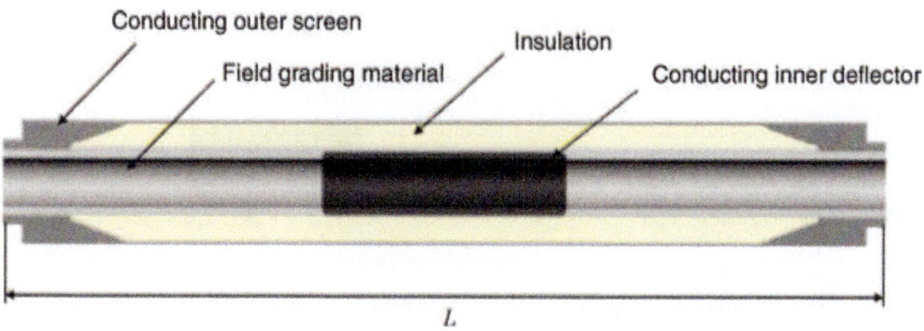

Figure 5.25 Prefabricated joint design (courtesy of ABB).

semi-conducting rubber electrically shields the connector. Through the whole length of the joint body there is a continuous layer of FGM, which even covers the inner deflector. A layer of insulating rubber outside the FGM represents the actual joint insulation. Finally, an outermost layer of the joint body is made up of a semi-conducting rubber screen.

Besides the material properties, the detailed geometry and the dimensions of the four layers are of paramount importance for the appropriate function of the cable joint. There are thermal, mechanical and electrical aspects to consider. The design must be simple enough to facilitate production, and to keep the total production cost at a reasonable level. The joint body geometry depends on numerous aspects. First, the inner deflector must safely enclose the connector, i.e. it must overlap the connector at both sides (see Fig. 5.26, L_1), because of a required installation robustness that permits the joint to be not perfectly centred. The length, L_2, between the inner deflector and the joint outer triple point (see below), is a critical parameter which determines the average tangential electrical field along the joint-cable interface. It is the task of the nonlinear resistive FGM to equalize this field and to suppress field enhancements in the region of deflector edge and outer triple point. It is vital that the insulation screen transition (semicon edge) of the cable is positioned inside the joint but behind the joint outer triple point (see below). Thus the end of the joint, L_3, must be made long enough to ensure a tolerable robustness in the field at installation. The total length (L) of the 150 kV and 320 kV joints are approximately 900 mm and 1400 mm, respectively.

One of the most important geometry parameters in the joint is the compound radii of curvature (Rid1, Rid2) of the inner deflector (Fig. 5.27), which determine the electric field in the FGM. A sufficiently large minimum radius of curvature is particularly relevant at capacitive or transient fields. Furthermore, the thickness of the FGM ($t$fgm) depends on the thickness of the inner deflector ($t$id), and on Rid,1,2, which affect the thin FGM layer thickness on top of the inner deflector, because this FGM layer must control the field enhancement of the deflector edge in the joint insulation. The joint insulation, with thickness (ti), is a further crucial design parameter. The outer screen (thickness $t$os) of the joint controls and confines the electric field to the inside of the joint body. It also ascertains the mechanical strength and handling robustness of the joint body [84].

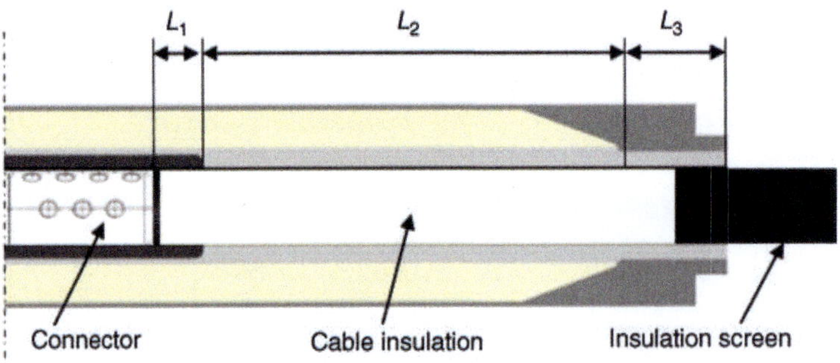

Figure 5.26 Critical joint design lengths (courtesy of ABB).

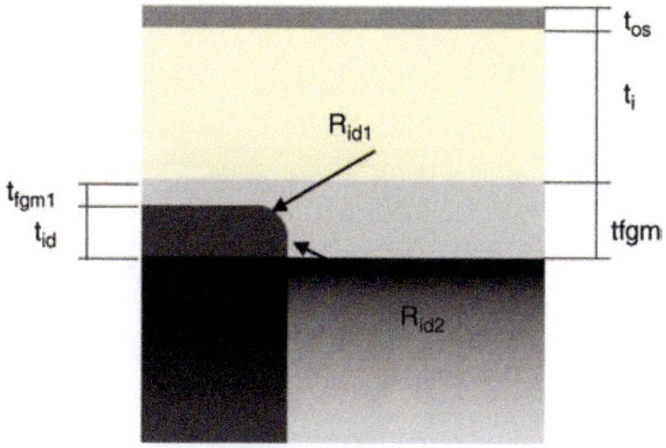

Figure 5.27 Important joint design dimensions (courtesy of ABB).

Interfaces between different materials are generally known as crucial areas in all electrical insulation systems. Of particular importance is a point or interface between three different materials. Such a point or interface is commonly known as a triple point. At each side of the joint body, an outer triple point occurs where the outer conducting screen meets the insulation and the FGM. It is well known that sharp edges result in high electric field strengths. To achieve a tolerable electric field, there is a radius (*R*otp) at the joint outer triple point. There are different ways to obtain a proper radius at the joint outer triple point. Moreover, the insulation forms an angle α with the FGM ([84], pp. 2–3).

5.7.3.2 Terminations

The HVDC [extruded cable termination illustrated in [84] is a true combination of a conventional AC termination based on geometric field control using a stress relief cone and a continuous layer of FGM connecting the live electrode to ground. The termination for 150 kV HVDC [extruded cable mentioned in [84] is a flexible dry type made up of separate rubber units, i.e. top sealing, weather-sheds, field grading adapters (FGA) and stress relief cone. Thanks to its design, the termination can basically be applied to any size of cable. Also, the creepage distance can easily be changed by varying the number of weather-sheds and FGAs. The heart of the 320 kV HVDC termination for extruded] cable [mentioned in [84]] is similar to the 150 kV design (FGA's), but is complemented with an oil filled porcelain body (Fig. 5.28) ([84], page 3).

5.7.3.3 Electric Fields in Accessories

As to the electric fields in accessories, in addition to the nominal operating DC voltage [the accessory] must also withstand switching impulses (SI) and in few cases lightning impulses (LI). The SI and LI requirements are determined by the system operation conditions. The accessories must have the ability to control all these inherently different voltage stresses. [Sörovist et al. [84] claim to have developed] an advanced computer

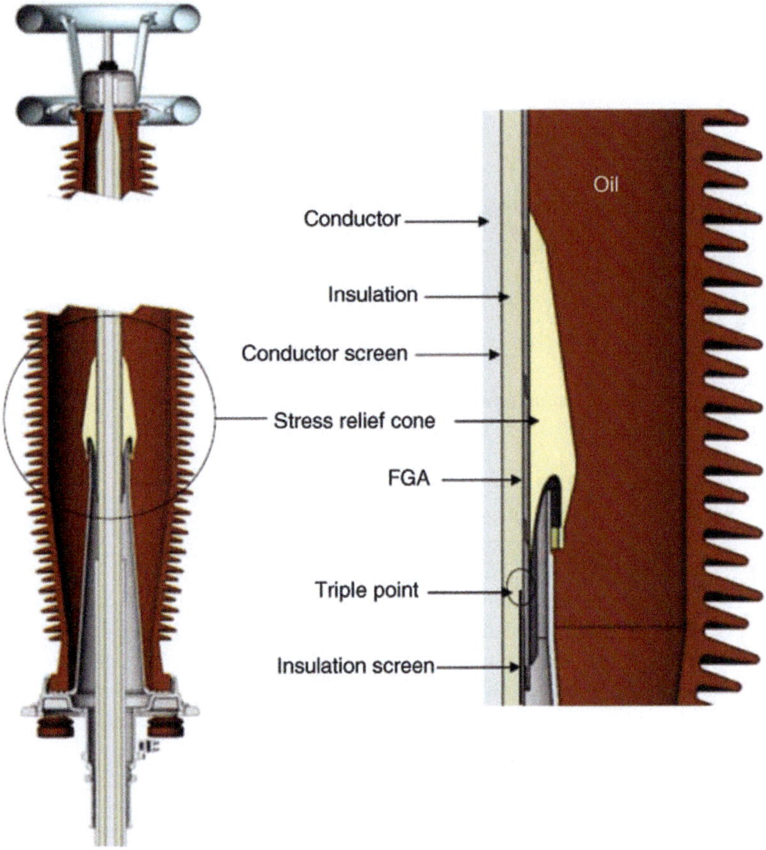

Figure 5.28 Schematic cut-away view of 320-kV HVDC extruded cable termination (courtesy of ABB).

aided tool to simulate (calculate) the electric fields and temperatures in the cable and accessories. Besides the full geometry, all relevant material data and ambient conditions must be known.

Examples of calculated equipotential lines, electric fields, and temperatures in the joint are shown in Fig. 5.29*a*, Fig. 5.29*b*, and Fig. 5.29*c*, respectively. Figure 5.29*a* indicates a smooth continuous potential drop along the FGM. The DC field in the joint is considerably smaller [low field: dark regions] than in the cable [high field: gray regions; see Fig. 5.29*b*]. Figure 5.29*c* reflects the fact that the temperature increase (gray: hot; dark: cold] inside the joint is uncritical. Electric full-scale testing, e.g. withstand tests, gives valuable information on the design limits ([84], page 4).

5.7.3.4 Installation of Accessories As far as the installation of accessories is concerned, in [84] the considerations already made at the beginning of this section are confirmed and completed, highlighting the great importance of the installation of accessories as far as cable system operation and reliability are concerned. Indeed,

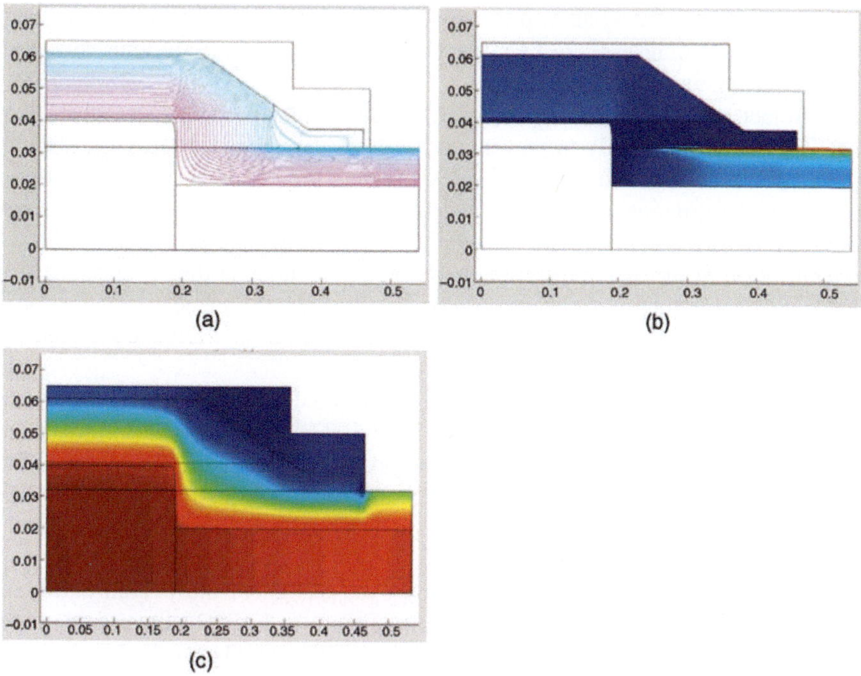

Figure 5.29 (*a*) Example of equipotential calculation, (*b*) example of electric field calculation, and (*c*) example of temperature calculation (courtesy of ABB).

most of the rare failures in service occur because of installation mistakes. The vital requirements for a proper installation are cleanliness and an adequate preparation of the cable insulation surface, which should be smooth and free from contaminants after the outer layers up to and including the insulation semicon are removed. The accessories are installed inside a jointing house, which provides a controlled environment from the viewpoint of both temperature and contamination, as well as a safe storage location for all the needed jointing tools. The quality control process of the installation requires that each single installation stage is verified and signed in ad hoc record sheets [84].

5.8 IMPROVED CABLE SYSTEM DESIGN

As far as the improved design of an extruded HVDC cable system as a whole is concerned, a major issue is bonding and grounding of the system, an important research subject that is analyzed in only a few studies to date and merits more extensive investigations [1].

Bonding and grounding systems of HVDC cable systems in general require a proper insulation coordination strategy relevant to the transient overvoltages that can be experienced by the system, primarily switching impulses, lightning impulses, polarity reversal transients (in LCC systems), commutation failure transients, and high-frequency harmonic ripple.

In fact, as pointed out in Chapter 3, the thermoplastic sheath (or jacket) of a cable is designed primarily to prevent corrosion of the underlying metallic sheath, which in turn among its functions ensures a sealing protection of the cable core. If insulation coordination is improperly dealt with, the above-mentioned transient overvoltages might puncture the thermoplastic sheath, thereby giving rise to localized corrosion of the metallic sheath and eventually damaging the cable insulation. Hence, regular testing of the thermoplastic sheath is a recommended practice for a prudent and safe operation and maintenance strategy to ensure reliable service and longevity to an extruded HVDC cable system.

Since HVDC transmission systems are best suited for realizing long links with massive power transmission capability, it is advisable to arrange sheath breaks at proper and regular intervals for isolation purposes, as well as for after-installation and in-service diagnostic tests on the system. Sheath break link boxes are also the place where sheath voltage limiters could be installed to reduce the risk of sheath injury during transient overvoltages.

One more possible issue for the improved design of the overall extruded HVDC cable system is represented by harmonics from HVDC converter stations that might affect the HVDC cable conductor. Actually these harmonics are a small fraction of the HVDC system ampacity, but nevertheless—if disregarded at the system design stage—they can disturb the proper operation of adjacent communications equipment by creating interferences. First of all, harmonics can be reduced by using highly conductive materials, such as aluminum or copper for the metallic sheath, and solidly grounding the cable sheath at both ends of the circuit, thereby attaining a low-resistance induction path in the metallic sheath itself: In this way, the magnetic field generated by harmonics outside the power cable decreases remarkably [1]. In addition, filters for filtering harmonics on the DC side to be installed at converter stations have also been studied and designed (see Chapter 2) [104].

5.9 TESTING OF HVDC EXTRUDED CABLE SYSTEMS

The first document containing recommendations for testing of HVDC extruded cable systems was CIGRÉ Technical Brochure 219 (TB 219) issued by CIGRÉ Working Group 21.01 [105]. Since commercial projects relevant to HVDC extruded cable systems above the 300-kV level are being undertaken (see Chapter 7), recently CIGRÉ Working Group 21.01 has replaced TB 219 by the new CIGRÉ Technical Brochure 496 (TB 496), thereby extending up to the 500-kV level the recommendations contained in the former TB 219 [106].

The tests in TB 496 "follow the same principles as in TB 219" and "the philosophy adopted is that the tests recommended should apply to the complete HVDC cable system as installed and as intended to function" [106].

Along with TB 219, TB 496 defines the following main tests to be performed on HVDC extruded cable systems [105,106]:

1. *Development tests*, performed during the development of the cable system
2. *Prequalification test*, performed before supplying on a general commercial basis a type of cable system covered by the TB in order to demonstrate satisfactory long-term performance of the complete cable system
3. *Type tests*, performed before supplying on a general commercial basis a type of cable system covered by the TB, in order to demonstrate satisfactory performance characteristics to meet the intended application
4. *Routine tests*, performed by the manufacturer on each manufactured component (length of cable or accessory) to check that the component meets the specified requirements
5. *Sample tests*, performed by the manufacturer on samples of complete cable or components taken from a complete cable or accessory, at a specified frequency, so as to verify that the finished product meets the specified requirements
6. *Tests after installation*, performed to demonstrate the integrity of the cable system as installed

An essential step in the qualification procedure according to both former TB 219 and latter TB 496 is the prequalification (PQ) test. Such a test has to be carried out on at least 100 m of cable, including all accessories (the so-called PQ loop) for about one year at a constant DC voltage but with time-varying loading. More precisely, a sequence of 360 "24-hour" load cycles has to be applied. The time duration of such load cycle tests is based (see Appendix A of [106]) on a *V*-*t* characteristic of the inverse power law type, with exponent n equal to 10, regarded as conservative, and on a design life of 40 years at constant rated voltage (see Chapter 6).

The 24-hour load cycles differ as to the load, that is, the current flowing in the conductor. Such periods—whose duration (in days), polarity, and magnitude of applied voltage (in pu of the rated DC voltage U_0) for cable systems using VSC[10] are listed in Table 5.4—are of three types:

1. "Zero-load" (ZL) periods, with no current flowing in the conductor

Table 5.4 Duration (in Days), Polarity, and Magnitude (in pu of U_0) of Applied Voltage During PQ Tests for Cable Systems Used with VSC

	LC	LC	HL	HL	ZL	LC	LC
Days	40	40	40	40	120	40	40
Polarity	+	−	+	−	−	+	−
Voltage ($\times U_0$)	1.45	1.45	1.45	1.45	1.45	1.45	1.45
Total days	40	80	120	160	280	320	360

Source: According to [106].

[10] Focus is here on PQ tests for cable systems using VSC, referred to as "VSC cables" for the sake of brevity.

2. "High-load" (HL) periods, with such a current in the conductor as to let the conductor reach a temperature no lower than the rated conductor temperature T_0
3. True "load cycle" (LC) periods, with 8 h heating and 16 h cooling, so that the conductor reaches a temperature $\geq T_0$ during the last 2 h of heating

According to [106], at the end of the PQ load cycles, impulse tests have to be performed, in order to verify that no major thermomechanical changes have taken place during the long-term testing. Typical loops for PQ tests are shown in Figures 5.30 and 5.31.

Figure 5.30 PQ test loop (courtesy of Prysmian).

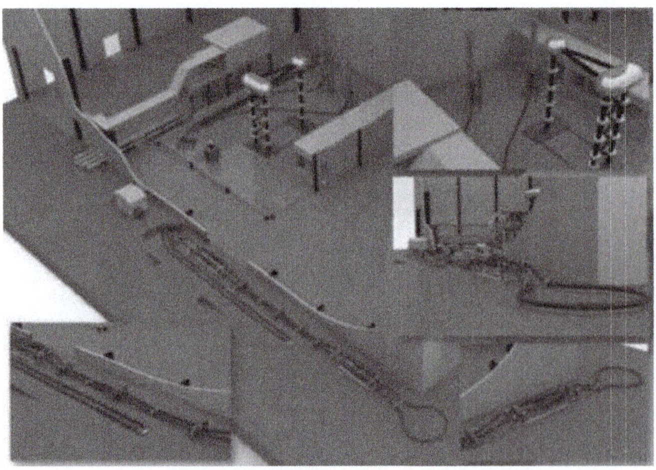

Figure 5.31 Sketch of PQ test loop (courtesy of Nexans).

Figure 5.32 Type test loop (courtesy of Nexans).

Also type tests involve load cycles, which are different for cable systems to be qualified for LCC operation with respect to cable systems to be qualified for VSC operation. Both 24-h and 48-h load cycles are defined for both cases. However, compared to PQ load cycle tests, type test load cycles are characterized by [106]:

- Shorter overall effective test duration (30 days instead of 360)
- Higher voltage levels ($U_T = 1.85 \times U_0$ instead of $U_{\mathrm{TP1}} = 1.45 \times U_0$ as constant type test voltage level for VSC operation)
- Reduced length loops (see Figs. 5.32 and 5.33)

More details about the nontrivial matter of testing HVDC extruded cable systems are omitted here for the sake of brevity. Interested readers should refer to [106]. Reference standards and recommendations are also [107–112] as well as [113–115] (although these latter strictly speaking are relevant to AC cables, nevertheless they provide useful indications also as far as HVDC extruded cable system testing is concerned).

Figure 5.33 Type test loop (courtesy of Silec).

REFERENCES

1. EPRI Report No. 1008720. *DC Cable Systems with Extruded Dielectrics*. Palo Alto, California, December 2004.
2. B. Jacobson, Y. Jiang-Häfner, P. Rey, G. Asplund, M. Jeroense, A. Gustafsson, and M. Bergkvist. "HVDC with Voltage Source Converters and Extruded Cables for up to ±300 kV and 1000 MW." Paper B4-105, CIGRE Session 2006.
3. J. Dorn, H. Huang, and D. Retzmann. "Novel Voltage-Sourced Converters for HVDC and FACTS Applications." Cigre Symposium, November 1–4, 2007, Osaka, Japan.
4. M. Salah Khalil. "International Research and Development Trends and Problems of HVDC Cables with Polymeric Insulation." *IEEE Electr. Insul. Mag.*, Vol. 13, No. 6, pp. 35–47, November–December 1997.
5. T.L. Hanley, R.P. Burford, R.J. Fleming, and K.W. Barber. "A General Review of Polymeric Insulation for Use in HVDC Cables." *IEEE Electr. Insul. Mag.*, Vol. 19, No. 1, pp. 14–24, January/February 2003.
6. R.N. Hampton. "Some of the Considerations for Materials Operating under High-Voltage, Direct-Current Stresses." *IEEE Electr. Insul. Mag.*, Vol. 24, No. 1, pp. 5–13, January–February 2008.

7. N. Ando, F. Numajiri, I. Tomori, I. Muraki, and T. Kumagai. "Development of ±250 kV DC Crosslinked Polyethylene Cable." *Hitachi Rev.*, Vol. 25, No. 10, pp. 355–360, 1976.
8. H. Fukagawa and Y. Sekii. "Insulation Properties of 250 kV DC XLPE Cables." IEEE PES Winter Meeting, Paper 81 WM 025-6, Atlanta, Georgia, February 1–6, 1981.
9. Y. Sakamoto, H. Fukagawa, K. Ninomiya, Y. Yamada, S. Yoshida, and N. Ando. "Investigation on Test Method of DC XLPE Cable." IEEE PES Summer Meeting, Paper 81 SM 386-2, Portland, OR, July 26–31, 1981.
10. Y. Sekii. "Electrical Characteristics of Extruded Insulation DC Cables." CIGRE Report WG22-02, pp. 1–9, 1982.
11. Y. Li and T. Takada. "Progress in Space Charge Measurement of Solid Insulating Materials in Japan." *IEEE Electr. Insul. Mag.*, Vol. 10, No. 5, pp. 16–28, Sept./Oct. 1994.
12. G. Asplund. "HVDC Using Voltage Source Converters. A New Way to Build Highly Controllable and Compact HVDC Substations." International Council on Large Electric Systems (CIGRÉ), 2000 Session, Panel 2, Paper 2–04, Paris, France, 2000.
13. Brochure CIGRÉ, 88. Overvoltages on HVDC Cables—Final Report, CIGRÉ Joint Working Group 33/21/14.16, August 1994.
14. M. Salah Khalil. "High Voltage Direct Current Cables with Polymeric Insulation: The State of the Art." Conference Record of the 1992 Nordic Insulation Symposium, Report No. 6.1.
15. N. Singh and A.C. Bruhin, "Application of Extrudable Insulants to DC Cables." CEIDP (Conf. on Electr. Insul. & Dielectric Phenomena) Ann. Report, pp. 60–69, 1973.
16. S.E. Kiersztyn and A.C. Bruhin. "Design Concept of Stress Inversion Free DC Cable." IEEE PES Summer Meeting, Paper A 77 540-6, Mexico City, July 17–22, 1977.
17. Extruded Dielectric DC Cable Development, EPRI Project 7828-1 Final Report, 1980.
18. A. Cherifi, A. Toureille, J.E. Reboul, and M. Salah Khalil. "Influence of Barium Titanate ($BaTiO_3$) Additive on Space Charge Formation in Polyethylene." *Conf. Electr. Insul. Dielectr. Phenom. Ann. Rep.*, pp. 140–145, 1993.
19. M. Salah Khalil, A. Cherifi, A. Toureille, and J. E Reboul. "On Effects of Polarizing Temperature and Additive Content on Space Charge Formation and Electric Field Distribution in Plain LDPE and LDPE Doped with $BaTiO_3$: Evidence from Measurements Using the Thermal Step Method." Proc. of the 5th International Conference on Conduction and Breakdown in Solid Dielectrics, pp. 288–292, United Kingdom, July 1995.
20. M. Salah Khalil, A. Cherifi, A. Toureille, and J.P. Reboul. "Influence of $BaTiO_3$ Additive and Electrode Material on Space Charge Formation in Polyethylene." *IEEE Trans. Dielectr. Electr. Insul.*, Vol. 3, pp. 743–746, 1996.
21. M. Salah Khalil and A.A. Zaky. "Influence of Iodine on the Distribution of Space Charge in LDPE." *Intl. J. Polymeric Mater.*, Vol. 11, pp. 1–8, 1985.
22. M. Salah Khalil, P.O. Henk, and M. Henriksen. "The Influence of Titanium Dioxide Additive on the Short Term DC Breakdown Strength of Polyethylene." Conference Record of the 1990 IEEE International Symposium on Electrical Insulation, pp. 268–271, Canada, June 1990.
23. M. Salah Khalil and P.O. Henk. "Effect of Barium Titanate ($BaTiO_3$) Additive on the Short-Term DC Breakdown Strength of Polyethylene." Conference Record of the 1996 IEEE International Symposium of Electrical Insulation, Vol. 2, pp. 654–656, Montreal, Canada, June 1996.

24. M. Salah Khalil. "On the Effect of Iodine on Space Charge in LDPE Using Field Probe Technique." Publication No. 8306, Department of Electric Power Engineering, Technical University of Denmark, November, 1983.

25. M. Salah Khalil. "The Use of Polyethylene as an Insulating Material for High Voltage DC Cables: Problems and Development Trends." Conference Record of the Second Arab International Conference on Materials Science (Polymeric Materials), Alexandria (Egypt), March 1990.

26. M. Salah Khalil, P.O. Henk, and M. Henriksen. "The Influence of Titanium Dioxide and Barium Titanate Additives on DC Conductivity of Low Density Polyethylene." Conference Record of the Nordic Insulation Symposium, NORDIS 88, Trondheim, Norway, June 1988.

27. M. Salah Khalil, P.O. Henk, and M. Henriksen. "Determination of Charge Carrier Mobility in Doped Low Density Polyethylene Using DC Transients." Proceedings of the 3rd International Conference on Conduction and Breakdown in Solid Dielectrics, Trondheim, Norway, pp. 192–195, 1989.

28. M. Salah Khalil, P.O. Henk, and M. Henriksen. "On the Effect of the Structural Properties of Polyethylene on the DC Conductivity in the Temperature Range from 40°C to 80°C." Conference on Electrical Insulation and Dielectric Phenomena Ann. Report, pp. 225–230, 1990.

29. S. Ogata, Y. Maekawa, K. Terkhima, R. Okiai, S. Yoshida, H. Yamanouchi, and S. Yokoya. "Study on the Electric Characteristics of DC XLPE Cables." *IEEE Trans. Power Del.*, Vol. 5, No. 3, pp. 1239–1247, July 1990.

30. Y. Maekawa, A. Yamaguchi, M. Haras, and Y. Sekii. "Research and Development of XLPE Insulated DC Cable." *Trans. IEE Jpn.*, Vol. 112-B, No. 10, pp. 905–913, Oct. 1992 [in Japanese].

31. H. Fukagawa, Y. Nitta, and Y. Sekii. "Development of a New Insulating Material for DC XLPE Cables." Conference Record of 1984 Jicable, pp. 283–287, 1984.

32. T. Suzuki, T. Niwa, S. Yoshida, T. Takahashi, and M. Hatada. "New Insulating Materials for HVDC Cables." Proceedings of the 3rd International Conference on Conduction and Breakdown in Solid Dielectrics, pp. 442–447, 1989.

33. T. Suzuki, T. Niwa, T. Takahashi, and H. Miyata. "Development of New Polymer Insulating Materials for HVDC Cable (Part 2)." *Trans. IEE Jpn.*, Vol. 112-B, No. 10, p. 914, Oct. 1992.

34. M. Pays, M. Louis, J. Perrep, C. Alquie, and J. Lewiner."Behaviour of Extruded HVDC Power Transmission Cables: Test on Materials and Cables." 21-07 *CIGRE Rept.*, pp. 1–11, 1988.

35. M. Pays and M. Louis. "Extruded Synthetic Insulations for Development of HVDC Energy Cables." Conference Record of 1987 Jicable, pp. 111–121, 1987.

36. M. Pays. "New Developments in the Field of High Voltage and Extra High Voltage Cables: The User Point of View." 1989 IEEE/PES Transmission and Distribution Conference, Paper No. 89 TD 356-7 PWRD, 1989.

37. B.S. Hansen. "High Voltage Direct Current Cables with Solid Insulation." Final Report of Erhversforskeruddanelse, EF74, Publication No. 8303, Electrical Power Engineering Dept., Technical University of Denmark, August 1983.

38. M.S. Khalil. "Development of New Polymeric Materials for HVDC Cables Using Additives."*Project ADDIS Report*, Lyngby, Denmark 1992.

39. M. Salah Khalil and A. Gastli. "Investigation of the Dependence of DC Insulation Resistivity of Ultra-Clean Polyethylene on Temperature and Electric Field." *IEEE Trans. Power Del.*, Vol. 14, No. 3, pp. 699–704, July 1999.
40. M. Salah Khalil. "On the Use of Doped Polyethylene as an Insulating Material for HVDC Cables." Conference Record of the 1996 IEEE International Symposium of Electrical Insulation, Vol. 2, pp. 650–653, Montreal, Canada, June 1996.
41. T. Ditchi, C. Alquie, J.E. Favrie, and R. Jocteur. "Electrical Properties of Electrode/Polyethylene/Electrode Structures." *IEEE Trans. Electr. Insul.*, Vol. 24, pp. 403–408, 1989.
42. T. Tanaka. "Charge Injection by Voltage Application into Polymer Dielectrics—New Proposal for DC Cable Polymer Insulation System." IEEE PES Summer Meeting, Portland, OR, July 18–23, 1976.
43. A. Cherifi, M. Salah Khalil, C. Meunir, J.E. Reboul, and A. Toureille. "Influence of the Polarizing Electrode Material on Space Charge Storage in Low Density Polyethylene." Proceedings of the 8th. International Symposium on Electrets (ISE-8), pp. 224–229, Paris, France, September 7–9, 1994.
44. S. Ou, A. Fujita, M. Okashita, and T. Shiono. Showa ElectricWire and Cable Co. Ltd., Japanese Patent JP11260158, 1999.
45. S. Ou, H. Tanimoto, and T. Shiono. Showa ElectricWire and Cable Co. Ltd., Japanese Patent JP10255561, 1998.
46. S. Ou, M. Fujita, H. Tanimoto, and F. Aida. Showa Electric Wire and Cable Co. Ltd., Japanese Patent JP9245521, 1997.
47. Y. Zhang, F. Zheng; Z. An, and X. Zhang. "Research on Space Charge in Insulating and Functional Dielectrics at Tongji University." *IEEE Electr. Insul. Mag.*, Vol. 28, No. 3, pp. 14–25, 2012.
48. Z. An, Q. Yang, C. Xie, Y. Jiang, F. Zheng, and Y. Zhang. "Suppression Effect of Surface Fluorination on Charge Injection into Linear Low Density Polyethylene." *J. Appl. Phys.*, Vol. 105, No. 6, art. no. 064102, 2009.
49. Y. Jiang, Z. An, C. Liu, F. Zheng, and Y. Zhang. "Influence of Oxyfluorination Time on Space Charge Behavior in Polyethylene." *IEEE Trans. Dielectr. Electr. Insul.*, Vol. 17, No. 6, pp. 1814–823, 2010.
50. Space Charge Measurement in DC Cable Materials. EPRI Report EL-7302, Research Project 7897, 1992.
51. P. Carstensen, A. Gustafsson, A. Farkas, A. Ericsson, J. Boström, B. Gustafsson, U. Nilsson, and F. Campus. Asea Brown Boveri AB, International Patent WO9933069, 1999.
52. P. Carstensen. Asea Brown Boveri AB, International Patent WO9940589 (1999).
53. B. Gustafsson, J. Boström, U. Nilsson, P. Nylander, P. Carstensen, A. Farkas, A. Gustafsson, and K. Hannesson. Asea Brown Boveri AB, International Patent WO0008655, 2000.
54. K. Ogawa, T. Suzuki, S. Katakai, M. Kanaoka, Y. Sekii, C. Ikeda, and H. Shirai. Tokyo Electric Power Co., Hitachi Cable Ltd., French Patent Fr 2629626, 1989.
55. T. Tanaka and K. Terajima. Furukawa Electric Co. Ltd., Electric Power Dev. Co. Ltd., Japanese Patent JP10289619, 1998.
56. H. Nomura. Furukawa Electric Co. Ltd., Japanese Patent JP9097522, 1997.
57. A. Yokoyama and H. Hiroyuki. Fujikura Ltd., Japanese Patent JP10162651, 1998.

58. X. Li, Y. Cao, Q. Du, Y. Yin, and D. Tu. "Charge distribution and crystalline structure in polyethylene nucleated with sorbitol." *J. Appl. Polym. Sci.*, Vol. 82, pp. 611–619, 2001.

59. T. Uozumi and K. Hirotsu. Sumitomo Electric Ind. Ltd., Japanese Patent JP11224544, 1999.

60. T. Uozumi and K. Hirotsu. Sumitomo Electric Ind. Ltd., Japanese Patent JP11224545, 1999.

61. T. Uozumi and K. Hirotsu. Sumitomo Electric Ind. Ltd., Japanese Patent JP11224543, 1999.

62. L.H. Sperling. *Polymeric Multicomponent Materials: An Introduction.* Wiley, New York, pp. 180–195, 256–273, 1997.

63. J. Duvall, C. Selliti, C. Meyers, A. Hiltner, and E. Baer. *J. Appl. Polym. Sci.*, (a) Vol. 52, pp. 207–216, 1994; (b) Vol. 52, pp. 195–206, 1994.

64. F. Ide and A. Hasegawa, *J. Appl. Polym. Sci.*, Vol. 18, p. 963, 1974.

65. F.M. Sweeney, Ed., *Polymer Blends and Alloys: Guidebook to Commercial Products.* Technomic, Lancaster, PA, pp. 206–220, 1988.

66. M. Salah Khalil. "The Role of $BaTiO_3$ in Modifying the DC Breakdown Strength of LDPE." *IEEE Trans. Dielectr. Electr. Insul.*, Vol. 7, pp. 261–268, 2000.

67. M.S. Khalil and J.A. Jervase. "Development of Polymeric Insulating Materials for HVDC Cables Using Additives: Evidence from a Multitude of Experiments Using Different Techniques." Conference Record 2000 IEEE Int. Symp. Electr. Insul., pp. 485–488, 2000.

68. M.M. Ueki and M. Zanin. "Influence of Additives on the Dielectric Strength of High-Density Polyethylene." *IEEE Trans. Dielectr. Electr. Insul.*, Vol. 6, pp. 876–881, 1999.

69. K. Nagata, S. Kodama, H. Kawasaki, S. Deki, and M. Mizuhata. "Influence of Polarity of Polymer on Inorganic Particle Dispersion in Dielectric Particle/Polymer Composite Systems." *J. Appl. Polym. Sci.*, Vol. 56, p. 1313; 1995.

70. M. Takaoka, T. Motai, M. Ono, S. Yoshida, T. Niwa, T. Takahashi, K. Kataoka, and H. Yamanouchi. Fujikura Ltd., U.S. Patent US4626618, 1986.

71. A. Yokoyama, A. Ono, and H. Miyata. Fujikura Ltd., Japanese Patent JP10269852, 1998.

72. T. Katagai, T. Yamazaki, and Y. Murata. Hitachi Cable Ltd., Japanese Patent JP11086636, 1999.

73. T. Katagai, Y. Murata, and H. Shirai. Hitachi Cable Ltd., Japanese Patent JP11086634, 1999.

74. T. Yoshizawa, H. Nomura, and T. Tanaka. Furukawa Electric Co. Ltd., Japanese Patent JP10134648, 1998.

75. J.A. Brydson. *Plastics Materials*, 7th ed. Butterworth Heinemann, Oxford, UK, p. 229, 1997.

76. T. Tanaka. "Dielectric Nanocomposites with Insulating Properties." *IEEE Trans. Dielectr. Electr. Insul.*, Vol. 12, No. 5, pp. 914–928, October 2005.

77. IEC60287. "Electric Cables—Calculation of the Current Rating—Part 1-1: Current Rating Equations (100% Load Factor) and Calculation of Losses—General." IEC 60287-1-1, 2006-12.

78. T.B. Worzyk, M. Jeroense, T. Sörqvist, and M. Farr. "The Murraylink Project—The First Commercial 150 kV Extruded HVDC Cable System." International Conference on Insulated Power Cables, Paper A7.4, pp. 217–222 (JiCable'03), Versailles, France, 22–24, 2003.

79. M. Jeroense, M. Bergkvist, A. Gustafsson, H. Rapp, T. Sörqvist, and J. Svahn. "Increased Voltage for the HVDC Light® Product Range—A Complete Solution." Paper B4.7, pp. 425–430, 2007 International Conference on Insulated Power Cables (Jicable'07), Versailles, France, June 24–28, 2007.

80. M. Byggeth, K. Johannesson, L. Palmqvist, J. Jonsson, C. Törnkvist, C. Liljegren, and U. Axelsson. "The Development of an Extruded HVDC Cable System and Its First Application in the Gotland HVDC Light Project." Paper B7.5, pp. 538–542, 5th International Conference on Insulated Power Cables (JiCable'99), Versailles, France, June 20–24, 1999.

81. M. Byggeth, K. Johannesson, C. Liljegren, and U. Axelsson. "Gotland HVDC Light—The World's First Commercial Extruded HVDC Cable System." Presented at CIGRE', Paris, France, August 2000.

82. J.O. Boström, A. Campus, U. Nilsson, P. Carstensen, A. Gustafsson, A. Farkas, and K. Johannesson. "Cross-Linked Polyethylene Materials for DC Power Cables." Paper C2.12, 5th International Conference on Insulated Power Cables (JiCable'99), Versailles, France, June 20–24, 1999.

83. M. Jeroense. "HVDC, the Next Generation of Transmission. Highlights with Focus on Extruded Cable Systems." Proc. 2008 International Symposium on Electrical Insulating Materials (ISEIM 2008), pp. 10–15, Yokkaichi, Japan, Sept. 7–11, 2008.

84. T. Sörqvist, T. Christen, M. Jeroense, V. Mondiet, and R. Papazyan. "HVDC-Light® Cable Systems—Highlighting the Accessories." 21st Nordic Insulation Symposium (NORDIS 2009), Gothenburg (Sweden), June 15–17, 2009.

85. F. Pittelli, M. Marelli, and E. Zaccone. "Cavi in Corrente Continua per uso terrestre." Atti del Convegno Nazionale AEIT 2008, Milano, June 27–29, 2011 (in Italian).

86. N. Kuljaca, M. Marelli, A. Micheletti, and E. Zaccone. "Cavi sottomarini per impianti di generazione off-shore—stato dell'arte e sviluppi futuri." Atti del Convegno Nazionale AEIT 2008, Milano, June 27–29, 2011 (in Italian).

87. M. Bacchini R. Grampa, M. Marelli, T. Westerweller, and N. Kelley. "200 kV DC Extruded Cables Crossing the San Francisco Bay." CIGRE Session paper B1–105, Paris, 2010.

88. T. Westerweller, K. Friedrich, U. Armonies, A. Orini, D. Parquet, and S. Wehn. "Trans Bay Cable—World's First HVDC System Using Multilevel Voltage-Sourced Converter." CIGRE Session Paper B4–101, Paris, 2010.

89. K. Terashima, H. Suzuki, M. Hara, and K. Watanabe. "Research and Development of ±250 kV DC XLPE Cables." *IEEE Trans. Power Del.*, Vol. 13, pp. 7–16, 1998.

90. K. Fujii, K. Terashima, C. Watanabe, H. Suzuki, N. Hozumi, M. Yoshida, K. Kataoka, K. Watanabe, A. Watanabe, T. Takahashi, H. Miyata, H. Yamanouchi, S. Yokoyama, N. Shiseki, H. Ikeda, and H. Sakaguchi "+/-250kV DC XLPE Cable and Factory Joint –Long-Term Performance Verification Test." *Fujikura Tech. Rev.*, No. 27, 1998.

91. S. Hirano, J. Haraguchi, T. Mizuno, and Y. Nagoya. "Development of DC +/– 250 kV coaxially-Integrated Return Conductor Extruded Cable." Paper B.7.3, 2007 International Conference on Insulated Power Cables (Jicable'07), Versailles, France, June 24–28, 2007.

92. Y. Sakai, H. Niinobe, and T. Hirasawa, "Development of Pre-molded Accessories for HVDC Extruded Cable System." Paper A.2.3, 8th International Conference on Insulated Power Cables (Jicable'11), Versailles, France, June 19–23, 2011.

93. T. Mizuno, T. Takahashi, H. Harada, N. Hayashi, Y. Tanaka, and T. Maeno. "Effect of Conductive Inorganic Filler on Space Charge Characteristics in XLPE as a HVDC Insulating Material." Paper C.5.2, 8th International Conference on Insulated Power Cables (Jicable'11), Versailles, France, June 19–23, 2011.

94. H. Niinobe, N. Ishii, T. Nakajima, A. Watanabe, H. Tanaka, and H. Kon. "HVDC and Power Electronic Systems for Overhead Line and Insulated Cable Applications." Paper B1-9, CIGRE 2012 San Francisco Colloquium, San Francisco, 2012.

95. K. Terashima, M. Asano, K. Watanabe, M. Yoshida, and H. Kon. "Research and Development of DC XLPE Cable Associated Factory Joint." Paper B7-6, 5th International Conference on Insulated Power Cables (JiCable'99), Versailles, France, June 20–24, 1999.

96. Y. Maekawa, K. Watanabe, S. Maruyama, Y. Murata, and H. Hirota. "Research and Development of DC +/−500 kv Extruded Cables." Paper 21-203, CIGRE Session 2002, Paris, France, 2002.

97. T. Worzyk. *Submarine Power Cables*. Spinger, Berlin, 2009.

98. S. Maruyama, N. Ishii, M. Shimada, S. Kojima, H. Tanaka, M. Asano, T. Yamanaka, and S. Kawakami. "Development of a 500-kV DC XLPE Cable System." *Furukawa Rev.*, No. 25, pp. 47–52, 2004.

99. T. Yamanaka, S. Maruyama, and T. Tanaka. "The Development of DC +/-500 kV XLPE Cable in Consideration of the Space Charge Accumulation." Proceedings of the 7th Intemational Conference on Properties and Applications of Dielectric Materials P3-44, Nagoya, pp. 689–694, June 1–5, 2003.

100. IEC, 62155. "Hollow Pressurized and Unpressurized Ceramic and Glass Insulators for Use in Electrical Equipment with Rated Voltages Greater than 1000 V." May 2003.

101. IEC, 61462. "Composite Hollow Insulators—Pressurized and Unpressurized Insulators for Use in Electrical Equipment with Rated Voltage Greater than 1000 V—Definitions, Test Methods, Acceptance Criteria and Design Recommendations." February 2007.

102. S. Delpino, D. Fabiani, G.C. Montanari, C. Laurent, G. Teyssedre, P.H.F. Morshuis, R. Bodega, and L.A. Dissado. "Polymeric HVDC Cable Design and Space Charge Accumulation. Part 2: Insulation Interfaces." *IEEE Electr. Insul. Mag.*, Vol. 24, No. 1, pp. 14–24, January–February 2008.

103. A. Ericsson, M. Jeroense, J. Miller, L. Palmqvist, B. Railing, and P. Riffon. "HVDC-Light Cable Systems—The Latest Projects." Nordic Insulation Symposium, Tampere, Finland, 2003.

104. Y. Jun, L. Zehong, and S. Yan. "Studies on Design of DC Filters of Three Gorges to Changzhou ±500 kV HVDC Link." Proceedings of 1998 International Conference on Power System Technology (POWERCON'98), Vol. 1, pp. 460–465, August 18–21, 1998.

105. Brochure CIGRÉ 219. "Recommendations for Testing DC Extruded Cable Systems for Power Transmission at a Rated Voltage up to 250kV." CIGRE, February 2003; also *Electra*, No. 206, pp. 46–53, February 2003.

106. Brochure CIGRÉ 496. "Recommendations for Testing DC Extruded Cable Systems for Power Transmission at a Rated Voltage up to 500kV." CIGRÉ Working Group B1-32, April 2012; *Electra*, No. 261, pp. 88–95, April 2012.

107. Electra 141. "Guidelines for Tests on High Voltage Cables with Extruded Insulation and Laminated Protective Coverings." 1992.

108. Electra 151. "Recommendations for Electrical Tests Type, Sample and Routine on Extruded Cables and Accessories at Voltages $>$ 150 (170) kV and $\leq$ 400 (420) kV." 1993.

109. Electra 173. "After Laying Tests on High Voltage Extruded Insulation Cable Systems." 1997.

110. Electra n. 189. "Recommendations for Tests of Power Transmission DC Cables for a Rated Voltage up to 800 kV (ELECTRA 72, 1980—Revision). 2000.

111. IEC 60229. "Electric Cables—Tests on Extruded Oversheaths with a Special Protective Function." Third edition, 2007.

112. Electra n. 171. "Recommendations for Mechanical Tests on Sub-marine Cables." 1997.

113. IEC 62067. "Power Cables with Extruded Insulation and Their Accessories for Rated Voltages above 150 kV ($U_m = 170$ kV) up to 500 kV ($U_m = 550$ kV)—Test Method and Requirements." Second edition, November 2011.

114. Electra 189. "Recommendations for Testing of Long AC Submarine Cables with Extruded Insulation for System Voltage above 30 (36) to 150 (170) kV." 2000.

115. IEC 60840. "Power Cables with Extruded Insulation and Their Accessories for Rated Voltages above 30 kV ($U_m = 36$ kV) up to 150 kV ($U_m = 170$ kV)—Test Method and Requirements." Fourth edition, November 2011.

CHAPTER 6

LIFE MODELING OF HVDC EXTRUDED CABLE INSULATION

This chapter tackles the subject of HVDC extruded cable insulation life modeling, a term that is fairly popular in the literature on dielectrics and electrical insulation. However, it could be misinterpreted when taking it literally outside of this context. Thus, it must be pointed out that by life modeling we mean "modeling the life expectancy" or "modeling the expected life" of the "non-self-healing" insulation of HVDC extruded cables. This issue is critical in HVDC cable systems—and more generally in high-voltage systems—since those insulations that by their nature are not self-healing are often also the "weakest link" of such systems, being prone to irreversible degradation (i.e., aging) and failure under applied stresses (i.e., voltage, temperature, mechanical load, environmental stresses, etc.). The contents of chapter 6 are as follows:

- A review of the basics of life modeling and reliability estimates of power cables (Section 6.1), which consists of the traditional approach to insulation life modeling and of the probabilistic framework of HVDC extruded insulation
- A detailed description of the theory and applications of the life models conceived for extruded HVDC cables, which are either based on space-charge effects (i.e., the field-limited space-charge model and the space-charge DMM model; see Section 6.2) or on the interaction between space charges and partial discharges (i.e., the life model based on damage growth from microvoids; see Section 6.3)
- Some hints about the long-debated question of whether space charge is a cause or an effect of aging (Section 6.4), based on life models that do not regard space charge as a direct cause of aging.

It must be emphasized that, although life and reliability of a cable system is determined more often by the behavior of the accessories—particularly of the joints—rather than by the cables, nevertheless life modeling of HVDC extruded cable insulation has a relevance to the life and reliability of the whole cable system. Indeed, cable insulation life models, often developed starting from tests on small-size specimens, provide precious indications for characterizing, comparing, and choosing the insulation of the various components of the cable system, what is essential for a

Extruded Cables for High-Voltage Direct-Current Transmission: Advances in Research and Development, First Edition. Giovanni Mazzanti and Massimo Marzinotto.

not yet fully established technology as HVDC extruded cable systems. Moreover, life models similar to those developed for cables could be set up in principle also for the insulation of accessories, by using some of the modelistic tools reported in this chapter (e.g., the enlargement law, the combined electro-thermal-mechanical model, etc.), thereby justifying the existence and the scope of this chapter.

6.1 FUNDAMENTALS OF LIFE MODELING AND RELIABILITY ESTIMATES OF POWER CABLES

Insulation life modeling consists in finding *the relationship between applied stresses (the independent variables) and time to failure—or life—of the insulation (the dependent variable).* Such a relationship can be expressed mathematically by an equation, the so-called *life model* of the insulation under consideration, which is characterized by a certain mathematical form as well as a certain number of model parameters. The more stresses are considered, the more complex is the life model and the lower is its practical applicability. The parameters of the model are in general derived by applying best-fitting techniques to the results of the so-called accelerated life tests (ALT) [1], that is, tests performed at various levels of stresses higher than the service stress of the insulation, in order to reduce the test time to acceptable time lengths—typically hours, days, or months (depending on stress levels)—much lower than the design life of full-size insulation (which for cables is around, say, 40 years).

For cable insulation (as in the case of the insulation of most power components), the purpose of life modeling is essentially threefold [2–4], namely:

1. To infer parameters useful for *insulating material characterization*, in order to *compare the different endurance characteristics of compounds* that are candidates for a given insulation system, that is, their ability to withstand the applied stresses during service life
2. To predict life at stresses lower than the test ones in order to estimate the life in service conditions, thereby supporting the selection of the main design parameters of cable insulation, that is, insulation thickness and the levels of applied stresses
3. To select test stress levels and times in the so-called voltage withstand test during routine tests, or during prequalification and type tests [5, 6], on the basis of design stress levels and design life of the power cable

While the first goal can be generally reached, the second and the third are much more critical. Indeed, the practice of performing accelerated life tests at stress values considerably larger than those considered for insulation design does not allow proper evaluation of the life model behavior at low stresses to be carried out. In fact, changes in aging mechanisms can invalidate the model holding at high stresses. Hence, the extrapolation usually needed to estimate life at design stresses requires great care, since it is based on the unverified assumption of validity of the model (derived on the basis of experimental results) over the whole range of stresses from test to design stress levels. Moreover, tests are carried out on small-size specimens, while life estimation at the design level is relevant to full-size insulation, which is sometimes different from tested objects both in dimensions and in geometry [7–10].[1] Hence an extrapolation also as to the geometry and size is often—if not always—needed when a life model has to be applied to power cable life estimation [7, 8].

A few more remarks are needed about the second and third goals above. The third goal is somewhat easier to achieve since both design stress levels and voltage withstand test stress levels are in fact constant with time—at least in principle[2]—thereby enabling the use of a "classic" constant-stress life model. On the contrary, the second goal, that is, life estimation in actual service conditions is made more cumbersome by the fact that no stress is constant during "real-world" operation of power cables (and of electrical devices in general) [11, 12]. However, life modeling can be helpful in a broader sense for the insulation characterization in that the knowledge of model parameters may provide insights into aging mechanisms; thus, fitting life data by a model could give an indication on the aging processes [2].

Life prediction via life models, being based upon extrapolation, requires that the simplest model be used, a linear one if possible. This involves that the most appropriate coordinate system be chosen for the time-to-failure versus stress relationship. In general, as pointed out above, the lower the number of parameters of a model, the higher the accuracy in estimating parameters and extrapolated quantities (life, stress). Of course, a compromise is always needed between model complexity and quality of fitting experimental results [2].

Several mechanisms have been proposed for the aging of HVDC extruded cable insulation in order to achieve practical life models. These proposals follow essentially two approaches:

1. A macroscopic—or "phenomenological"—approach, based on a global description of aging
2. A microscopic—or "physical"—approach, based on the hypothesis that the main cause of electrical and mechanical aging is accelerated localized degradation triggered by microdefects [13]

The macroscopic approach, historically the first, has provided simple life models employed mainly to fit the results of accelerated electrical, thermal, and mechanical life tests: for example, the Arrhenius thermal life model (see Section 6.1.2.3), suggested by Dakin also for electrical aging, and the thermodynamic model of Eyring and Zhurkov, applied to electrical and mechanical aging [14–16].

The microscopic approach assumes that bulk degradation of a homogeneous dielectric at typical design stresses would require extremely long times, thus the time to

[1] Tests for cable development are performed mostly on cable models of (considerably) smaller size than the ultimate power cable, as well as on flat specimens, on Rogowsky objects [9], and so on. Only in the final development stage of the cable are tests performed on full-size prototypes of cable, but of smaller than delivery lengths [3, 4, 10].

[2] Tests at constant voltage are often performed under load cycling (see, e.g., prequalification tests and electrical type tests after [5]) so the constant stress models strictly speaking are not applicable to such tests. However, different from real service conditions, the temperature variation during load cycle tests is known and regular, so that constant stress models can be applied easily in an approximated way in the framework of Miner's law, as shown in [11, 12].

failure of practical insulation is associated with micro- or macrodefects, for example, voids, conducting particles, and protrusions. Hence, microscopic models [17–19] consist in relationships describing the growth of damage from such defects up to either failure or the onset a different aging mechanism—for instance, electrical treeing—which will eventually lead insulation fail [13]. Being conceived for providing a physical explanation of aging, microscopic models appear as an advance with respect to phenomenological models, which are based on fitting parameters derived through life tests only and whose number increases with the number of applied stresses [3]: for example, the phenomenological electro-thermo-mechanical model of Eq. (6.51) features 10 parameters!

The rest of Section 6.1 first recalls the traditional "inverse power model (IPM)-based" approach to insulation life modeling, then outlines the correct probabilistic framework of modern insulation life modeling, and subsequently presents some fundamental life models of the phenomenological type available in the literature and that can be employed for time-to-failure estimation of HVDC cable insulation. A few life models of the physical type, conceived specifically for HVDC cable insulation and focused mainly on space charge, are illustrated in Sections 6.2 and 6.3. Finally, a discussion on whether space charge is a cause or an effect of aging is carried out in Section 6.4 resorting to a few more "physical" life models.

6.1.1 Traditional Approach to Insulation Life Modeling

Quite a lot of knowledge on the physics of aging has been gathered in the last century, thereby giving rise to fairly complex "physical models" that, as far as HVDC cable insulation is concerned, are described, for example, in [13,20–23] (see Sections 6.2 and 6.3 for a detailed review). These models predict the life of cable insulation, that is, the time to failure [10], qualitatively—and on laboratory sample scale sometimes even quantitatively. From a theoretical point of view these models condense our knowledge of aging physics. For practical engineering purposes relevant to power cable insulation, a more pragmatic approach is often motivated and commonly followed, based on "phenomenological models", such as the inverse power model (IPM) [10,24], also referred to as *V-t* characteristic(s) [3–5]. This well-known and often used empirical law can be expressed and applied in many different ways (see, e.g., [24] for a review and Section 6.1.2 for a thorough probabilistic framework); for instance, according to [10] the IPM is given by

$$t_2 = t_1 \left(\frac{U_2}{U_1} \right)^{-n} \tag{6.1}$$

where n is referred to as life exponent [3–5], or voltage coefficient [10] or—more properly—as the voltage endurance coefficient (VEC) [24]; t_2 stands for the time to failure at voltage U_2; and t_1 stands for the time to failure at voltage U_1, typically the service voltage.[3] It must be pointed out here that it is a common practice to replace the voltage U in Eq. (6.1) with the mean electric field stress E_m [10].

[3] Note that in [10] t_2 is misleadingly defined as the "*time during which the test object is put on voltage* U_2" and t_1 as the "*time during which the test object is put on voltage* U_1" ([10], page 13). Similar misleading or even wrong definitions, often encountered throughout the empirical literature about electrical insulation (see, e.g. [3]), are focused only on the achievement of the third goal mentioned at Section 6.1.

Otherwise, the inverse power model can be expressed even more simply via the so-called *V-t* characteristic (in practice, a life model with one single stress applied, i.e., voltage), namely [3, 4]:

$$U^n t = \text{constant} \tag{6.2}$$

where the variables have the same meaning as above, that is, U is the applied voltage; t is the time to failure (note that in [3] t is rudely defined as "duration of voltage" [3], page 14); and n is the life exponent. Due to its mathematical expression [see Eqs. (6.1) and (6.2)], the IPM is plotted at best in $\log(U)$ versus $\log(t)$ coordinates, that is, in a log–log or "bilog" plot, where the life curve becomes a straight line.

The IPM is the model used traditionally in the determination of voltage levels for DC withstand voltage tests [3, 4]. In that case—and only in that case—can the time to failure be replaced simply with the "time during which the test object is put on voltage" ([10], page 13). Indeed, by postulating the validity of the IPM as the law that governs cable insulation aging, an equivalence is set between the factor $U^n t$ at rated voltage U_0 (then $U^n t = U_0^n t_0$, where t_0 is the design life of the cable) and the same factor at a much higher test voltage U_{DC} (then $U^n t = U_{DC}^n t_{DC}$, where t_{DC} now means time duration of DC test voltage, set by test protocols, standard recommendations, or laboratory constraints) [3–6].

For example, the DC withstand voltage performances of a prototype 250-kV class DC undersea cable was evaluated in [3] by applying a DC withstand voltage U_{DC} for a fixed time duration $t_{DC} = 3\,\text{h}$. U_{DC} was calculated as follows [3]:

$$U_{DC} = U_0 K_1 K_2 K_3 \tag{6.3}$$

where

- U_0 = rated "pole-to-ground" voltage (250 kV, relevant to the whole DC cable system)
- K_1 = deterioration coefficient, namely the value that must multiply U_0 for obtaining the voltage U_{DC} corresponding to a fixed time duration $t_{DC} = 3\,\text{h}$. A value of $K_1 = 3.54$ is calculated from *V-t* characteristics according to a life exponent $n = 9$ (see below)
- K_2 = temperature coefficient, assumed equal to 1.0
- K_3 = safety factor for uncertainties, assumed equal to 1.1

The value of K_3 is explained in [3] as a safety factor with respect to the withstand voltage level adopted for AC XLPE or DC fluid-filled cables, since the design parameters of the DC cables were not sufficiently known when those test were performed (around 1998), while the value of temperature coefficient $K_2 = 1.0$ is explained as adopted "because we evaluate performance at high temperature similar to real operation" ([3], page 10).

As emphasized above, in [3] the deterioration coefficient K_1 was determined based on the *V-t* characteristic of Eq. (6.2), as follows. From the above definitions of

U_0, t_0, and U_{DC}, t_{DC}, the following equalities hold:

$$U^n t = U_0^n t_0 = U_{DC}^n t_{DC} = \text{constant} \tag{6.4}$$

Then

$$\frac{U_{DC}^n}{U_0^n} = \frac{t_0}{t_{DC}} \Rightarrow U_{DC} = U_0 \left(\frac{t_0}{t_{DC}}\right)^{1/n} \tag{6.5}$$

Now, if $K_2 = 1.0$, by assuming for the sake of simplicity also $K_3 \sim 1.0$ (in fact, it is $K_3 = 1.1$), we can rewrite Eq. (6.3) as follows:

$$U_{DC} \sim U_0 K_1 \tag{6.6}$$

A comparison between (6.6) and (6.5) yields straightforwardly:

$$K_1 = \left(\frac{t_0}{t_{DC}}\right)^{1/n} \tag{6.7}$$

By setting in (6.7) $n = 9$ (a value derived conservatively in [3] from HVAC XLPE-insulated cables), $t_0 = 30$ years[4] (design life of the cable after [3]) and $t_{DC} = 3$ h, one obtains immediately

$$K_1 = \left(\frac{30 \times 8760}{3}\right)^{1/n} = 3.541 \quad QED \tag{6.8}$$

In general, the life exponent n for DC cables is considered to be greater than for AC cables [3,10]. In fact, in [3]—dating back to 1998—a value of $n \cong 15$ was estimated after long-term aging tests with load cycles. It must be pointed out that the value of n is, of course, strongly dependent on the type of compound, additives, and manufacturing procedures—which in turn are strongly dependent on the manufacturer. Therefore, such a high estimate has to be considered carefully since the latest improvements in the insulation of HVDC extruded cables might have implied different and peculiar values of n for the various cable designs.[5]

Following the same procedure, the IPM is used also by CIGRÉ Brochure 496 [5] in the derivation of DC voltage factors as test parameters to be employed in constant DC voltage tests at a voltage higher than design voltage U_0. In [5] the multiplication factors for the test voltages and periods are determined based on the same considerations about the V–t characteristics available in the literature. Although acknowledging that the precise nature of the V–t characteristic is not yet determined for DC operation, however, the WG B1-32 of CIGRÉ that wrote the

[4] It is important to state that the design life of a HVDC extruded cable is usually 40 years as recognized by CIGRÉ WG B1-32 [5].

[5] In [4]—dating back to 2004—a prototype (referred to as "subject") DC-XLPE was applied to develop, design and fabricate 500-kV DC XLPE cables and factory joints. A prequalification test [5] was then carried out to assess the 500-kV DC XLPE cable system, which was found successful in all tests. However, neither an updated estimate of n was reported in [4] nor a practical 500-kV HVDC cable system has been realized after the tests in [4] yet.

Table 6.1 IPM Applied to Determination of Test Factor for Load Cycle Tests at the Prequalification Stage and at the Type Test Stage

Parameter	Prequalification Test	Type Test
Design Life, t_0(years)	40	40
Test Duration, t_1 (days)	360	30
Deterioration Factor, k_1	$\sqrt[10]{40 \times 365/360}$	$\sqrt[10]{40 \times 365/30}$
Test factor	1.45	1.85

Source: After [5].

CIGRÉ Brochure 496 was of the opinion that the inverse power model provided a conservative basis for the work. The approach is exactly as in previous Eqs. (6.3)–(6.7), apart coefficients K_2 and K_3 that are omitted [5].

Using this approach it is possible to determine the test voltage factors that are equivalent to a prescribed system life when testing for a shorter time. WG B1-32 used a system life of 40 years. The approach requires knowledge of the life exponent n, which is determined empirically from endurance tests on cables. The knowledge of n under DC was not sufficient for the WG to identify a precise value, but they estimated a lower limit value of $n = 10$, which was used to determine the test voltage factor. As a consequence, Table 6.1 is derived, which shows how to calculate the deterioration coefficient K_1 and the relevant test factor (in practice, the numerical value of K_1), which should multiply the rated voltage U_0 for obtaining the constant absolute value of the DC voltage[6] to be applied to cables during load cycle tests as a function of the test duration, which is 360 days at the prequalification test stage, 30 days at the type test stage: Thus, the test factor turns out to be 1.45 and 1.85, respectively. On this basis the WG identified the DC voltage $U_{\text{TP1}} = 1.45U_0$ as equivalent to approximately 40 years operation at rated voltage when applied for 1 year, and the DC voltage $U_T = 1.85U_0$ as equivalent to approximately 40 years of operation at rated voltage when applied for 30 days [5, 25].

A better understanding of the traditional approach to life modeling can be gained from a deeper insight into the value of the stress coefficient K_1 selected in constant voltage DC and AC tests. Hence, starting from the above hypothesis that prequalification test voltage applied for 1 year should be equivalent to rated voltage applied over a lifetime of 40 years, let us:

Compare DC prequalification tests on HVDC extruded cables after CIGRÉ TB 496 [5] with AC prequalification tests on HVAC extruded cables after IEC 62067 [6], and the underlying IPMs.

Consider these underlying IPMs in terms of (mean) electric field instead of voltage.

[6] Note that the load cycle scheme for prequalification tests and (electrical) type tests according to [5] is quite complex. For VSC cables, [5] prescribes a certain polarity (positive or negative) of the DC voltage to be applied for selected subsequent periods at different load levels (LC = load cycle, HL = high load, ZL = zero load). However, the absolute value of DC voltage is constant throughout the cycle (different from LCC cables, which should be subjected to polarity reversals) and is, respectively, $U_T = 1.85U_0$ and $U_{\text{TP1}} = 1.45U_0$ for type tests and prequalification tests (see Section 5.9). Due to the complexity of test schemes, refer to [5] for details.

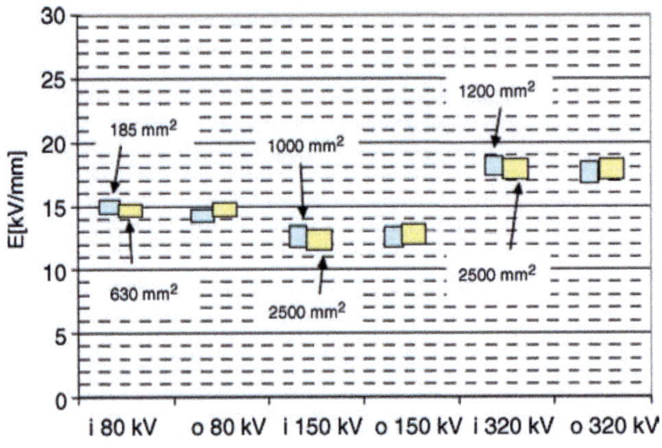

Figure 6.1 DC electric fields for 80-, 150-, and 320-kV HVDC cables: For every voltage, the DC fields both at the inner semicon (e.g., i80 kV) and the outer semicon (e.g., o80 kV) are given for a smaller and a larger conductor cross section. The fields have been calculated for a temperature drop ΔT across the insulation of 0 and of 10°C. Values of the temperature drop ΔT in between 0 and 10°C are located inside the boxes (courtesy of ABB).

First of all, one can easily show that, while the value of $K_1 = 1.45$ selected in DC prequalification tests after [5] comes from an IPM with voltage coefficient $n_{DC} = 10$ (see above), the value of $K_1 = 1.7$ selected in constant AC voltage tests after [6] comes from an IPM with voltage coefficient $n_{AC} = 7$ (both n_{DC} and n_{AC} are conservative according to the relevant standards and literature).

Further on, the IPMs relevant to the mean electric field in the cable can be derived trivially from the IPMs relevant to the rated voltage provided that insulation thickness is known, which means chosing a specific cable type and voltage rating. Starting from HVDC extruded cables, for the sake of illustration let us take a 320-kV class cable as a reference, whose insulation thickness is 18 mm according to [26]. This involves a mean electric field of ~18 kV/mm for this cable, which is in agreement with Figure 6.1 (after [10]); the figure also shows that the electric field in the cable is quite constant all over the insulation wall, as well as for different temperature drops (hence different load conditions of the cable). Thus the adoption of an IPM relevant to the mean electric field—rather than to the maximum electric field, whose value and location vary with load conditions—is more than justified for this and for other HVDC extruded cables.[7]

As to HVAC extruded cables, let us take a $\sqrt{3} \times 320\,\text{kV} \sim 500\,\text{kV}$ (or equivalent) class cable as a reference in order to make the comparison with the HVDC cable consistent.[8] For a 500-kV HVAC cable, an insulation thickness of 25 mm is reported in [27,28], that implies a mean electric field of 12.8 kV/mm. Therefore, an

[7] It must be pointed out that in an AC cable the field distribution is Laplacian at all loads (see Chapter 3), thus the field might be more nonuniform than in a DC cable.

[8] The pole-to-ground voltage of the 320-kV HVDC cable is the same as the phase to ground of the 500-kV HVAC cable, if one takes $U_m = 550$ kV for the HVAC cable [6].

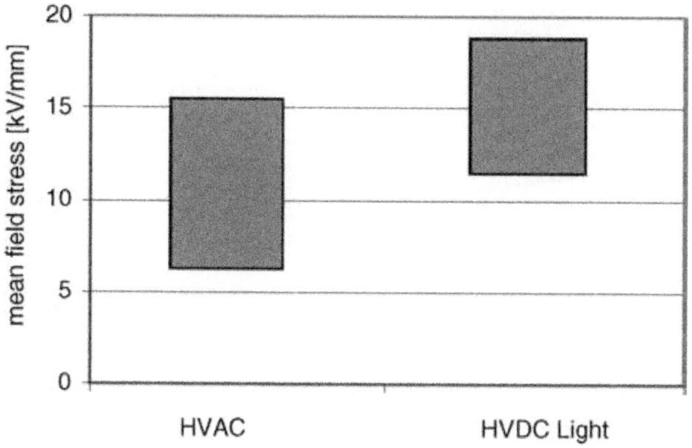

Figure 6.2 AC and DC cable comparison as to the mean field stress (courtesy of ABB).

insulation thickness of 25 mm and a mean HVAC field of ~13 kV/mm will be hypothesized for the 500-kV HVAC cable.

It must be pointed out that the mean electric fields assumed here for the reference HVDC and HVAC extruded cables are consistent with those reported in Figure 6.2 after [10]. Indeed, the figure shows that the upper bound of the HVDC cable mean field stress is at ~19 kV/mm, very close to the value of 18 kV/mm assumed for the reference HVDC cable. Hence, a mean field of ~14 kV/mm (also close to the ~15-kV/mm upper bound of HVAC cable mean field stress) should be reasonable for the 500-kV HVAC cable. However, according to both the experience of the authors and to the wider spread of HVAC versus HVDC mean fields in Figure 6.2, the value of ~13 kV/mm assumed for the reference HVAC cable seems more appropriate.

The IPM of mean electric field versus life in years (with a design life of 40 years; see above) for the 320-kV HVDC and the 500-kV HVAC cable based on the previous hypotheses are shown in Figure 6.3, together with the following constant values of mean electric field:

1. $E_{m,\mathrm{DC}}$, that is, the design mean electric field in the DC cable
2. $E_{m,\mathrm{AC}}$, that is, the design mean electric field in the AC cable
3. $E_{m,\mathrm{DC},1.45}$, that is, the mean electric field in the DC cable at a constant voltage $1.45U_0$ (DC voltage in prequalification test for VSC cables after [5])
4. $E_{m,\mathrm{AC},1.70}$, that is, the mean electric field in the AC cable at a constant root-mean-square (rms) voltage of $1.70U_0$ (AC rms voltage in prequalification test after [6]).

It can be argued that the mean electric field is significantly higher in the DC cable during the test at $1.45U_0$ than in the AC cable during the test at $1.7U_0$, since $E_{m,\mathrm{DC},1.45} = 25.8$ kV/mm and $E_{m,\mathrm{AC},1.70} = 21.8$ kV/mm, respectively.

Of course, the above life estimations have to be regarded with care. Indeed, some authors claim that the electrical aging of insulating materials is prohibited

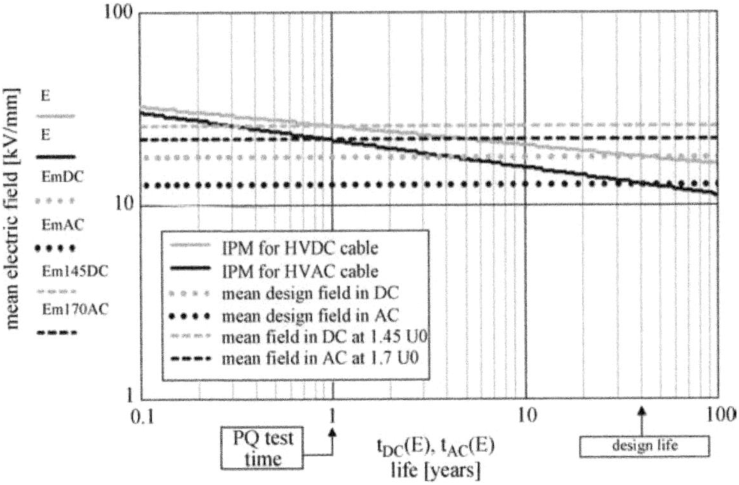

Figure 6.3 IPM of mean electric field versus life in years for the 320-kV HVDC and the 500-kV HVAC cable, together with $E_{m,\mathrm{DC}}$ (design mean field in the DC cable), $E_{m,\mathrm{AC}}$ (design mean field in the AC cable), $E_{m,\mathrm{DC},1.45}$ (mean field in the DC cable at a constant voltage $1.45U_0$), $E_{m,\mathrm{AC},1.70}$ (mean electric field in the AC cable at a constant rms voltage $1.70U_0$).

under a certain stress level, known as the aging electrical threshold (see, e.g., [2,10,29,30]). For (XL)PE this threshold is said in [10,29,30] to be ~10 kV/mm, although it must be pointed out that the threshold is strongly dependent on temperature (see Fig. 6.4 after [30], relevant to XLPE-insulated minicables tested under AC voltage), and it must be checked—depending on the chosen cable

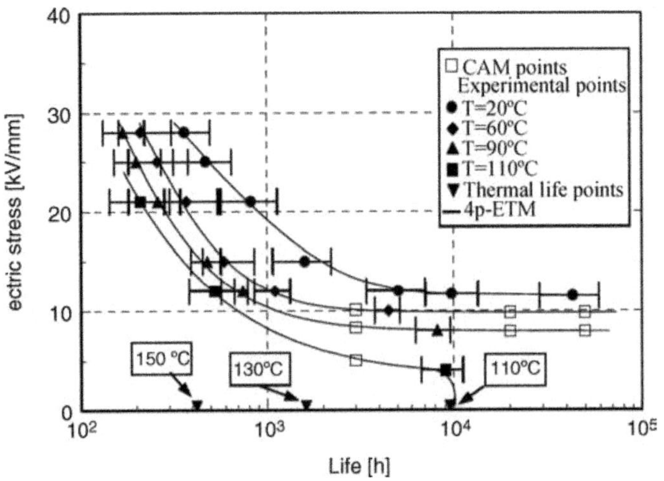

Figure 6.4 Electrical life lines at different temperatures under AC voltage, derived fitting experimental points at 50% failure probability (with confidence intervals at probability 95%). XLPE cable models. Semilog plot. Electrical thresholds, strongly dependent on temperature, are clearly observed, which disappear above 90°C (after [30]).

insulation—if it tends to disappear as temperature approaches the operating temperature (90°C for extruded HVAC cables, 70°C for extruded HVDC cables; see previous chapters). Since DC prequalification tests after CIGRE 496 are performed under load cycling, neglecting a possible threshold might be conservative in any case.

Some modified versions of Eq. (6.2) have been proposed that cater to a possible electrical aging threshold. For example [31],

$$(U - U_T)^n t = \text{constant} \tag{6.9}$$

where U_T is the voltage threshold below which no electrical aging takes place.

Threshold or not, it can also be argued that different aging regimes exist at different ranges of applied stresses. A higher rate of aging is expected at higher stresses with respect to lower stresses, and this may imply different values of n. The theory that the higher the electric stress, the lower the voltage coefficient is supported in [10], by commenting on Figure 6.5 as follows:

> Imagine that the test object does not break down after time t_1. One can then motivate with the IPM that the object probably would last at least during a time t_2 at service stress E_2. It would probably hold much longer, e.g. at least t_2'. For this reason the estimated time t_2 appears conservative using this method, under the above hypothesis of decreasing n as voltage (field) increases. This is exactly the reasoning that the CIGRÉ Recommendations in Brochure 496 have used. This CIGRÉ document recommends the use of an aging factor n of 10 which tends to be a conservative estimation ([10], page 13).

In [10] the theory "the higher the electric stress, the lower the voltage coefficient" is supported by showing aging test results of XLPE plate samples stressed at 80°C and very high electric fields, resulting in an estimated voltage coefficient $n \sim 10$, consistently with [5], although in [10] it is argued that the aging factor is likely to be lower at lower temperature and electric stress. Thus the voltage coefficient likely to be higher. Provided that is true (and, of course, this should be demonstrated case by case—i.e., insulation by insulation—on experimental grounds), the use of the simple IPM that

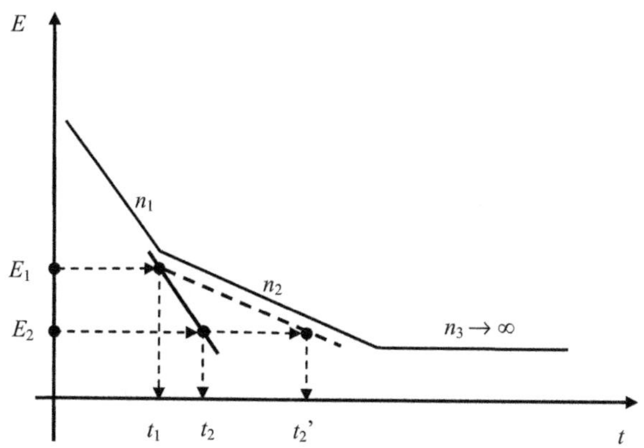

Figure 6.5 Inverse power model and aging curves (reprocessed from [10]).

takes into account neither thresholds nor different values of n would then be automatically conservative when used correctly.

However, it is a matter of fact that different insulating materials exhibit different behaviors as to the voltage coefficient. For instance in Figure 6.6, electrical life lines at different temperatures derived by graphically fitting experimental points at 50% failure probability relevant to EPR cable models are displayed in a log–log plot. A sharp downward curvature is exhibited by life lines in the Figure 6.6, thereby suggesting that in this case the voltage coefficient increases with the electric stress.

It is true that EPR is not used as insulation for HVDC cables; however, as to extruded HVDC insulation—though being essentially based on the same polymeric material, that is, PE—strong differences are encountered among the different manufacturers and the various R&D efforts, as highlighted in Chapter 5. For this reason, a behavior as shown in Figure 6.6 cannot be ruled out a priori and one should carefully check to be on the safe side when performing life estimates or test voltage level selection on the basis of a given life model (be it of the IPM kind or not).

Finally, it is worthwhile highlighting that Figure 6.6 clearly proves that "increasing aging rate" does not necessarily mean "decreasing voltage coefficient," as tentatively concluded in [10]: indeed, aging is surely increased (it would be better to say "accelerated") when raising the temperature at a constant electric field (and in this case n decreases), but aging is surely decreased when lowering the electric stress at a given temperature, nevertheless the curves at temperatures >60°C do show a decrease in the voltage coefficient as the electric stress is lowered at constant temperature.

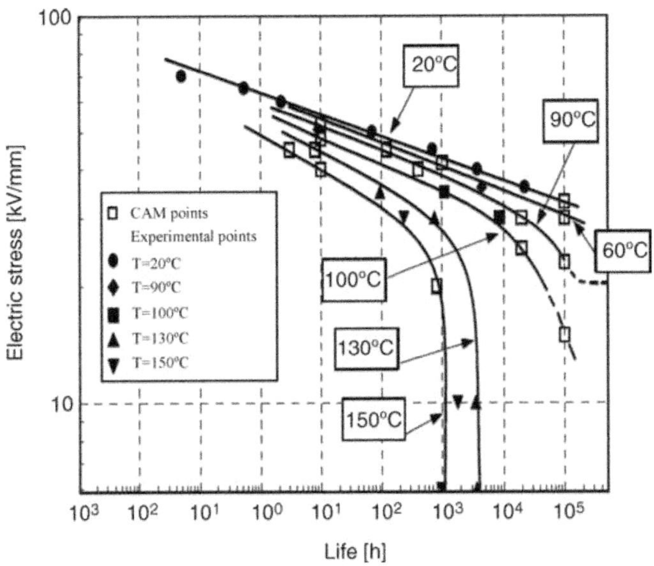

Figure 6.6 Electrical life lines at different temperatures under AC voltage, derived by graphically fitting experimental times to failure at 50% failure probability relevant to EPR cable models (after [30]).

In conclusion, one must be very careful in estimating the trend of n as the electric stress level is varied (diminished, in particular), and such an estimate must be based on solid experimental grounds and on a sound understanding of the aging processes that are active at the various stress levels.

6.1.2 Probabilistic Framework of HVDC Extruded Insulation Life Modeling

6.1.2.1 Basic Aspects of Probabilistic Life Modeling As can be argued from Section 6.1.1, the "practical" traditional approach to insulation life modeling misses completely a fundamental aspect of life modeling, which is not only quite obvious but also strictly connected to the statistical processing of ALT data (see Figs. 6.4 and 6.6) and discussed for a long time (see, e.g. [29, 30, 32–34]): the *life or time to failure of an insulation*, or of a component of which the insulation is a part, *is a random variable*, due to the intrinsic randomness of the breakdown processes. This holds for electrical insulation in general and a fortiori in an inherently nonhomogeneous material such as extruded insulation.

Thus, *time to failure* is *always associated with the relevant failure probability*, that is, the probability of failing under the action of applied stresses, or, conversely, *with the corresponding reliability*, that is, the probability of withstanding the applied stresses and surviving. In fact, aging and failure processes are regulated by stochastic laws, as demonstrated by the fact that identical specimens manufactured with the same procedure and material, subjected to the same levels of stresses, exhibit different failure times, because of the intrinsic inhomogeneities of the materials, the uncertainties in the manufacturing processes, the imperfect control of the test conditions, and so on. These concepts are not emphasized in the literature on cable design and testing, though some articles make extensive use of life modeling (see, e.g., [10, 27, 28]).

In the particular case of electrical breakdown of solid insulation subjected to applied voltage, a huge series of experimental tests have been conducted over many decades. The results of such tests, combined with sound mathematical reasons related to the "extreme value" theory and the advances in physical knowledge of the mechanisms of wear and stresses acting on insulation, have shown that the Weibull distribution is the best probability distribution for reproducing the relationship between failure probability and times to failure at a constant value of applied voltage (or electric field), as well as between failure probability and breakdown voltage (or dielectric strength) under ramp-voltage tests [31, 35, 36].

Under the Weibull hypothesis, the relationship between failure probability and life at a constant value of applied voltage (or electric field) is the so-called Weibull distribution of times to failure (or times to breakdown, or lives), which relates the cumulative failure probability, F, to time to failure, t_F, and can be expressed as follows [1, 33, 34, 36]:

$$F(t_F) = 1 - \exp\left[-\left(\frac{t_{\rm f}}{\alpha_{\rm t}}\right)^{\beta_t}\right] \tag{6.10}$$

where α_t is the scale parameter and β_t is the shape parameter of the distribution, and time to failure is indicated as t_F in order to emphasize that it is the generic 100 Fth percentile of failure time. β_t is linked to the dispersion of the times to failures, while α_t coincides with the breakdown time at 63.2% failure probability, that is, with the 63.2th percentile of time to failure. It must be highlighted that both parameters are functions of the stresses applied to the insulation, though the dependence of β_t on stresses is usually weaker and is mostly if not always neglected in practice [2,24]. Hence, by indicating with $S_1, \ldots, S_N$ the values of the N stresses applied to the insulation (assumed as constant with time), α_t can be written as $\alpha_t = \alpha_t(S_1, \ldots, S_N)$. As a consequence, Eq. (6.10) can be rewritten as follows:

$$F(t_F; S_1, \ldots, S_N) = 1 - \exp\left\{-\left[\frac{t_F}{\alpha_t(S_1, \ldots, S_N)}\right]^{\beta_t}\right\} \tag{6.11}$$

From (6.11), by virtue of the meaning of α_t and of t_F, the so-called probabilistic life model of the considered insulating system can be derived, namely a relationship between life, stress levels, and failure probability (or, conversely, reliability) [37]. In order to do that, Eq. (6.11) should be expressed as a function of t_F:

$$t_F(S_1, \ldots, S_N) = [-\ln(1-F)]^{1/\beta_t} \alpha_t(S_1, \ldots, S_N) \tag{6.12}$$

which is the expression of the probabilistic life model implicit in terms of the stresses. Notice that it enables the derivation, for any value of stresses $S_1, \ldots, S_N$, of the relevant 100 Fth percentile of breakdown time, $t_F(S_1, \ldots, S_N)$.

Note that, for its statistical significance, the scale parameter $\alpha_t = \alpha_t(S_1, \ldots, S_N)$ of the Weibull distribution is commonly chosen as the reference percentile of the distribution of failure times coming from breakdown tests on electrical insulation. Any other failure time percentile can thus be derived from $\alpha_t = \alpha_t(S_1, \ldots, S_N)$ and β_t resorting to Eq. (6.12).

Probabilistic life models are the fundamental tool for carrying out a reliability analysis on the basis of laboratory test results, before the component is put in service. Indeed, under the hypothesis that stress levels are constant and fixed, probabilistic life models enable the estimation of key reliability parameters and functions, such as the mean time to failure (MTTF), the hazard function, and the reliability function of the insulation, thus of the electrical device of which the insulation is the weakest part (e.g., a power cable). In fact, from Eq. (6.11) reliability at mission time t_F can be evaluated trivially via the following reliability function $R(t_F; S_1, \ldots, S_N)$:

$$R(t_F; S_1, \ldots, S_N) = 1 - F(t_F; S_1, \ldots, S_N) = \exp\left\{-\left[\frac{t_F}{\alpha_t(S_1, \ldots, S_N)}\right]^{\beta_t}\right\} \tag{6.13}$$

Thus, failure rate at the same time t_F can be estimated via the following hazard function $h(t_F; S_1, \ldots, S_N)$ [33, 34]:

$$h(t_F; S_1, \ldots, S_N) = \frac{\beta_t}{\alpha_t(S_1, \ldots, S_N)}\left[\frac{t_F}{\alpha_t(S_1, \ldots, S_N)}\right]^{\beta_t - 1} = \frac{\beta_t {t_F}^{\beta_t - 1}}{[\alpha_t(S_1, \ldots, S_N)]^{\beta_t}} \tag{6.14}$$

Actually, Eqs. (6.11)–(6.14) are not fully explained versus applied stresses; thus the functional dependence of the relevant reliability model is not fully assessed until the functional dependence of α_t on $S_1, \ldots, S_N$ is not fully assessed as well. This is needed for the estimation of reliability and related quantities. Such functional dependence can be explained provided that the life model holding for the considered insulation (or component) has been singled out.

6.1.2.2 Probabilistic Life Models for HVDC Extruded Cable Insulation

Life modeling of power cable insulation in HVDC regime has the goal of determining the most appropriate mathematical relationship (model) between the time-to-failure (life) of HVDC cable insulation and the levels of the various stresses applied to such insulation [2, 24, 29]. Referring to the 63.2% failure probability a life model in its most general form can be expressed as follows:

$$\alpha_t(S_1,\ldots,S_N) = f(S_1,\ldots,S_N;p_1,p_2,\ldots,p_M) \tag{6.15}$$

where $p_1, \ldots, p_M$ are the model parameters and $f(S_1, \ldots, S_N; p_1, \ldots, p_M)$ is a proper mathematical function of model parameters and applied stresses. Moreover, the functional dependence of the model on applied stresses and model parameters is generally such that relationship (6.15) can be recast trivially in the following form:

$$\begin{aligned}\alpha_t(S_1,\ldots,S_N) &= f(S_{1,0},\ldots,S_{N,0};p_1,\ldots,p_M)\frac{f(S_1,\ldots,S_N;p_1,\ldots,p_M)}{f(S_{1,0},\ldots,S_{N,0};p_1,\ldots,p_M)}\\ &= \alpha_t(S_{1,0},\ldots,S_{N,0})f'(S_1,\ldots,S_N;p_2,\ldots,p_M)\\ &= \alpha_0 f'(S_1,\ldots,S_N;p_2,\ldots,p_M)\end{aligned} \tag{6.16}$$

where $\alpha_0 = \alpha_t(S_{1,0}, \ldots, S_{N,0}) = p_1$ is the 63.2th failure time percentile at reference values of stresses $S_{1,0}, \ldots, S_{N,0}$ and is sometimes referred to as "scale parameter" of the life model [11, 32]; $f'(S_1, \ldots, S_N; p_2, \ldots, p_M) = f(S_1, \ldots, S_N; p_1, \ldots, p_M)/f(S_{1,0}, \ldots, S_{N,0}; p_1, \ldots, p_M)$ is a dimensionless function that encompasses the whole dependence of the model on applied stresses.

Although the 63.2th percentile is usually chosen as the reference one of the failure time distribution obtained from accelerated life tests on electrical insulation [33, 34], other life percentiles than the 63.2th could be considered, for example, the 50th at the processing stage of experimental results coming from laboratory tests on specimens, or extreme percentiles such as the 5th and the 95th for deriving confidence bounds, or again extreme low percentiles such as the 1th and the 5th at the design stage of the power cable [see Eq. (6.12)] [2, 24, 29]. Hence, when a life percentile other than the 63.2th is chosen, from (6.12) Eq. (6.16) becomes

$$t_F(S_1,\ldots,S_N) = [-\ln(1-F)]^{1/\beta_t}\alpha_0 f'(S_1,\ldots,S_N;p_2,\ldots,p_M) \tag{6.17}$$

As can be argued from (6.15)–(6.16), a life model valid for HVDC cable insulation provides life estimates for that insulation at selected levels of the applied stresses on condition that the values of the model parameters are known.

Life models for cable insulation are commonly used first of all for characterizing and comparing the endurance properties of various materials candidates for the realization of the cable insulation; this was pointed out in Section 6.1, highlighting that for a given insulating material, the life model parameters are usually derived via laboratory tests performed on small-size specimens (typically flat, Rogowski-like, minicables or prototype cables), thereby achieving considerable time and cost savings with respect to tests on full-size insulation systems. However, it was also pointed out that the second goal of cable insulation life models is inferring the service life of the power cable, but this requires an extrapolation of test results and relevant model parameter values to the full-size cable, thereby introducing a further degree of uncertainty in power cable life estimation.

The extrapolation can be performed, for example, via the statistical "enlargement law." The enlargement law is a relationship that expresses the enlargement in terms of the scale parameter of the Weibull distribution with respect to volume. By denoting the "large" cable as cable 1, and the small cable as cable 2, in the case of a HVAC cable—where the electric field distribution within insulation thickness is always of the capacitive or "Laplacian" type (see Section 3.1.3.1)—the enlargement law can be written as follows [7, 8, 36, 38–40]:

$$\frac{\alpha_{E,1}}{\alpha_{E,2}} = \left(\frac{l_2}{l_1}\right)^{1/\beta_E} \left(\frac{r_{i2}}{r_{i1}}\right)^{2/\beta_E} H_{\mathrm{AC}}(r_{i1}, r_{i2}, r_{o1}, r_{o2}) \tag{6.18.a}$$

$$H_{\mathrm{AC}}(r_{i1}, r_{i2}, r_{o1}, r_{o2}) = \left[\frac{(r_{o2}/r_{i2})^{2-\beta_E} - 1}{(r_{o1}/r_{i1})^{2-\beta_E} - 1}\right]^{1/\beta_E} \tag{6.18.b}$$

where

$\alpha_{E,1}$ = scale parameter (63.2% failure probability percentile) of the Weibull distribution of the AC or impulse dielectric strength for the "small" cable, or alternatively of the Weibull distribution of the constant electric field corresponding to a fixed time to failure for the "small" cable

$\alpha_{E,2}$ = scale parameter of the Weibull distribution of the AC or impulse dielectric strength for the large cable, or alternatively of the Weibull distribution of the constant electric field corresponding to a fixed time to failure for the large cable

β_E = shape parameter of the dielectric strength distributions for both minicables and full-size cables (β_E is not affected by the scaling process [7])

l_1, r_{i1}, r_{o1} = respectively, length and inner and outer insulation layer radii of the small cable

l_2, r_{i2}, r_{o2} = respectively, length and inner and outer insulation layer radii of a large cable.

Note that the expression of H_{AC} holds for $\beta_E \neq 2$.

Once the Weibull distribution parameters relevant to electric breakdown tests[9] performed on cable models have been estimated, the enlargement law enables the evaluation of the Weibull scale parameter of AC or impulse dielectric strength for a real (full-size) cable, manufactured using the same materials and the same manufacturing technologies as the minicables or prototype cables.

By the same formula, once the constant electric field corresponding to a fixed time to failure has been estimated according to a certain life model (e.g., the IPM) valid for minicables, the constant electric field corresponding to the same fixed time to failure (e.g., design life) of a large full-size cable, manufactured as the small minicables, can be evaluated as well [29, 38].

After the evaluation of $\alpha_{E,2}$ via Eq. (6.18), the 63.2th percentile of the breakdown voltage for full-size cables, V_2, can be estimated using the following equation:

$$V_2 = \alpha_{E,2} r_{i2} \ln \frac{r_{e2}}{r_{i2}} \tag{6.19}$$

The same was shown to be valid for impulsive stress in [41], where experimental data from 1.2/50-μs impulse tests carried out on EPR-insulated minicables were plotted in $\log(E_i)$ versus $\log(\alpha_n)$ coordinates—E_i being the maximum impulsive electric field and α_n the number of impulses that correspond to a 63.2% breakdown probability—giving rise to a straight line. Tests performed with specimens at both room temperature and 90°C well fit the IPM in the following form:

$$\alpha_n = k_E E^{-n_{imp}} \tag{6.20}$$

where k_E is a constant, depending on the insulating material and the geometry of the specimens, and n_{imp} the life exponent, depending on the insulating material only. Figure 6.7 reports the experimental data after [41] in which the life exponent n_{imp} is equal to 20.2 and 11.6 when the insulation temperature is room and 90°C, respectively.

As mentioned above, the enlargement law expressed by Eq. (6.18) is based on the "Laplacian" or "capacitive" electric field distribution within the insulation wall of an AC coaxial cable. A problem arises in the application of this formula to HVDC cables, since under constant DC voltage the field distribution within the dielectric is of the "Poisson" or "resistive" type and depends strongly on temperature drop across the insulation, thus on load conditions, as broadly discussed throughout this book and particularly in Section 3.1.3.2. Hence, if one is interested in extrapolating the constant DC voltage or electric field corresponding to a fixed time to failure from a small to a large HVDC cable, Eq. (6.18) should be properly modified via a rigorous theoretical treatment.

An innovative enlargement law for HVDC cables was proposed for the first time in [42,43], by using a procedure similar to that employed in [35,38] for AC cables. The starting point is the resistive field distribution holding under constant DC

[9] Here electric breakdown tests include ramped AC or DC voltage, constant AC or DC voltage, or impulse voltage tests.

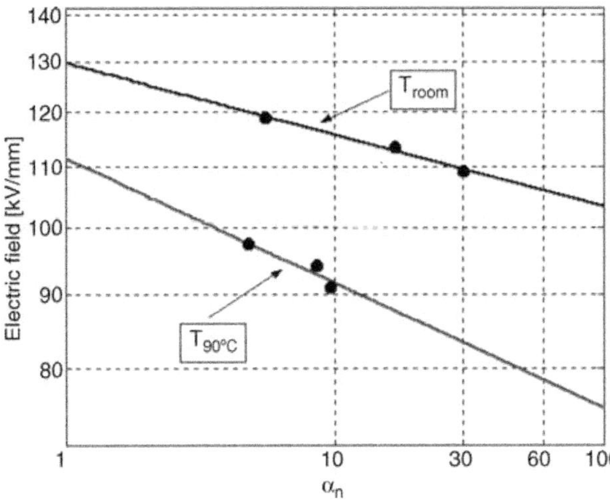

Figure 6.7 Electrical lifetime of an EPR compound under impulsive (1.2/50 μs—positive polarity) voltage stress at room temperature and 90°C (after [41]).

voltage [see Eq. (3.29) in Section 3.1.3.2]. After several manipulations reported in detail in [42], the enlargement law for HVDC cables can be obtained, namely:

$$\frac{\alpha_{E,1}}{\alpha_{E,2}} = \left(\frac{l_2}{l_1}\right)^{1/\beta_E} \left(\frac{r_{i2}}{r_{i1}}\right)^{2/\beta_E} H_{\mathrm{DC}}(r_{i1}, r_{i2}, r_{o1}, r_{o2}, V_{\mathrm{DC},1}, V_{\mathrm{DC},2}, W_{C1}, W_{C2}, \beta_E, a, b, \lambda_{T,d}) \tag{6.21}$$

$$H_{\mathrm{DC}}(r_{i1}, r_{i2}, r_{o1}, r_{o2}, V_{\mathrm{DC},1}, V_{\mathrm{DC},2}, W_{C1}, W_{C2}, \beta_E, a, b, \lambda_{T,d}) = \left[\frac{(r_{o2}/r_{i2})^{2-\eta_2} - 1}{(r_{o1}/r_{i1})^{2-\eta_1} - 1}\right]^{1/\beta_E} \tag{6.22}$$

where, consistent with Section 3.1.3.2, $V_{\mathrm{DC},1}$ = DC voltage applied to the small cable, $V_{\mathrm{DC},2}$ = DC voltage applied to the large cable, W_{C1} = conductor losses (W/m) in the small cable, W_{C2} = conductor losses (W/m) in the large cable, $\lambda_{T,d}$ = thermal conductivity of insulation (W m^{-1} K^{-1}), a = temperature coefficient of electrical resistivity (K^{-1} or °C^{-1}), b = electrical stress coefficient of electrical resistivity (m/MV or mm/kV). η_j (j being either 1 or 2) has the following expression:

$$\eta_j = (1 - \delta_j)\beta_E \tag{6.23}$$

with δ_j given by

$$\delta_j = \left[\frac{aW_{Cj}}{2\pi\lambda_{T,d}} + \frac{bV_{\mathrm{DC}j}}{(r_{oj} - r_{ij})}\right] \Big/ \left[\frac{bV_{\mathrm{DC}j}}{(r_{oj} - r_{ij})} + 1\right] \tag{6.24}$$

Since in the DC case the maximum electric field in the insulation wall depends on the load, the 63.2th percentile of dielectric strength on the conductor is not so interesting as in the AC case. Consequently, the 63.2th percentile of DC breakdown voltage ($V_{\mathrm{DC},1}$ for the cable in the reduced state, $V_{\mathrm{DC},2}$ for the cable in the enlarged state) is of much greater significance and the following relationship should be considered instead of (6.21):

$$\frac{V_{\mathrm{DC},1}}{V_{\mathrm{DC},2}} = \frac{\alpha_{E,1}}{\alpha_{E,2}} \cdot \frac{k_2}{k_1} \cdot \frac{\delta_2}{\delta_1} \tag{6.25}$$

where k_j (j being either 1 or 2) is given by

$$k_j = \frac{(r_{ij}/r_{oj})^{\delta_j - 1}}{r_{oj}[1 - (r_{ij}/r_{oj})^{\delta_j}]} \tag{6.26}$$

An alternative, but even more cumbersome approach to the enlargement law was followed in [44] for HVDC cables in conjunction with the application of the so-called Dissado–Mazzanti–Montanari (DMM) model [20,21] to power cables and is omitted here for the sake of brevity (interested readers should refer to [44] for details).

However, in [29] a simpler—though approximate—approach is proposed that consists in neglecting the dependence of the enlargement formula on insulation thickness (which involves the dependence on field distribution, thus on inner and outer radius of model and power cables), thereby applying it only to cable length (in [32] a "dual" or complementary approximate approach is followed since the dependence of the enlargement formula on cable length is neglected, focusing only on insulation thickness). Of course this involves negligible errors only if the insulation thickness and radii of cable models are not too small with respect to power cable. By doing so, the enlargement formula can be rewritten, dropping the term associated with the radii, as follows:

$$\frac{\alpha_{E,1}}{\alpha_{E,2}} \cong \left(\frac{l_2}{l_1}\right)^{1/\beta_E} = N_l^{1/\beta_E} \tag{6.27}$$

where N_l is the so-called enlargement factor, relevant to enlargement over the length only.

As a general rule, it can be argued that if insulation thickness is neglected, the enlargement is governed by cable length only. Then, also the enlargement law for HVAC cables, that is, Eq. (6.18), reduces to (6.27), which becomes the unique enlargement law holding for both DC and AC cables, irrespectively, whether it refers to constant, ramped, or impulse voltage tests.

Under this hypothesis, a relationship complementary to (6.27), namely relevant to the 63.2th percentiles of time to failure of the small (i.e., short) and big (i.e., long) cable, can be obtained. Indeed, if the role of insulation thickness in the volume enlargement is negligible, the extruded insulation of a long cable subjected to a fixed field/voltage level behaves as a chain made of N_l series-linked rings, that is, short cables: The chain fails as soon as one of its N_l rings fails. If $R_{\text{chain}}(t_F)$ and $R_i(t_F)$ are the reliabilities of the chain and of the ith ring at time t_F, respectively, by assuming that all rings in the chain have the same reliability, $R(t_F)$, it is derived straightforwardly:

$$R_{\text{chain}}(t_F) = \prod_{i=1}^{N_l} R_i(t_F) = [R(t_F)]^{N_l} \tag{6.28}$$

Hence, the cable reliability at time t_F will be smaller than that of every single ring. When rewriting relationship (6.28) according to the Weibull hypothesis for the rings, that is, the short cables [see Eq. (6.10)], the following expression for the reliability of the whole chain, that is, the long cable, is attained:

$$R_{\text{chain}}(t_F) = \prod_{i=1}^{N_l} \exp\left[-\left(\frac{t_F}{\alpha_{t,1}}\right)^{\beta_t}\right] = \exp\left[-\left(\frac{t_F}{\alpha_{t,1}/N_l^{1/\beta_t}}\right)^{\beta_t}\right] = \exp\left[-\left(\frac{t_F}{\alpha_{t,2}}\right)^{\beta_t}\right] \tag{6.29}$$

where $\alpha_{t,1}$ is the scale parameter of the Weibull distribution of times to failure for the short cables, $\alpha_{t,2}$ is the scale parameter of the Weibull distribution of times to failure for the long cable, and β_t is the shape parameter of the Weibull distribution of times to failure for both the short and the long cable (β_t is not affected by the scaling process like β_E [7]).

Thus, it can be concluded that it holds:

$$\frac{\alpha_{t,1}}{\alpha_{t,2}} = N_l^{1/\beta_t} \cong \left(\frac{l_2}{l_1}\right)^{1/\beta_t} \tag{6.30}$$

By applying Eq. (6.30) to Eq. (6.12), one obtains the relationship between $t_{F,2}(S_1, \ldots, S_N)$—full-size cable life at applied stresses $S_1, \ldots, S_N$ and failure probability F—and $\alpha_{t,1}(S_1, \ldots, S_N)$—short cable life at the same values of stresses, but at failure probability 63.2%, that is usually the reference probability for processing test results relevant to extruded insulation cables—namely:

$$t_{F,2}(S_1, \ldots, S_N) = \left[\frac{-\ln(1-F)}{N_l}\right]^{1/\beta_t} \alpha_{t,1}(S_1, \ldots, S_N) \tag{6.31}$$

Let us consider now the design values of applied stresses, $S_{1,0}, \ldots, S_{N,0}$, and failure probability, P_0[10]: From (6.31), the relevant expression of full-size cable life, $t_{P_0,2}$ $(S_{1,0}, \ldots, S_{N,0})$, versus $\alpha_{t,1}(S_{1,0}, \ldots, S_{N,0})$—short cable life at the same values of stresses, but at failure probability 63.2%—can be achieved, namely:

$$t_{P_0,2}(S_{1,0}, \ldots, S_{N,0}) = \left[-\frac{\ln(1-P_0)}{N_l}\right]^{1/\beta_t} \alpha_{t,1}(S_{1,0}, \ldots, S_{N,0}) \tag{6.32}$$

[10] Design failure probability P_0 is much lower than 63.2%. Typically, it ranges from 1 to 5%; see above.

By combining Eq. (6.32) with (6.12) written for a generic probability P_T, an even more general relationship can be found between $t_{P_0,2}$ ($S_{1,0}, \ldots, S_{N,0}$) and $t_{P_T,1}$ ($S_{1,0}, \ldots, S_{N,0}$)—short cable life at the same values of $S_{1,0}, \ldots, S_{N,0}$, but at a failure probability P_T different from 63.2%—namely [45]:

$$t_{P_0,2}(S_{1,0},\ldots,S_{N,0}) = \left\{\frac{1}{N_l}\frac{[-\ln(1-P_0)]}{[-\ln(1-P_T)]}\right\}^{1/\beta_t} t_{P_T,1}(S_{1,0},\ldots,S_{N,0}) \qquad (6.33)$$

Remember that the applicability of the enlargement process from minicable to full-size cables instead of short cables (of the same cross section of full-size cables) shall be extended carefully since the largest is the radial enlargement, the greatest shall be the care with the application of (6.33).

During the last three decades, the understanding of aging mechanisms for cable insulating materials has grown continuously, leading to significant achievements. As hinted at the beginning of the present Section 6.1, such achievements go from the development of "phenomenological" life models—able to fit failure time data for various stresses (singly or simultaneously applied) and useful for deriving parameters for material evaluation (see, e.g., [29])—to "physical" models—which describe different physical-chemical mechanisms responsible for insulation degradation under different types and/or ranges of applied stress (see, e.g., [13, 20–23] for extruded HVDC cable insulation). All this information provides considerable help at the design stage of a full-size cable. The introduction of normative references in IEC and IEEE publications (see, e.g., [46,47]) regarding aging under thermal, electrical, mechanical, environmental, and multiple stresses support the maturity and the progresses obtained on this topic, especially as far as polymeric insulation for HVAC cables is concerned, whereas it must be acknowledged that as far as polymeric insulation for HVDC cables is concerned the modelistic achievements are much less well established. However, most of the newly acquired knowledge is associated with the diffusion of polymeric cable insulation and the requirement of increasing design stresses, in order to reach more compact devices and reduce costs without affecting insulation system reliability, as required by the deregulated electricity market.

In what remains of Section 6.1, a few fundamental life models of the phenomenological type available in the literature and that can be employed for time-to-failure estimation of HVDC cable insulation are presented in Section 6.1.2.3. For the sake of brevity, the main focus is on electrical and thermal stresses, singly or simultaneously applied—in the latter case giving rise to the so-called thermoelectrical or electrothermal stress—these being the stresses that mostly age and cause failure of HVDC cable insulation. However, some hints at electro-thermo-mechanical stress are also given. Indeed, often cables are subjected to mechanical stress superimposed on electrical and thermal stresses during their service life: for example, cables affected by the mechanical strain associated with thermal expansion/contraction during load cycles; underground cables laid close to a motorway route and exposed to mechanical vibrations associated with the vehicles (particularly heavy trucks); or undersea cables exposed to waves and tides in shallow water, to sea currents in deep water [31].

When such models are inserted in the probabilistic framework outlined at Section 6.1.2.1, reliability can be evaluated first of all in aprioristic terms with respect to real operating conditions of insulating materials and systems—which usually include time-varying stresses—that is, on the basis of "rated" or "design" stress levels, which are assumed as constant. This kind of indirect reliability evaluation, illustrated in what follows, provides fundamental indications for the design of the insulating systems of power cables.

6.1.2.3 Life Models for HVDC Extruded Cables under Single and Combined Stress The problem of insulation life modeling under the combination of electrical and thermal stress (i.e., when both voltage and temperature are applied), referred to as electrothermal stress, was solved for HVAC cables by combining two single-stress life models that hold, respectively, when only either temperature or voltage is applied [29,48]. This approach is valid also for extruded HVDC cable insulation subjected to electrothermal stress. It is, therefore, convenient to review briefly these single-stress life models.

Single-Stress Thermal Life Model When only a constant temperature (thermal stress) is applied to insulation, electrical breakdown does not occur, of course. Thus, failure is said to take place conventionally when a selected diagnostic property that has to be monotonous in time and correlated to thermal degradation (e.g., dielectric strength, yield strength, weight, density, etc.) reaches a proper endpoint, that is, a fixed limit beyond which the insulation is no more able to perform satisfactorily [29, 46]. The initial studies on insulation aging regarded mainly endurance to thermal stress. A fundamental approach was Dakin's theory, dating back to 1948 [14], according to which temperature speeds up the rate of thermally activated degradation reactions (e.g., oxidation, cross-linking, etc.), thereby accelerating the chemical aging of insulation. As a consequence, it can be shown that the logarithm of the time to end point, that is, "thermal life," is inversely proportional to absolute temperature, T [46]. For this reason, every percentile of thermal life, for example, the 63.2th percentile[11] $\alpha_t(T)$, can be expressed as [compare with Eq. (6.16)] [24, 29]

$$\alpha_t(T) = \alpha_0 \exp\left[-B\left(\frac{1}{T_0} - \frac{1}{T}\right)\right] = \alpha_0 \exp(-BT') \qquad (6.34)$$

where T' is the so-called conventional thermal stress, defined as

$$T' = \frac{1}{T_0} - \frac{1}{T} \qquad (6.35)$$

where T_0 is a reference value of absolute temperature (e.g., design temperature, or sometimes even room temperature), α_0 is the 63.2th percentile of thermal life at such a reference temperature, $B = \Delta W/k_B$ is a thermal endurance constant typical of the

[11] Actually, the probability distribution best suited for expressing the relationship between failure probability and test temperature in thermal life tests is not the Weibull but the log-normal distribution [1, 33, 34, 46]. Reference is made here to the 63.2th percentile (scale parameter of the Weibull distribution) with the ultimate purpose of combining electrical and thermal life models into one single electrothermal model based on electric breakdown tests at various temperatures.

material, ΔW is the activation energy of the main thermal degradation reaction involved, and $k_B = 1.38 \times 10^{-23}$ J/K the Boltzmann constant. Equation (6.34) is referred to as the Arrhenius model: Based on the chemical reaction rate theory, it was proposed also for electrical aging by Dakin [13]. The Arrhenius model is thus characterized by two parameters, namely α_0 and B.

Some alternatives are available for the selection of the reference temperature, as well as for expressing the Arrhenius model. For instance, it is common practice to choose as the reference temperature a value T_H higher than the operating temperature of the insulation. In that case, the Arrhenius model can be rewritten trivially from (6.34) as follows:

$$\alpha_t(T) = \alpha_H \exp\left[-B\left(\frac{1}{T_H} - \frac{1}{T}\right)\right] \tag{6.36}$$

where α_H is the 63.2th percentile of thermal life at reference temperature T_H. A value often chosen for T_H is the temperature that corresponds to a mean life of 20,000 h (such a value in °C is referred to as the "temperature Index," TI) [46]. In this way, α_H is known and fixed a priori, and the Arrhenius model is once more characterized by two parameters, that is, T_H and B.

The Arrhenius model is usually represented in the so-called Arrhenius graph, having coordinates log(life) versus $-1/T$, thereby giving rise to a straight line of slope $-B$ that enables the extrapolation from test to service temperatures. In particular, B is very important: An insulation that exhibits a higher value of B (for the same value of α_H and T_H, with T_H higher than the service temperature as pointed out above) features a longer life at temperatures lower than T_H, that is, down to the service temperature of the insulation. Then, the higher is B, the better is the insulation [24, 29] (see Fig. 6.8*a*). International standards have been established for evaluating thermal endurance capabilities of insulating materials via indices such as the above-mentioned TI and the HIC (halving interval in celsius), that is, the temperature difference giving rise to the halving of life starting from the temperature of TI [46].

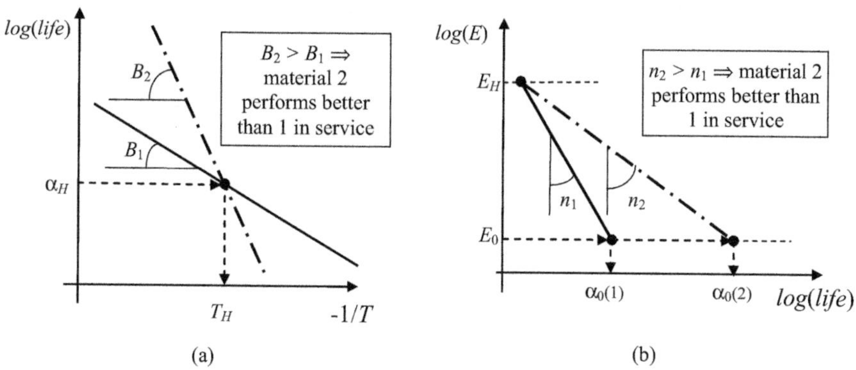

Figure 6.8 Single-stress electrical life models: (*a*) Arrhenius model in log(life) versus –1/T coordinates (Arrhenius graph) and (*b*) inverse power model in log(*E*) versus log(life) coordinates.

Single-Stress Electrical Life Model(s) As is well known, when a fast increasing voltage is applied to an insulation system, electrical breakdown (i.e., the discharge of the whole insulation thickness) takes place as the applied voltage exceeds a value typical of the considered insulation, the breakdown voltage; the relevant electric field is referred to as dielectric strength and depends on several quantities [24,29]. When a constant (in the rms sense) voltage only is applied to an insulation system, the above-mentioned inverse power model (IPM) or the so-called exponential model (EM) is mostly used for expressing the relationship between applied voltage/electric field and time to breakdown.

In addition to the expressions already presented in Section 6.1.1, a further expression of the IPM often used in the literature is as follows:

$$\alpha_t(E) = \alpha_0 \left(\frac{E}{E_0}\right)^{-n} \qquad \text{(IPM)} \qquad (6.37)$$

where $\alpha_t(E)$ is the 63.2th percentile of time to failure, also referred to as "electrical life," E is the magnitude of electric field, also referred to as "electrical stress" (proportional to the applied voltage via trivial geometrical factors; see Chapter 3 and Section 6.1.1), E_0 is a reference value of electric field (e.g., design electric field of cable insulation), α_0 is the value of the 63.2th percentile of time to breakdown corresponding to E_0 [i.e., $\alpha(E=E_0)=\alpha_0$], and n is the above-mentioned voltage coefficient, which should be more properly referred to as the voltage endurance coefficient (VEC), as suggested since 1983 and before (see, e.g., [29]).

As to the EM, it can be expressed as follows:

$$\alpha_t(E) = \alpha_0 \exp[-h_{EM}(E - E_0)] \qquad \text{(EM)} \qquad (6.38)$$

where the meaning of the quantities involved is the same as in Eq. (6.37); here coefficient h_{EM} plays the same role as the VEC n in Eq. (6.37), and for this reason is sometimes—but not always—called VEC as well. Note that Eq. (6.38) becomes linear in E versus log(life) coordinates, that is, in a semilog plot; for this reason the EM is better plotted in such semilog coordinates, as done, e.g., in Figure 6.4.

Similar to the Arrhenius model, the IPM and EM models can be written in alternative ways to Eqs. (6.37) and (6.38) by choosing as the reference value of the electric field a value E_H much higher than the design field, for example, the initial (or "as manufactured") electric strength of the insulation, thereby attaining trivially the following expressions from Eqs. (6.37) and (6.38):

$$\alpha_t(E) = \alpha_H \left(\frac{E}{E_H}\right)^{-n} \qquad \text{(IPM)} \qquad (6.39)$$

$$\alpha_t(E) = \alpha_H \exp[-h_{EM}(E - E_H)] \qquad \text{(EM)} \qquad (6.40)$$

Equations (6.37) and (6.38)—as well as Eqs. (6.39) and (6.40)—provide straight lines in log–log and semilog coordinate systems, respectively, with slopes $-1/n$ and $-1/h_{EM}$, if—as usual—E is in ordinate and α_t (≡life) in abscissa (see Section 6.1.1). The VEC, together with dielectric strength, is a fundamental parameter for insulation characterization and design: Indeed, for the same values of initial dielectric strength E_H, an insulation that exhibits a higher value of n (or h_{EM}) features a longer life for electric field values below E_H. Then, the larger the VEC, the better the insulation endurance, that is, its ability to endure electrical stress, as highlighted in Figure 6.8*b*.

The IPM and the EM have essentially an empirical background because most of the accelerated life test data can be fitted by straight lines in log–log or semilog plots (the linearization of the stress–life relationship is needed to extract coefficients for material characterization, as well as to derive design field estimation through extrapolation from the results of accelerated life tests, carried out at stresses considerably larger than the service one, as repeated many times above). However, both models can acquire a theoretical background; in particular, the IPM was associated with a statistical approach based on the Weibull distribution and was applied plenty of times to power cable insulation, as shown also in Section 6.1.1 (see, e.g., [1, 8, 29, 31, 35]).

Combined Electrothermal Life Model By combining the Arrhenius thermal model in the form of Eq. (6.34) with the IPM in the form of Eq. (6.37), the following "electrothermal life model" was obtained [29]:

$$\alpha_t(E,T) = \alpha_0 \left(\frac{E}{E_0}\right)^{-(n_0 - b_{\mathrm{ET}}T')} \exp(-BT') \tag{6.41}$$

In Eq. (6.41) $\alpha_t(E,T)$ is the 63.2th percentile of electrothermal life, $\alpha_0 = \alpha_t(E_0,T_0)$, n_0 is the value of the VEC at reference temperature T_0,[12] b_{ET} accounts for the synergism between electrical and thermal stress [49], and all the other quantities have the same meaning as in Eqs. (6.34) and (6.37).

Model (6.41), which features four parameters (i.e., B and n_0 for thermal and electrical endurance, respectively, b_{ET} for the extent of stress synergism, and α_0 as the scale parameter) was fitted satisfactorily to several sets of data, relevant to different materials [2,24,29,30], and can be used also for extruded HVDC cable insulation subjected to the combination of electrical and thermal stresses.

From these considerations—and by comparing Eqs. (6.15) and (6.16) with (6.41)—it can be concluded that, in the presence of electrical and thermal stresses only, $\alpha_t(S_1, \ldots, S_N)$ reduces to $\alpha(E,T)$, which can be expressed through Eq. (6.41); moreover, $\alpha_0 = \alpha_t(S_{1,0}, \ldots, S_{N,0})$ reduces to $\alpha_0 = \alpha_t(E_0,T_0)$. Hence, by inserting Eq. (6.41) into Eq. (6.12), one gets the following probabilistic electrothermal life model that relates the 100 Fth percentile of cable time to failure under electrical and thermal stress to the values of applied electrical stress E, applied temperature T, and cumulative failure probability F:

$$\begin{aligned} t_F(E,T) &= [-\ln(1-F)]^{1/\beta_t} \alpha_0 \left(\frac{E}{E_0}\right)^{-(n_0 - b_{\mathrm{ET}}T')} \exp(-BT') \\ &= t_F(E_0,T_0) \left(\frac{E}{E_0}\right)^{-(n_0 - b_{\mathrm{ET}}T')} \exp(-BT') \end{aligned} \tag{6.42}$$

[12] The VEC is temperature dependent, as proven by Figure. 6.6 in Section 6.1.1.

while from Eqs. (6.11), (6.13), and (6.14), respectively, one obtains the relevant cumulative failure probability, reliability, and hazard functions:

$$F(t_F; E, T) = 1 - \exp\left\{-\left[\frac{t_F}{\alpha_0 (E/E_0)^{-(n_0 - b_{\mathrm{ET}} T')} \exp(-BT')}\right]^{\beta_t}\right\} \tag{6.43}$$

$$R(t_F; E, T) = 1 - F(t_F; E, T) = \exp\left\{-\left[\frac{t_F}{\alpha_0 (E/E_0)^{-(n_0 - b_{\mathrm{ET}} T')} \exp(-BT')}\right]^{\beta_t}\right\} \tag{6.44}$$

$$h(t_F; E, T) = \frac{\beta_t t_F{}^{\beta_t - 1}}{[\alpha(E, T)]^{\beta_t}} = \frac{\beta_t t_F{}^{\beta_t - 1}}{\left[\alpha_0 (E/E_0)^{-(n_0 - b_{\mathrm{ET}} T')} \exp(-BT')\right]^{\beta_t}} \tag{6.45}$$

Equation (6.44) represents a "phenomenological reliability model" that can be used for reliability estimation in the case of a typical extruded HVDC cable insulation subjected to the combination of electrical and thermal stresses. It can be argued that it is a Weibull reliability function, thus it is characterized by the relevant mathematical properties [33, 34].

A comparison of Eqs. (6.41) and (6.42) with Eq. (6.31) provides the following relationship between the 100 Fth percentile of full-size cable time to failure under electrical field E and temperature T, $t_{F,2}(E, T)$, and the values of:

1. Applied electrical field E
2. Applied temperature T
3. Failure probability F
4. Dimensional factor of the enlargement law from tested short cable (i.e., full-size cable length tested in laboratory) to long cable (i.e., full-size cable installed in field), N_l
5. 63.2th percentile of short cable time to failure under design electrical stress E_0 and design temperature T_0, $\alpha_{0,1}$

$$t_{F,2}(E, T) = \left[-\frac{\ln(1 - F)}{N_l}\right]^{1/\beta_t} \alpha_{0,1} \left(\frac{E}{E_0}\right)^{-(n_0 - b_{\mathrm{ET}} T')} \exp(-BT') \tag{6.46}$$

From Eq. (6.46) with $F = P_0$ [or by comparing Eqs. (6.41) and (6.42) with Eq. (6.32)], one gets the following relationship between the 100 P_0th percentile of full-size cable time to failure under electrical field E and temperature T, $t_{P_0,2}(E, T)$, and the values of:

1. Applied electrical field E
2. Applied temperature T
3. Design failure probability of power cable P_0

4. Dimensional factor of the enlargement law from tested short cable (full-size cable length tested in the laboratory) to long cable (full-size cable installed in the field), N_l
5. 63.2th percentile of short cable time to failure under design electrical stress E_0 and design temperature T_0, $\alpha_{0,1}$

$$t_{P_0,2}(E,T) = \left[-\frac{\ln(1-P_0)}{N_l}\right]^{1/\beta_t} \alpha_{0,1}\left(\frac{E}{E_0}\right)^{-(n_0-b_{\mathrm{ET}}T')} \exp(-BT') \quad (6.47)$$

Note that Eq. (6.47) rewritten for $E=E_0$ and design temperature $T=T_0$ provides life at design electrical stress E_0, design temperature T_0, and design failure probability of full-size cable P_0, that is, the design life t_0 of the power cable according to the modern probabilistic design philosophy, namely:

$$t_0 = t_{P_0,2}(E_0,T_0) = \left[-\frac{\ln(1-P_0)}{N_l}\right]^{1/\beta_t} \alpha_{0,1} \quad (6.48)$$

A comparison of Eq. (6.42) with Eq. (6.33) provides the following relationship between the $100P_0$th percentile of full-size cable time to failure under electrical field E and temperature T, $t_{P_0,2}(E,T)$, and the values of:

1. Applied electrical field E
2. Applied temperature T
3. Design failure probability of power cable P_0
4. Dimensional factor of the enlargement law from tested short cable (full-size cable length tested in the laboratory) to long cable (full-size installed in the field), N_l
5. 100 P_Tth percentile (different from 63.2th percentile) of short cable time to failure under design electrical stress E_0 and design temperature T_0, $t_{P_T,1}(E_0,T_0)$;

$$t_{P_0,2}(E,T) = \left\{\frac{1}{N_l}\frac{[-\ln(1-P_0)]}{[-\ln(1-P_T)]}\right\}^{1/\beta_t} t_{P_T,1}(E_0,T_0)(E/E_0)^{-(n_0-b_{\mathrm{ET}}T')}\exp(-BT') \quad (6.49)$$

Note that from the ratio between Eq. (6.46) and (6.48), one obtains the following relationship between the $100F$th percentile of full-size cable time to failure under electrical field E and temperature T, $t_{F,2}(E,T)$, and the values of:

1. Applied electrical field E
2. Applied temperature T
3. Design failure probability of power cable P_0

4. Life at design electrical stress E_0, design temperature T_0, and design failure probability of power cable P_0, that is, the design life t_0 of the power cable according to the modern probabilistic design philosophy:

$$t_{F,2}(E,T) = \left[-\frac{\ln(1-F)}{\ln(1-P_0)}\right]^{1/\beta_t} t_0 \left(\frac{E}{E_0}\right)^{-(n_0 - b_{\mathrm{ET}}T')} \exp(-BT') \qquad (6.50)$$

Combined Electro-Thermo-Mechanical Life Model Often, a mechanical stress, due, for example, to vibration, tension/compression, bending, thermomechanical fatigue, and so on is superimposed to the electrothermal stress. To describe this complex situation, the following electro-thermo-mechanical life model was proposed [50], analogous to Eq. (6.41), but with further terms devoted to mechanical stress M and to the thermomechanical and electromechanical synergism:

$$\alpha_t(E,T,M) = \alpha_0 \exp(-BT' - n_0E' - m_0M' + b_{\mathrm{ET}}E'T' + b_{\mathrm{MT}}M'T' + b_{\mathrm{EM}}E'M') \qquad (6.51)$$

In Eq. (6.51) $\alpha_t(E,T,M)$ is the 63.2th percentile of electro-thermo-mechanical life, M being mechanical load (its dimensions are usually $\mathrm{N/m^2 = Pa}$); $\alpha_0 = \alpha_t(E_0,T_0,M_0)$, M_0 being the value of mechanical load at design conditions; m_0 is the value of the so-called mechanical endurance coefficient (MEC) m at reference temperature T_0 and reference electric field E_0; and b_{MT} and b_{EM} are parameters that account for the thermomechanical and electromechanical synergism, respectively, analogously to b_{ET} for electrothermal synergism. The other parameters appearing in Eq. (6.51) are the same previously defined when dealing with Eq. (6.41). As to M' and E', they are referred to here as conventional electrical stress and conventional mechanical stress, respectively, defined for making Eq. (6.51) more compact, namely:

$$E' = \ln\left(\frac{E}{E_0}\right) \qquad (6.52)$$

$$M' = \ln\left(\frac{M}{M_0}\right) \qquad (6.53)$$

Like model (6.41), model (6.51) derives from the combination of three single-stress models, each holding when one stress is applied, and the other two are inactive. However, if model (6.41) is more complex than the two single-stress models from which it derives, the electro-thermo-mechanical model of Eq. (6.51) is far more complicated than the three "parent" single-stress models since Eq. (6.51) is virtually characterized by 10 parameters: α_0, E_0, M_0, T_0, n_0, m_0, B, b_{ET}, b_{MT}, b_{EM}. The determination of so many parameters all together from multistress life test data is not simple since it requires the use of complex iterative numerical algorithms, which sometimes are unstable and can diverge, especially in case of strongly nonlinear models. So, this model falls beyond the compromise mentioned at the beginning of Section 6.1 between model complexity and quality of fitting experimental results in the direction of model complexity.

Actually, the electro-thermo-mechanical model of Eq. (6.51) was not applied to extruded HVDC insulation failure data under electro-thermo-mechanical stresses because such data are not easily available in the literature. In the literature, test results relevant to the behavior of XLPE insulated HVDC minicables or laboratory full-size cables subjected to electrical and mechanical stresses are rather uncommon and only relevant to static tension such as bending (see, e.g., [51]). In general, the test description is not so accurate as to enable data to be used for fitting purposes via Eq. (6.51). To the best of our knowledge, the only application of Eq. (6.51) was related to HV generator stator insulation made of epoxy-bonded mica splitting, subjected to 50-Hz AC voltage (from 22 to 26 kV rms), high temperature (from 155 to 210°C), and mechanical vibration (from 2.4 to 66.4 N/cm^2, 120 Hz frequency). The application is illustrated in more detail in [50].

However, it must be pointed out that the short-term and long-term load cycle tests, prescribed in [5] at the prequalification and type test stage of HVDC extruded cables, try and account not only for the electrothermal stress on the insulation of cables and accessories but also for the thermomechanical fatigue due to contraction and expansion of cable system components as the load varies. This type of stress is particularly active at the joints and terminations.

6.2 SPACE-CHARGE-BASED LIFE MODELS FOR EXTRUDED HVDC CABLES

Taking advantage of the increase in amount and accuracy of information provided by the recent advances in space-charge measurements (see Chapter 4), this section illustrates those life models that emphasize the contribution of space charge in DC regime only. Since the aim of [13] is quite similar, although encompassing both DC and AC regime, this work is extensively quoted in the present section, referring to both macroscopic and microscopic models. Focus is essentially on dry aging and on phenomena that are precursors of the electrical treeing.

Storage centers of space charge—either inherent to the dielectric or injected and trapped at defect sites—can be related to microscopic defects (see Chapter 4): for instance, protrusion/conductive contaminants that can inject space charge or voids where space charge can be trapped at the void–insulation interface. In both cases space charge can modify remarkably the local field, raise the aging reaction rate, or even give rise to new degradation processes [13], thereby playing a major role in the aging of extruded HVDC cable insulation, as discussed extensively in Chapter 4.

Although the Eyring model mentioned in Section 6.1.1 was developed when space-charge measurement techniques were not available—thus it does not cater explicitly for the presence of space charge—it can be profitably recalled for the subsequent reasoning, in the following form [15]:

$$L = C_E\left(\frac{h_P}{k_B T}\right)\exp\left(\frac{\Delta G}{kT}\right) \tag{6.54}$$

where L is time to failure (or lifetime, or life), k_B and h_P are the Boltzmann and Planck constants, C_E is a model parameter, which depends on material and, in most cases, electric field and temperature (E and T), and ΔG is the Gibbs (free) energy of the electrothermal degradation process. Endicott et al. [48] rewrote the Eyring model (6.54) to include explicitly electrical and thermal stresses:

$$L(E,T) = C_E T^{-w} \exp\left(\frac{-\Delta G}{k_B T}\right) \exp\left[-\left(k_1 + \frac{k_2}{T}\right) E\right] \tag{6.55}$$

where w, k_1, and k_2 are coefficients, discussed in detail in [52]. The Eyring model has been recast similarly in order to account for mechanical stress only [16] and, later on, electrical, thermal, and mechanical stresses, both in the framework of space-charge models and in the presence of mechanical strain/fatigue, as illustrated in Section 6.2.2 and in [13].

As far as the space-charge behavior is concerned, the Eyring model has two main characteristics [13]:

1. It assumes implicitly a geometrical (Laplacian) field distribution in the dielectric.
2. Its activation energy is affected (generally reduced) by the presence of space charges.

However, provided that the space-charge distribution in the insulation is known, the local field can be evaluated and the Eyring model can be employed using local (Poissonian) field values instead of the Laplacian (applied) field. As an alternative, the electric field dependence of the activation energy can be accounted for thoroughly by including the space-charge contribution to the local Poissonian field. In this framework, a macroscopic model as the Eyring model can be applied locally to defect sites, because—consistent with the extreme value probability distributions used for electrical failure data [53]—electric breakdown is associated with the growth of damage from the weakest point: In fact, this is the approach followed when developing the space-charge DMM model (see Section 6.2.2).

6.2.1 Field-Limited Space-Charge Model

At very high fields (say, above 100–200 kV/mm) the electrical conductivity of insulating polymers can increase by several orders of magnitude because of the strong dependence of its activation energy on the electric field. Thus large current densities can occur at particular spots within the insulation, thereby giving rise to the formation and growth of local conductive paths: Once they reach a critical length, breakdown ensues [54–58]. In practice, these phenomena usually happen [13]:

1. At interfaces between bulk insulation and defects (e.g., protrusions and metallic contaminants) capable of greatly enhancing the local electric field compared to the design field
2. At locations of field amplification due to huge bipolar space-charge accumulation, which can take place nearby strongly injecting electrodes and ionic contaminants

Life models were developed by Zeller [54–56] and Boggs [57, 58] based on these premises, as well as on the so-called field-limited space-charge current (FLSC) theory (see Chapter 4). The FLSC theory states that a threshold field, E_{mc}, exists below which the mobility of injected charges is low but finite. Hence, at long times the equilibrium situation according to the space-charge-limited conduction (SCLC) model [59] is achieved; above E_{mc} the strongly nonlinear dependence of mobility on electric field gives rise to space-charge "packets" [60, 61] that penetrate quickly into the insulation bulk, thereby reducing the electric field at the interface between insulation and electrode/defect in a quite brief time. Above such thresholds, dielectric heating (under AC voltage) and Maxwell stresses may take place, and highly energetic local damage events may start, particularly if filamentary currents are produced [54]; this damage process can be described by an electric life model that holds for $E > E_{mc}$ and has the following form [13]:

$$L \propto \frac{1}{r_{\mathrm{FLSC}}^2 E_l (E/E_{mc} - 1)^2} \tag{6.56}$$

where L is the time to failure relevant to the above mechanism of local damage, assumed to cause the precursor of an electrical tree, E_l is the Laplacian field, E is the local (Poissonian) field, E_{mc} is the Poissonian threshold field for the onset of the FLSC, and r_{FLSC} is the radius of the charge injecting tip. Life is infinite below the threshold field for the activation of FLSC; thus, for $E \leq E_{mc}$, aging must be addressed to other mechanisms (see below).

Therefore, depending on the ratio E_l/E_{mc}, the local damage will:

1. Either evolve abruptly into an electrical tree, from which a fast breakdown will ensue
2. Or undergo an incubation period during which a cavity of growing size is formed from the initial defect, followed by partial discharges and electrical treeing onset, causing breakdown in longer times.

As argued in [13], under AC conditions:

> according to [58] the FLSC conductivity is equal to the product $\omega\varepsilon$, being ω angular frequency and ε dielectric permittivity. When this condition is reached, the resistive current density is equal to the capacitive current density and space charge is generated to limit the field to close to E_{mc}. This has the practical effect of reducing the Laplacian field at the defect tip to a near constant Poissonian value of $\sim E_{mc}$ for a length inside the dielectric that is established by the extent of the difference between the Laplacian and FLSC threshold fields (see Fig. 6.9) ([13], page 878).

In [13], after estimating approximately as $\sim 10000\,\mathrm{C/m^3}$ the space-charge density formed by this process at the tip of the defect for $\varepsilon = 2.3\varepsilon_0$, $E_{mc} = 250\,\mathrm{kV/mm}$, and $r = 1\,\mu\mathrm{m}$, it is argued that:

> Under AC a large part of this very large amount of charge will be injected and extracted each half cycle, causing massive localised degradation, either through chemical modification from carrier recombination and photon emission followed by relevant photochemistry, or from kinetic energy transferred to the polymer by ballistic impact during electron acceleration . . . A critical length of 1.5 μm becomes a failure criterion for an insulating material at such high fields as those dealt with by these models.

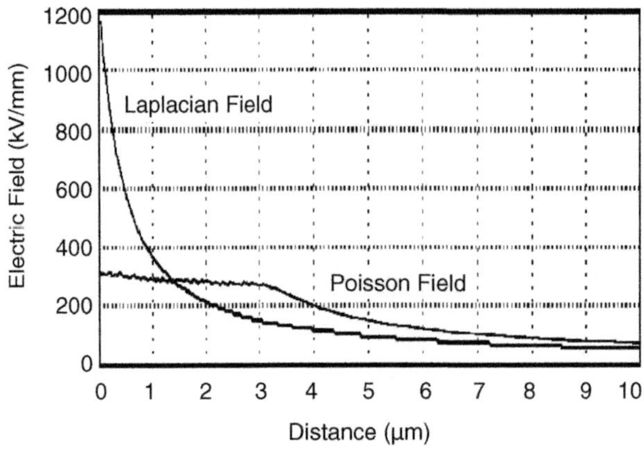

Figure 6.9 Example of a plot of the Laplacian (without accounting for space charge) and Poissonian (with space charge) electric field on the axis of symmetry versus distance from a defect tip when the Laplacian field at the tip is appreciably above the threshold field for the FLSC mechanism. In this case, the space-charge-limited field region extends from the defect tip for approximately 3 μm (after [58]).

> Degradation can proceed, obviously, also under DC, but in this case it is speculated that its rate is much slower than under AC. Damage can penetrate into the polymer from the defect tip through electron avalanches generated by the very high local field, but the electrons available for such a process must be extracted from the defect/polymer interface, with a rate that depends on defect work function and polymer electron affinity . . . Therefore, degradation times much longer than those involved in a half cycle of AC supply voltage would be involved.
>
> Summarizing, under both AC and DC fields, space charge seems to play a predominant role in polymer degradation, due to massive high-field injection, electron acceleration to gain high energy and, thus, ability to break polymer bonds, and penetration of space charge clouds into the polymer bulk. Even if the mechanism of charge penetration and its dependence on temperature needs to be better modeled, and the failure criterion does not correspond to insulation breakdown, but to the onset of a degradation mechanism based on partial discharge along the damage pit, this model based on high-field space charge effects can account for the very fast breakdown caused by conductive or semiconductive sharp defects which may be present in insulation systems due to manufacturing processes ([13], page 878).

6.2.2 Space-Charge DMM Model

The Dissado–Mazzanti–Montanari (DMM) model relies on a thermodynamic approach to the aging of insulation under electrical and thermal stress in the presence of accumulated space charge. According to this approach, the local stress enhancement associated with space-charge storage within a polymeric insulation subjected to an applied voltage produces aging since—as highlighted at the beginning of Section 6.2 while describing the Eyring model—the relevant electromechanical energy storage reduces the free energy barrier that should be overcome for degradation to occur [13]. At first the model was conceived for insulation subjected to DC voltage [62], but it was extended later on to the AC voltage regime [63]. Only the DC version of the DMM model is reviewed and discussed below since focus is here on extruded HVDC cables.

6.2.2.1 The DC DMM Model The DC DMM model [20, 21, 62] is based on a series of hypotheses, which can be summarized as follows [13]:

1. Since a DC voltage is applied to a polymeric insulation that is homogeneous on a macroscale, local "charge centers" (i.e., space-charge concentrations) are generated within small voids of size from a few nano- to a few micrometers, formed unavoidably during manufacturing.[13]
2. Charge stored at charge centers is injected from electrodes and comes into equilibrium in a much shorter time than the insulation time to failure.
3. Near the charge centers, the applied (Laplacian) field, E, is negligible compared to the space-charge (Poissonian) field.
4. Charge centers are spherical, of radius r_0, and have uniform charge density.
5. All charge centers have the same amount of charge, q_C.
6. Charge q_C depends on applied field E via a power law relationship, namely $q_C = C_q E^{b_q}$, where constant C_q and exponent b_q depend on the space charge storage properties of the insulation (see, e.g., Section 4.5.2.2, Fig. 4.50).
7. A number N_m of microstructural units of the polymer that undergo degradation (e.g., weakening/breaking of interchain cohesion and/or intrachain bonds, cross-linking bonds, etc.), referred to as "moieties" in the model, are influenced on average by each microvoid.
8. These "moieties" lie within a thin shell of thickness λ on the void wall.

The thermodynamic approach followed when developing the DMM model implies that insulation aging is the effect of the net progress of local "reactions" that modify the local polymer morphology so as to promote a faster degradation process, essentially partial discharges (PDs) and electrical treeing, from which rapid failure ensues. Both a "forward" reaction and a "reverse" reaction are allowed that carry reacting moieties from reactant (as manufactured) state 1 to product (degraded) state 2 and vice versa, respectively, through an intermediate transition state at a higher free energy. The rates of forward and reverse reactions depend on the relevant concentrations of moieties in the two local states and the magnitude of the free energy barrier in between. The application of an external electric field puts this system in a global nonequilibrium state, thus the system changes in time until it either comes to a thermodynamic equilibrium or it reaches an amount of local "reactions" sufficient to initiate a faster aging mechanism that eventually leads the system to failure: In this latter case, the time to trigger the faster aging mechanism is the time to failure [13].

From the above hypotheses (see in particular items 4 and 5), it follows that trapped space charges raise the local electric field up to a maximum at the boundary of each space-charge center, yielding elemental electromechanical stresses and

[13] The shape, size, and density of such microvoids differ by much for different insulation compounds and designs.

strains associated with storage of electromechanical energy per moiety in this region. This electromechanical energy lowers the free energy barrier for the forward degradation reaction and raises that for the reverse reaction by an amount proportional to the stored energy, thereby increasing the net transfer into the degraded state. Failure is said to take place whenever the fraction of degraded moieties, A, exceeds locally a critical fraction A^*, which involves the enlargement of the charge center up to a cavity large enough for the inception of PDs (thus of electrical treeing). Hence, the DMM model does not predict the time to breakdown but the time to the attainment of critical void size [13].

It must be pointed out that an aging threshold may, in fact, arise from the need to concentrate, at a space-charge center, a minimum amount of energy required to cause a local plastic deformation. Such an energy concentration is required since energy cannot be transferred continuously to the polymer in this mechanism. Balancing processes ensure an equilibrium unless a critical amount of local deformation is achieved.

A lengthy analytical treatment based on all the above hypotheses gives eventually the expression of the DC DMM model, that is [13, 21],

$$L(E,T) = \frac{h_P}{2k_BT}\exp\left(\frac{\Delta H/k_B - C'E^{2b_q}/2}{T} - \frac{\Delta S}{k_B}\right) \times \ln\left[\frac{A_{\text{eq}}(E)}{A_{\text{eq}}(E) - A^*}\right]\left[\cosh\left(\frac{\Delta/k_B - C'E^{2b_q}}{2T}\right)\right]^{-1} \quad (6.57)$$

$$A_{\text{eq}}(E) = \left[1 + \exp\left(\frac{\Delta/k_B - C'E^{2b_q}}{T}\right)\right]^{-1} \quad (6.58)$$

where $A_{\text{eq}}(E)$ is the value of A at the equilibrium between forward and backward reactions, T is the absolute temperature, ΔH and ΔS are the activation enthalpy and entropy per moiety respectively, Δ is the free energy difference between degraded and product state, C' is a constant depending on material properties, namely [13, 21]

$$C' = \frac{B_q \alpha_E \delta_S \lambda C_q^2}{16\pi\epsilon^2 r_0^2 N_m k_B} \quad (6.59)$$

where B_q is the proportionality constant between the stored electromechanical energy per moiety and its contribution to the degradation free energy barrier, δ_S is elemental strain [20, 21], and α_E is the electrostriction coefficient. The above-mentioned constant C_q and exponent b_q appear in Eqs. (6.57)–(6.59) due to the relationship between stored charge per charge center and applied electric field [13].

6.2.2.2 *Application of the DC DMM Model to HVDC Extruded Insulation*

The DC space-charge life model of Eqs. (6.57)–(6.59) was used to fit times to failure obtained at room temperature in an accelerated life test performed under DC voltage on press-molded plaques made of an XLPE material for HVDC cable insulation (the mean thickness of the plaques was 0.15 mm) [21]. When plotted in

Table 6.2 Values of Parameters of DC DMM Model [Eq. (6.57)] Obtained from DC Life Tests on XLPE Material for HVDC Cables

Parameter	Dimension	Value
Δ	J	3.40×10^{-21}
ΔG	J	2.1×10^{-19}
ΔG from OIT [65]	J	2.11×10^{-19}
C'	$K\,(\text{kV/mm})^{-2b}$	8×10^{-4}
b_q	Dimensionless	1.438
$A*$	Dimensionless	0.305

Source: After [21].

log E versus log L coordinates (as usual, see Section 6.1), these lifetimes fit well the IPM, with a value of $n \sim 20$ that is consistent with the value of $n = 15$ after [3]. By fitting the DMM model to the 63.2th percentiles of these times to failure via the Levenberg–Marquardt (LM) algorithm [64], the values of the parameters listed in Table 6.2 were calculated [21]. The table shows that the value of ΔG[14] is in good agreement with that obtained previously in [65] from oxidation induction time (OIT) measurements. However, it can be argued that the activation enthalpy ΔH in the two cases differs due to the difference in activation entropy ΔS between OIT measurements and life tests [21]. In addition, since such OIT measurements were performed between 130 and 180°C, a significantly different estimate of ΔG could be attained at temperatures <100°C because of the thermodynamic transition present in XLPE close to 110°C. Figure 6.10 displays the experimental life points and the electrical life line at room temperature obtained from Eqs. (6.57)–(6.59) with the parameter values reported in Table 6.2: The fitting is quite good.

Furthermore, the model predicts a value of ~20 kV/mm at room temperature for the DC field threshold below which no electrical aging takes place. Although life tests at these field levels are unfeasible since they would last too long, space-charge measurements via the PEA technique and electroluminescence observations performed on similar materials provide a very similar estimate for the field below which no space charge can be detected [66,67] (see also Figs. 6.15–6.16 in what follows). Since space charge is necessary for aging via the mechanism involved in this model, any threshold for aging should not be lower than that for the generation of space charge [21]. However, because test electrical fields are all much higher than the predicted threshold, accurate estimates for A^* and Δ could not be obtained. Since these parameters are strictly linked to the critical local concentration of energy, thus with the threshold value inferred by the DMM model, the above threshold estimate has to be regarded with care and is not conservative, being slightly above most present values of design electric field of extruded HVDC cables (see Fig. 6.1) [10], [21].

As a last remark about the DMM model applied to HVDC extruded insulation, it must be pointed out that the good fitting of Figure 6.10 should be regarded with care and the possibility that it is a consequence of "phenomenological" rather than "physical

[14] Since the life test under exam is relevant to one single temperature, only $\Delta G = \Delta H - \Delta S/T$ can be estimated [20, 21].

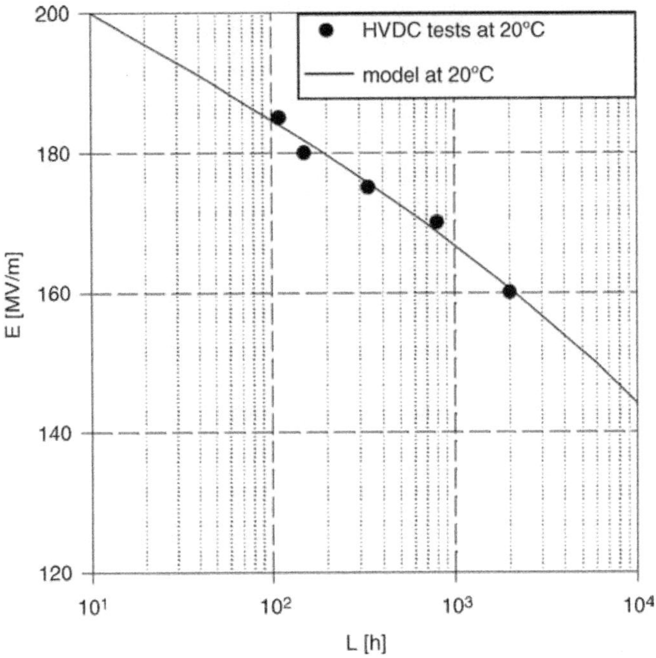

Figure 6.10 Electrical life line at room temperature derived by means of the DMM model [Eqs. (6.57)–(6.59)] with parameter values listed in Table 6.2 when applied to XLPE for HVDC data after [21]. Experimental life points (at 63.2% failure probability) are also displayed (after [21]).

effects" should not be ruled out a priori. Indeed, the DMM model features a large number of parameters, thus it is quite flexible and capable of fitting satisfactorily quite different data sets without a precise physical meaning of the best fitting parameters. On the other hand, it must be emphasized that, for the XLPE material for HVDC cable insulation considered in this application, a fairly good correspondence was found in [21] between the values of some parameters of the DMM model obtained from fitting accelerated life test data and those derived from PEA measurements made on the same material (details are omitted for the sake of brevity). This further supports a possible physical meaning of the good data fitting of the experimental data in Figure 6.10.

6.3 FROM SPACE CHARGES TO PARTIAL DISCHARGES: LIFE MODEL BASED ON DAMAGE GROWTH FROM MICROVOIDS

Aging caused by PDs in a cavity—which becomes particularly severe under AC voltage when discharges may occur many times on each half cycle—can be regarded as involving space-charge when considering that the space charge accumulation at the interface between the dielectric and the cavity controls the phenomenon of electron

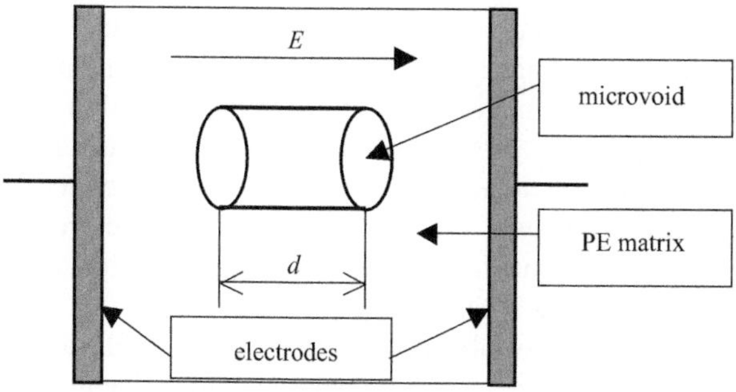

Figure 6.11 Sketch of a PE matrix with an embedded cylindrical void of size enlarged for the sake of illustration (after [22]).

injection in the void and therefore the repetition rate of the discharges under DC regime [68]. This is made clear in the following framework, outlined in [13, 22, 23].

As previously mentioned, polymeric insulation of power cables typically contains microvoids from the manufacturing stage. Therefore, if a gas (air)-filled microvoid is large enough (typically a few tens of microns at rated electric fields and temperatures [13,69]) to incept and support PDs under the applied electrical field, the system consisting of such a void plus the neighboring polymer (see Fig. 6.11) can be assumed to be in local breakdown conditions. The higher permittivity of the polymer implies a field concentration in the void; however, the discharge should be triggered by a charge carrier, primarily an electron: such a carrier, after being transported through the insulation due to conduction process and getting trapped as space charge at the polymer–void interface, is eventually injected into the void and starts hot-electron avalanches via impact ionization of gas molecules. Such avalanches collide with the opposite void–polymer interface damaging the adjacent polymer lattice and depositing space charge (essentially electrons and positive ions) on the cavity walls. The damage accumulation at the void–polymer interface—at a rate depending on applied voltage and temperature, micro-void size, electron energy distribution—may eventually result in the growth of a pit that causes electrical treeing and finally the breakdown of the whole insulation thickness [13].

These processes were carefully studied in [22, 23], having the polymer–void system sketched in Figure 6.11 as a reference, namely a homogeneous polyethylene (PE) matrix[15] lying between plane parallel metallic electrodes, with an embedded air-filled cylindrical void of radius *R*, circular base *S*, and height *d* along the field direction. These studies led to a PD aging model for the system of Figure 6.11, based on the following three-stage aging process [13]:

1. Charge storage at the PE–void interface and injection into the void
2. Hot-electron avalanche formation inside the void
3. Damage accumulation at the void–PE interface and growth into the PE matrix

[15] A homogeneous PE matrix is considered as the reference insulation in [22, 23]. Of course, the theory illustrated here can be applied in a quite similar way—*mutatis mutandis*—to any PE-based insulation of interest in this book.

Thereafter, similar to the FLSC approach, the occurrence of failure was associated with the attainment of a damaged zone large enough to intercept an electrical tree, and insulation life was estimated as the time to the formation of a damaged zone of a critical size. As in the case of the DMM, it can be argued that, strictly speaking, this life is lower than the actual time to failure—that is, the time to breakdown of the whole insulation thickness—but not by much, since tree growth (described by different models) will take much less time than the prior degradation process at usual service fields of extruded HVDC cables [13].

Now, let us recall in sequence the three above degradation stages of the aging and life model based on damage growth from air-filled voids, so as to derive the life model expression.

6.3.1 Charge Storage at PE–Void Interface and Injection into the Void

In [22, 23] focus is on electrons as precursors of the proposed damage mechanism. In fact, as discussed in Chapter 4, electrons—already within the insulation bulk or injected from the electrodes—are transported in the dielectric, for example, via thermally activated hopping and may eventually reach the PE–void interface, where the negative electron affinity of PE favors electron trapping and subsequent space-charge storage. Then, these trapped electrons can be injected into the void through the Schottky effect because both the Laplacian field and the space-charge Poissonian field contribute to a Coulombic repulsion between stored charges, and as a consequence the activation energy for electron injection, ϕ, is lower than the actual charge trap depth, ΔG_t. The time derivative of the number of electrons stored at time t at the PE–void interface, $n(t)$, is derived from a balance between electrons carried to the interface by the conduction current density, J_C, and electrons removed from the interface by Schottky injection into the void [13, 22, 23, 63]:

$$\frac{dn(t)}{dt} = \int_0^R \frac{J_C}{q_e}(1 - e^{-\phi(r,t)/k_B T})2\pi r\, dr \tag{6.60}$$

where q_e is the electron charge, k_B is the Boltzmann constant, T is the absolute temperature, and r is the radial coordinate over the void circular surface S. Under the following additional assumptions:

Chemical traps at the interface can be neglected.

Electrons arriving at the PE–void interface are all trapped.

Trapped electrons are uniformly distributed over S.

De-trapping rates are much slower than trapping ones.

Relationship (6.60) can be rewritten as follows [13]:

$$\frac{dn(t)}{dt} = \frac{J_C S}{q_e}\left\{1 - \exp\left[-\frac{\Delta G_t - q_e(q_e \varepsilon_r E/4\pi\varepsilon_o)^{1/2} - (2q_e^2/4\pi\varepsilon_0 R)(J_C S t/q_e)}{k_B T}\right]\right\} \tag{6.61}$$

where ε_0 is the void (air) permittivity, ε_r is the relative permittivity of PE, E is the local electric field in the PE matrix.[16] After solving Eq. (6.61) for $n(t)$, one obtains the number of injected electrons at time t, $n_{\text{inj}}(t)$, via the following relationship:

$$n_{\text{inj}}(t) = \frac{J_C St}{q_e} - n(t) \tag{6.62}$$

as well as the time to the injection of the first electron, t_{inj} (see [22, 23, 70] for details). Finally, t_{inj} yields the injection rate $R_{\text{inj}} = 1/t_{\text{inj}}$ [13]:

$$R_{\text{inj}} = \frac{J_C S}{q_e k_B T} \frac{2q_e^2}{4\pi\varepsilon_0 R} \left\{ \ln \left[1 + \frac{2q_e^2}{4\pi\varepsilon_0 R k_B T} \exp\left(\frac{\Delta G_t - q_e (q_e \varepsilon_r E / 4\pi\varepsilon_0)^{1/2}}{k_B T} \right) \right] \right\}^{-1} \tag{6.63}$$

6.3.2 Hot-Electron Avalanche Formation inside the Void

At typical design fields of extruded HVDC insulation, the electric field within the void, $E_V = \varepsilon_r E$, can be as high as, say, 40 kV/mm, that is, well above the critical value for avalanche inception (the dielectric strength of dry air at atmospheric pressure and room temperature is ~2 kV/mm). Therefore, at usual service fields electrons injected into the void and accelerated by the field can ionize gas molecules and generate electron avalanches formed mostly of hot electrons, that is, electrons with kinetic energy above 8 eV [13]. Since hot electrons are capable of damaging the polymer, modeling the degradation process involved by the successive collisions between avalanches and the void–PE interface requires the knowledge of the energy (w_e) distribution of electrons produced by ionization in the void (at a given time and value of E_V), $n(w_e; E_V, t)$. Typically the electron energy spectrum reaches the equilibrium distribution $n(w_e; E_V)$ very shortly (say, in ~10^{-10}–10^{-11} s) and remains steady thereafter, while the total number of electrons increases exponentially according to the field-dependent multiplication rate $\beta(E_V)$. Therefore, $n(w_e; E_V, t)$ can be written as [13]

$$n(w_e; E_V, t) = n(w_e; E_V) \exp[\beta(E_V) t] \tag{6.64}$$

In the framework of the life model based on damage growth from air-filled voids, $n(w_e; E_V)$ and $\beta(E_V)$ are calculated as described in [70], together with the number of avalanche electrons produced across the void starting from one single injected electron, N_{el}, that is [13]:

$$N_{\text{el}} = \exp[\alpha(E_V) d] \tag{6.65}$$

where $\alpha(E_V)$ is the well-known first Townsend coefficient, for which semiempirical expressions are available in the literature; see for example [59]. E_V is assumed as constant inside the void, thereby neglecting the influence of the spatial distribution of electrons [13].

[16] See footnote 14: a homogeneous PE matrix is considered as the reference insulation in [22, 23], but this theory holds—*mutatis mutandis*—for any PE-based insulation of interest in this book.

6.3.3 Damage at Void–PE Interface Growing into the Polymer

Hot-electron avalanches generated inside the cavity hit the void–PE interface, where the electrons release their kinetic energy through high-energy scatterings inside PE layers close to the void surface. As pointed out at the beginning of Section 6.3, the electric field in PE is generally lower than that in the cavity, while the dielectric strength of PE is well above 100 kV/mm, thus much higher than service electric fields of HVDC extruded insulation, although very large field values might occur nearby conductive, sharp particles (see the FLSC theory above and in Chapter 4). Hence, electrons in PE are far from the breakdown conditions typical of the void and are subjected to a fast thermalization, essentially via two strongly inelastic scattering mechanisms that happen above ~8 eV, namely [13, 22, 23, 70]:

1. Impact ionization
2. Dissociative electron attachment (DEA) of C–H bonds

DEA is important since it causes chemical damage (hence aging), which accumulates in the polymer as successive avalanches cross the void and hit the void–PE interface. The role of DEA in the model based on damage growth from air-filled voids was investigated in particular in [70]; the investigation can be summarized in the following steps:

1. For a polyolefin like PE, chemical damage is thought of as the breaking of a percentage of CH bonds large enough to yield irreversible aging
2. From the mean free paths for electron scatterings in PE, the whole hot-electron cooling process is assumed to take place in a layer (or "slab") of thickness D_{dis} ~400 Å
3. By combining Eqs. (6.63) and (6.65), the rate of electrons impinging the void–PE interface after being injected into the void and multiplied by the avalanche mechanism, R_{el}, is estimated
4. By solving the electron diffusion equation in the energy–time domain by means of experimental scattering rates, the hot-fraction (energy >8 eV) of avalanche electrons hitting the void–PE interface, F_{hot}, and the fraction of avalanche hot electrons that are effective in generating chemical damage through DEA, F_{eff}, are derived versus electric field
5. By defining t_{dis} as the time to disruption (i.e., to severe and irreversible chemical degradation) of the slab affected by the aging process, t_{dis} is roughly evaluated as the time needed to dissociate half of the CH bonds in the slab, whose number is N_{CH}, namely [13]:

$$t_{dis} = \frac{N_{CH}}{2R_{el}F_{eff}F_{hot}} \tag{6.66}$$

6. The damage growth rate in PE, R_{dis}, is determined versus electric field E, temperature T, and void height d, as follows [13]:

$$R_{dis} = \frac{D_{dis}}{t_{dis}} \tag{6.67}$$

Finally, an expression for the life L of the system was attained under the assumption that failure happens as soon as the damaged zone (made of contiguous sequentially degraded slabs of thickness D_{dis}) reaches a critical size d_C large enough to start an electrical tree—for example, by enhancing the local field, or, simply, by increasing the volume of damaged PE. As a consequence, L can be estimated from Eqs. (6.66) and (6.67), as follows [13]:

$$L = \frac{d_C}{R_{\mathrm{dis}}} \tag{6.68}$$

As observed in [13], all the parameters of the model based on damage growth from air-filled voids are related to physical-chemical properties of PE and of the air filling the void, with the exception of d_C, the only phenomenological parameter. d_C can be either guessed a priori via theoretical hypotheses (see, e.g., [58]) or a posteriori by comparing life estimates derived from Eq. (6.68) with experimental lifetimes. However, the approach by Jiang et al. in the framework of the FLSC theory [58] can suggest a reference value of d_C, namely, the growth of a pit ~1.5 μm long [13].

6.3.4 Fitting the Model to Extruded HVDC Insulation Times to Failure

In [23] the life model based on damage growth from air-filled voids was compared with times to failure relevant to HVDC cable insulation, namely:

The results of DC life tests after [21,23], performed at 20°C on two kinds of XLPE flat specimens, denoted as XLPE 1[17] and XLPE 2.

Lifetimes derived from long-term aging tests relevant to the same prototype 250 kV class DC undersea cables already treated in Section 6.1.1, which consisted in a sequence of 260 daily loading cycles (8 h heating, 16 h cooling) at voltage levels from ±375 to ±500 kV, with and without polarity reversals (see [3] for details). By processing these test results, an IPM life model—whose parameters are reported in [3]—was attained and validated, thereby enabling the extrapolation of the breakdown times versus poling field relationships down to service stress levels.

The times to failure versus poling field at room temperature (circles) for XLPE 1 and XLPE 2 [23] are compared in Figures 6.12 and 6.13, respectively, with the best fitting line derived via the life model of Eq. (6.68) (solid line) taking a value of void height $d = 10$ and 13 μm, respectively; in both cases the other life model parameter values are $R = 10$ μm, $\Delta G_t = 0.7$ eV, critical size of damaged zone $d_C = 3$ μm, plus values of current density J_C from XLPE data fitting after [22]. Both figures show that the life model fits fairly well the experimental failure times relevant to flat specimens [23].

Coming to the prototype 250-kV class DC cable dealt with in [3], the relevant values of some extrapolated lifetimes (in hours) versus poling field at room temperature

[17] This data set was already fitted to the DMM model; see Figure 6.10.

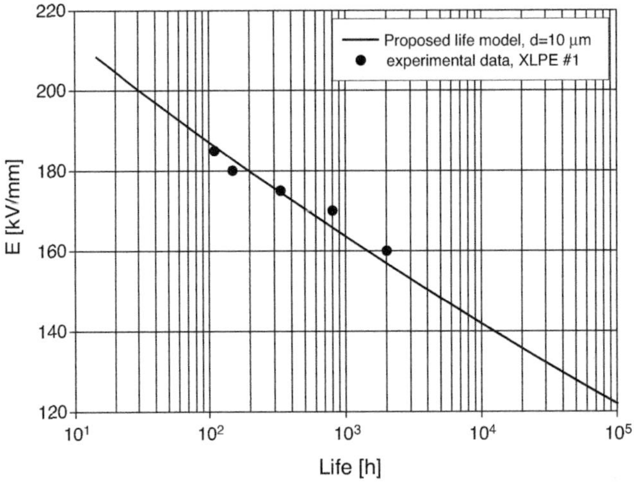

Figure 6.12 Comparison between experimental breakdown times (circles) obtained for XLPE 1 at room temperature (20°C) (after [21]) and the relevant best fitting values obtained at the same temperature by the life model of Eq. (6.68) (solid line) assuming a value of void height d = 10 μm and of critical size of damaged zone $d_C = 3$ μm. Other life model parameter values are $R = 10$ μm, $\Delta G_t = 0.7$ eV, plus current density J_C from [22] (after [23]).

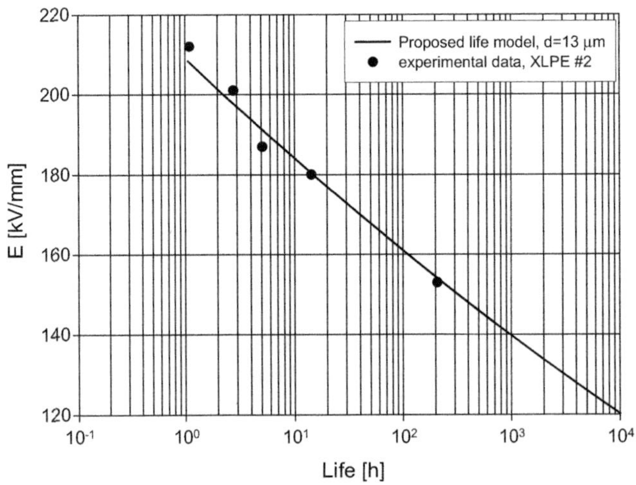

Figure 6.13 Comparison between experimental breakdown times (circles) obtained for XLPE 2 at room temperature (20°C) (after [23]) and the relevant best fitting values obtained at the same temperature by the life model of Eq. (6.68) (solid line) assuming a value of void height $d = 13$ μm and of critical size of damaged zone $d_C = 3$ μm. Other life model parameter values are $R = 10$ μm, $\Delta G_t = 0.7$ eV, plus current density J_C from [22] (after [23]).

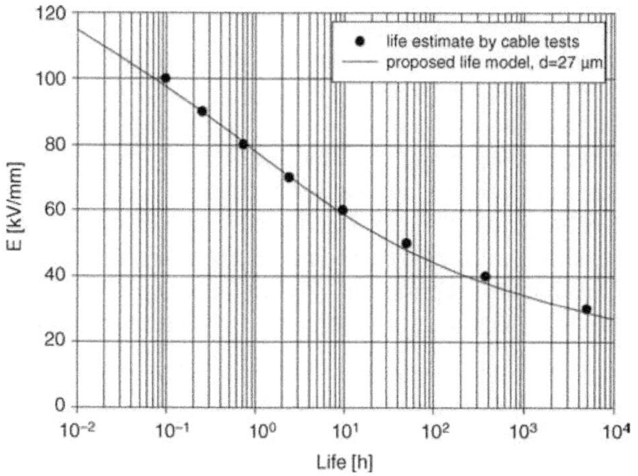

Figure 6.14 Comparison between some breakdown times obtained, at room temperature, by interpolating and extrapolating HVDC cable life test results via the IPM model (as done in [3]) and the relevant best fitting values obtained, at the same temperature, by the life model of Eq. (6.68) (solid line) under the following hypotheses: $d = 27\ \mu m$, $d_C = 3\ \mu m$, $R = 10\ \mu m$, $\Delta G_t = 0.7$ eV, plus current density J_C from [22] (after [23]).

are compared in Figure 6.14 with the best fitting line derived via the life model of Eq. (6.68) (solid line) under the following hypotheses: $d = 27\ \mu m$, $d_C = 3\ \mu m$, $R = 10\ \mu m$, $\Delta G_t = 0.7$ eV, and current density J_C from XLPE data fitting after [22]. Once more, the figure shows a good agreement between life model and test data [23].

Figures 6.12–6.14 confirm that void height d and critical size of damaged zone d_C are essential microstructural parameters for the quality of manufacturing and, ultimately, the reliability of cable insulation. Thus in [23] attention was focused on the values of d and d_C derived by fitting the life model of Eq. (6.68) to the experimental data displayed in Figures 6.12–6.14: These values are listed in Table 6.3.

The table shows that the best fitting is achieved with a larger void height for cable prototypes (Fig. 6.14) than for flat specimens (Figs. 6.12 and 6.13). This does make sense since it is likely that the cable manufacturing process (in particular the extrusion stage) can generate voids of much larger size than the press-molding process of flat specimens. Actually, a value of $d = 27\ \mu m$ is in the range of maximum

Table 6.3 Values of Void Height, d, and Critical Size of Damaged Zone, d_C, Derived by Fitting the Life Model of Eq. (6.68) to the Experimental Data Displayed in Figs. 6.12–6.14[a]

Data Set → / ↓ Parameter	XLPE 1 (Fig. 6.12)	XLPE 2 (Fig. 6.13)	HVDC Cable [3] (Fig. 6.14)
d [μm]	10	13	27
d_C [μm]	3	3	3

[a]All relative to room temperature. The values of the other life model parameter values are the same for all figures, i.e., $R = 10\ \mu m$, $\Delta G_t = 0.7$ eV, plus current density J_C from [22].

Source: After [23].

size of technological imperfections commonly encountered in up-to-date extra-clean XLPE compounds for EHV applications, since dry cured, extra clean XLPE cables are expected to have maximum void sizes below 30–40 μm [3,17–19,27,69]. As for d_C, as previously pointed out, its value is consistent with physical arguments about the critical size of damaged regions stemming from local defects [57, 58].

According to [23] all these considerations support a physical meaning of the good data fitting in Figures 6.12–6.14, although more experimental support is required through dedicated tests. On the other hand, it should be pointed out here that, particularly as far as HVDC extruded insulation is concerned, the physical meaning of the good fitting in Figures 6.12–6.14 has to be regarded with great care for some more reasons:

The role of charge carriers other than electrons is neglected in the aging and life model based on damage growth from air-filled voids, but other charge carriers may have an effect, for example, ions.

The field distribution within the discharging void is assumed as uniform, but this may be a quite rude hypothesis in the presence of avalanches and high density of stored charge on cavity walls [68].

Actually the PD repetition rate in DC is orders of magnitude lower than at AC voltage with the same amplitude [71].[18] Thus it is reasonable that partial discharge aging does not play a major role in HVDC extruded insulation at usual service fields. Also for this reason, the life model based on damage growth from air-filled voids—initially developed for DC regime—has been extended and focused on the AC regime thereafter [68].

The mechanism of degradation proposed in the aging and life model based on damage growth from air-filled voids is completely different and alternative to that proposed in the DMM model. Nevertheless, they both fit well the same data set of times to failure (compare Fig. 6.10 with 6.12). As highlighted in Section 6.2.2.2, this could be addressed to "phenomenological effects" related to these two life models since they are both characterized by a large number of parameters, thus with a large flexibility, this can accomplish a good fitting to different data sets without a precise physical meaning.

6.4 SPACE CHARGE: CAUSE OR EFFECT OF AGING?

Since more and more refined measurement tools of space charge in dielectrics became available (see Chapter 4), an ever enduring discussion started about the origin of space charges and their aging effects. Mainly two alternative hypotheses were supported by different authors [13]:

1. The onset and/or increase of space-charge distributions in insulation is a consequence of aging, but any direct effect of space charge on electrical aging can be neglected [65, 72–78];

[18] There is one exception to this rule, that is, relative high repetition rates can occur in DC insulation systems when the insulation is polarized or depolarized during, respectively, first application and turning off of the voltage [71].

2. Space charge can be both cause and consequence of aging. This is the approach followed by the model illustrated in Section 6.2 [20,21,62], as well as by models dealt with in other articles and described for the sake of completeness in the present section [79–83].

These two approaches differ substantially:

1. From a theoretical viewpoint: Indeed, if the first hypothesis above holds, no influence of space charge in insulation should be accounted for in the development of life models.
2. From a practical viewpoint. Indeed, if the first hypothesis above holds, insulation systems could be designed without taking into account space-charge behavior and regarding space charge only as an aging marker. This, however, is in contrast with research and development activities that ultimately led to the main HVDC cable system designs, as extensively illustrated in Chapter 5.

Two electrical life models, presented in the last two decades by Lewis and co-workers [72–75] and Crine [65,78], neglect space charges as a direct cause of aging and consider electromechanical stress only. Both these models share a common thermokinetic background with the DMM space-charge life model (see Section 6.2 [20,21,62]), regarding electrical aging as a thermally activated process that can be summarized as follows:

1. A free energy barrier ΔG shall be crossed for the system to pass from the unaged state (of energy G_1) to the aged state (of energy G_2).
2. When an electric field is applied, the barrier ΔG decreases with the square of the field due to an electromechanical modification of the intrinsic free energy barrier.

The model set up by Lewis et al. [72–74] is based on electromechanical stresses, at low and high fields, as the source of aging through microvoid enlargement in the incubation period of electrical trees. This model is similar, as hypotheses and development (thermodynamic basis, electromechanical energy causing local damage, partially reversible reactions, and the existence of electrical threshold), to the DMM space-charge model of Section 6.2, but initially it did not consider the presence of space charge as a cause of degradation. The final stage of aging, promoting the appearance of electrical trees, is explained by the Griffith crack propagation theory. The time to trigger crack growth, t (assumed to be the far longest period of the total failure process, thereby coinciding in practice with time to failure), is given by

$$t = \int_{N_b}^{\eta_C} \left\{ \frac{k_B T}{h_P} \left[\exp\left(\frac{-G_b(E)}{kT} \right) (N_b - \eta) - \exp\left(\frac{-G_f(E)}{kT} \right) \eta \right] \right\}^{-1} d\eta \qquad (6.69)$$

where h_P and k_B are, as usual, the Planck and Boltzmann constants, respectively, N_b is the number of breakable bonds, G_f and G_b are the energies for bond forming and breaking, respectively (correlated to the above energies G_1 and G_2), η is the number of broken bonds, and η_C is its critical value. This model shows a threshold field below which electrical degradation does not occur.

More room for space-charge-related effects was given in subsequent investigations carried out in the framework of this life model. Indeed, according to [76], "a combination of the Maxwell stress and electrostriction associated with a change in permittivity with strain will cause mechanical stress and consequent strain. For a fluid the electrically induced stress is given by the Helmholtz expression" ([76], page 721):

$$\mathbf{F} = \rho\mathbf{E} - \left(\frac{1}{2}\right)E^2\nabla\varepsilon + \left(\frac{1}{2}\right)\nabla\left[\rho_m \frac{d(\varepsilon E^2)}{d(\rho_m)}\right] \tag{6.70}$$

where ρ and ρ_m are charge and mass density, respectively. The terms in Eq. (6.70) are as follows [76]:

- [First term]: Simple Coulomb interaction between the field and charge in the system.
- [Second term]: A dielectrophoretic stress pulling regions of high permittivity into those of low permittivity. It accounts for non-uniform polarizability, for example in the interface regions of any dielectric boundaries.
- [Third term]: An electrostrictive stress, depending on the change in permittivity with density.

> [As pointed out in [76]] . . . strictly the Helmholtz version is valid only for fluids and in the case of . . . PE where a shearing stress can exist, the electrostrictive term in Equation (6.70) must be replaced by expressions for incremental increases corresponding to strains parallel and orthogonal to **E** . . . Two features are significant. The induced mechanical stress depends quadratically on the field, and a stress component orthogonal to the field direction also is generated, which will tend to expand the structure, increasing free volume and the likelihood of microvoids with major axes in the field direction ([76], page 721–722).

It has been shown that this stress orthogonal to the electrical field has magnitude [13, 74, 76]:

$$\sigma_m = -\varepsilon E^2 \tag{6.71}$$

As further highlighted in [76]:

> The field-generated stress will alter the potential energies G_1 and G_2 [of the two-state thermo-kinetic model] and so encourage conformational relaxations of the morphology . . . Over long periods of electrical stress, the morphological changes might be considerable, particularly in regions such as interfaces where the field is likely to be enhanced. There is evidence . . . that electrical stress, in altering the morphology, can change the shear modulus of the polymer ([76], page 722).

As far as σ_m is concerned, Crine has hypothesized in [78] that it acts against the van der Waals' attractions, straining polymer chains and enlarging nanocavities, thereby yielding insulation damage up to breakdown. In this framework, a life model can be obtained by considering partially reversible—that is, both direct and reverse—reactions and by inserting the electromechanical energy reduction term of Eq. (6.71) in the activation energy barrier ΔG. Then, insulation life L as a function of electric field E and temperature T can be attained in the following form [13, 78]:

$$L = \left(\frac{h_p}{2k_B T}\right) \exp\left(\frac{\Delta G}{k_B T}\right) \operatorname{csch}\left(\frac{1}{2}\frac{\varepsilon \Delta V E^2}{k_B T}\right) \tag{6.72}$$

where ΔV is the activation (or "strained") volume. Relationship (6.72) holds for DC electrical fields, while for AC fields a similar expression has been suggested, where the AC frequency appears at the denominator of the preexponential term [13, 78].

Crine's model has been successful in fitting experimental time-to-failure data [78], nevertheless—as it holds for the models dealt with in Sections 6.2 and 6.3—it must be emphasized that it features a large number of parameters, some of which affect strongly the estimated life, thereby fostering the achievement of good data fitting.

As discussed in [13], Crine's model is valid until the onset of PD phenomena since other degradation processes will become active as cavity size increases, for example, partial discharges (PD). Furthermore, in Crine's model the role of space charge is subordinate to the formation and enlargement of nanocavities: the higher the density of nanocavities as aging progresses, the higher the amount of charges trapped with respect to the unaged insulation [13]. On these grounds, space-charge density in the dielectric is expected to rise with aging, in agreement with most experimental results [78, 84–87].

Contrary to the models proposed by Crine and Lewis, other electrical life models—in addition to that treated in Section 6.2—consider space charge not only as an effect but also as a direct cause of aging. This is consistent with the many outcomes of the research and development activity on HVDC polymeric cables that shows how cable design has to account for the presence of space charge (see, e.g., [3–5, 10, 25, 65]). Since constant DC voltage tests with and without voltage polarity reversal, as well as impulse tests with superimposed DC voltage have shown that insulation life can drop even dramatically above a certain space charge threshold, a life model able to predict life under voltage polarity inversion has been developed on top of an extensive experimental campaign on PE-based materials. This model has the following expression [13, 80]:

$$\ln\left[\left(\frac{L}{L_c}\right) - 1\right] = A_1 + A_2 \ln(q) + A_3 \ln(E_M) + A_4 \ln(s) + A_5 \ln(f) \tag{6.73}$$

where L and L_i are life without and with polarity inversion, respectively, E_M is the maximum "Poissonian" field measured in the insulation, q is the space charge per unit volume, s is the exponent in the time dependence of the charge depletion relationship ($q \propto t^{-s}$), and f is the frequency of polarity inversion.

Life model (6.73) can be made simpler resorting to the power law dependence of space-charge density on poling (Laplacian) field, E, found above the field threshold for space-charge accumulation, E_T (described, e.g., in [88,89], and discussed herein in Section 4.5.2.2):

$$q = A_0 E^b \tag{6.74}$$

where b is the power law exponent, which can be determined as the slope in a log–log plot of the q versus E characteristic above the threshold E_T (see Fig. 6.15). When replacing q in Eq. (6.73) with relationship (6.74), after proper manipulations described in detail in [80] one gets (see also [68])

$$\ln\left[\left(\frac{L}{L_i}\right) - 1\right] = b' \ln(KE) + A_5 \ln(f) \tag{6.75}$$

where K and b' depend on the coefficients A_0, A_1, A_2, A_3, and A_4. Both the more complex equation (6.73) and the more compact equation (6.75) show that the larger the amount of space charge, the shorter is life under voltage polarity inversion with respect to life without inversion. Of course, space charge can also affect significantly the life without voltage polarity inversions [13], as discussed in previous chapters.

The above-mentioned field threshold for space charge accumulation can provide more arguments about the debated matter whether space charge is a cause or an effect of aging. Indeed in the early 2000s a few papers [90–92] have proposed that the threshold for electrical aging under DC field may be close to the threshold above which space charge becomes trapped in the insulation, under the assumption that electric fields are moderate and the insulation is free from macroscopic defects which might lead to premature failure through highly energetic phenomena (this is likely to be the case in the improved insulation for HVDC cable systems dealt with in Chapter 5). In [93], observing that the presence of space charge has consequences for other electrical properties of the material, these properties were used in conjunction with direct measurements of space charge to infer the absence or the existence of space charge as a function of applied DC electric field—thereby checking for the existence of a space charge threshold—by using complementary techniques, namely:

1. space charge measurements via the PEA technique;
2. conduction current measurements;[19]
3. electroluminescence (EL) measurements.[20]

[19] In the SCLC theory, the critical field separating Ohmic conduction from the nonlinear regime in the current–voltage characteristic corresponds to the onset of space charge accumulation, see Section 4.2.2.2 [93].

[20] Electroluminescence is the process by which light is emitted from the polymer under electrical stress. EL detection was selected as complementary technique providing information on space charge accumulation and charge dynamics since under moderate field conditions—when the existence of hot-carriers can be discarded—EL excitation needs the mediation of space charge and can therefore be used to infer space charge accumulation [93].

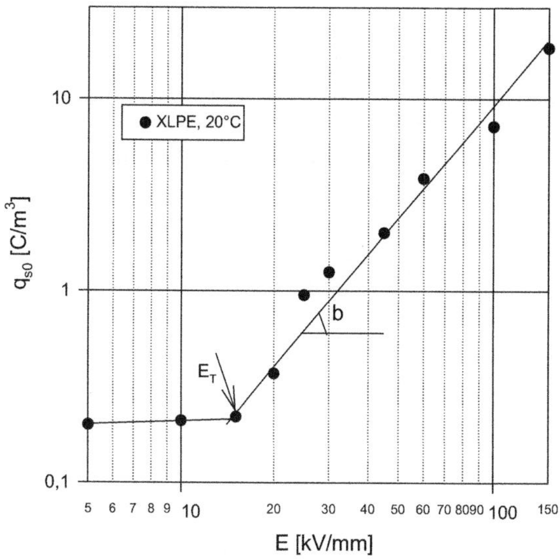

Figure 6.15 Example of threshold characteristic, relevant to XLPE, 20°C. The threshold for space-charge accumulation, E_T, is indicated by an arrow (after [84]).

Among the experimental results relevant to various polymers reported in [93], we quote those relevant to XLPE, considered here as a reference material for HVDC extruded cables due to the reasons illustrated in Chapter 5. The tested XLPE specimens were 150 μm thick films peeled from the insulation of a power cable and subjected to a thermal treatment (60°C, 48 h) in order to remove most crosslinking by-products. As Figure 6.16 shows clearly, similar values of space charge threshold for XLPE are found with the three above-mentioned techniques. This supports the physical explanation of a threshold for trapped space charge accumulation.

These results provide one more clue about the relationship between space charge and aging, especially when considering the results reported in [94], an up-to-date milestone paper that discusses very thoroughly the relationships between charge dynamics and electrical aging. In particular one of the goals of [94] is investigating whether the main processes involved in charge dynamics (i.e., charge injection/extraction, charge trapping/detrapping, and charge recombination) yield energetic features able to bring irreversible changes in the local physical/chemical nature of the polymers. In this framework, it is observed in [94] that "Charge trapping/de-trapping processes are primarily reversible processes by which thermal energy is exchanged between the charge and its surroundings. Conversely, charge recombination . . . involves the creation of electronic excited states that are of paramount importance in chemical reactions . . . The excited states can decay to their ground states but they can also decay by other routes where bond breaking and bond rearrangements, i.e. chemistry, are involved. Recombination is a natural link to aging processes" ([94], page 12).

Furthermore, it is shown in [94] after a broad and detailed illustration that in additive-free LDPE "the electroluminescence spectra match the electron beam-induced luminescence spectra advocating the existence of massive chemical/physical degradation in the electrical regime where recombination has been found to be a predominant mechanism" ([94], page 19). Of course additive-free LDPE cannot be compared directly with XLPE with additives used in HV applications (especially in DC applications, see Chapter 5). Nevertheless, in the perspective outlined in [94] the evidence after [93] that the threshold for the onset of space charge practically coincides with the onset of electroluminescence activity could be a clue that above such threshold the field is capable of activating irreversible aging processes associated with the presence of space charge and its dynamics.

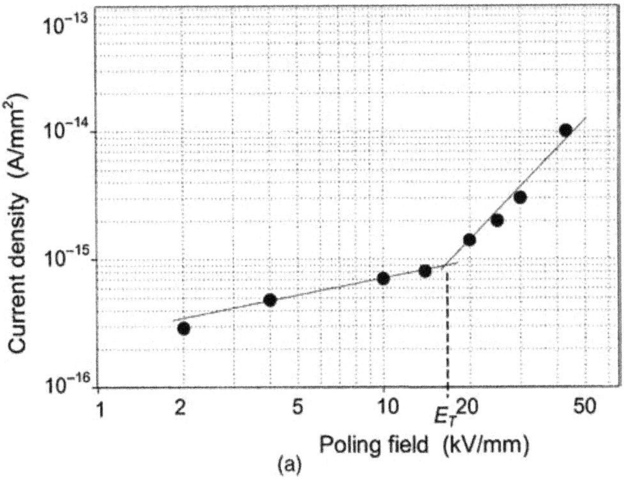

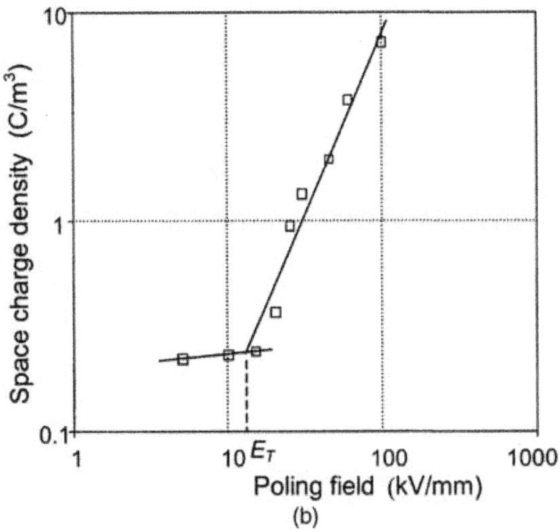

Figure 6.16 Threshold field for space charge accumulation, E_T (indicated by an arrow) estimated in XLPE sample by plotting vs. DC poling field: (*a*) the measured current density; (*b*) the space charge density measured using the PEA; (*c*) the number of EL pulses (integrated EL) at short-circuit (the polarization time, 1 h, was kept constant at each field level). After [93].

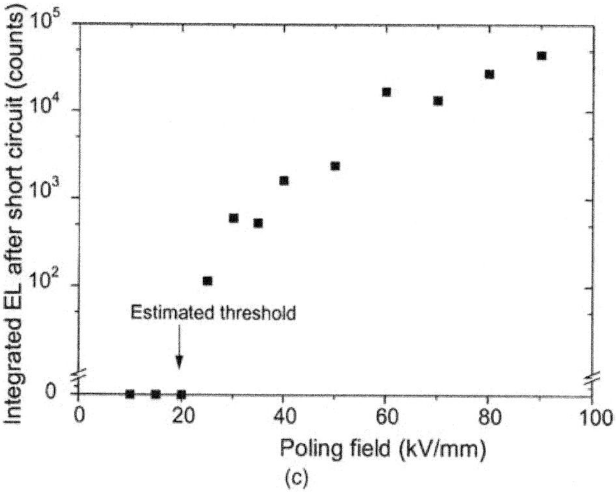

Figure 6.16 (*Continued*)

Summarizing, it can be concluded in agreement with [13] that "the above results and experimental observations seem definitely to support an aging acceleration effect under DC electrical fields caused by space charge. This might indicate that . . . even the simplest phenomenological life models can be applied with the consideration of the maximum field magnification in an insulation, and in such a way can provide life estimations which are based on the extreme value (weakest point) approach" [95] ([13], page 887).

REFERENCES

1. W. Nelson. *Accelerated Testing*. Wiley, New York, 1990.
2. G.C. Montanari, and G. Mazzanti. "Insulation Aging Models." In *Encyclopedia of Electrical and Electronics Engineering*, Vol. 10, J. Webster, Ed. Wiley, New York, pp. 308–319, 1999 (updated in 2007).
3. K. Terashima, H. Suzuki, M. Hara, and K. Watanabe. "Research and Development of ±250 kV DC XLPE Cables." *IEEE Trans. Power Del.*, Vol. 13, pp. 7–16, 1998.
4. S. Maruyama, N. Ishii, M. Shimada, S. Kojima, H. Tanaka, M. Asano, T. Yamanaka, and S. Kawakami. "Development of a 500-kV DC XLPE Cable System." *Furukawa Rev.*, No. 25, pp. 47–52, 2004.
5. Brochure CIGRÉ 496. Recommendations for Testing DC Extruded Cable Systems for Power Transmission at a Rated Voltage up to 500 kV. CIGRE WG B1–32, April 2012.
6. IEC 62067. Power Cables with Extruded Insulation and Their Accessories for Rated Voltages above 150 kV ($U_m = 170$ kV) up to 500 kV ($U_m = 550$ kV)—Test Methods and Requirements, 2nd ed., November 2011.
7. M. Marzinotto, G. Mazzanti, and C. Mazzetti. "A New Approach to the Statistical Enlargement Law for Comparing the Breakdown Performance of Power Cables—Part 1: Theory." *IEEE Trans. Dielectr. Electr. Insul.*, Vol. 14, No. 5, pp. 1232–1241, October 2007.

8. M. Marzinotto, G. Mazzanti, and C. Mazzetti. "A New Approach to the Statistical Enlargement Law for Comparing the Breakdown Performances of Power Cables—Part 2: Application." *IEEE Trans. Dielectr. Electr. Insul.*, Vol. 15, No. 3, pp. 792–799, June 2008.
9. E. Ildstad, J. Sletbak, B.R. Nyberg, and J.E. Larsen."Factors Affecting the Choice of Insulation System for Extruded HVDC Power Cables." Paper D1-203, CIGRE Session 2004.
10. M. Jeroense. "HVDC, the Next Generation of Transmission. Highlights with Focus on Extruded Cable Systems." Proc. 2008 International Symposium on Electrical Insulating Materials (ISEIM 2008), pp. 10–15, Yokkaichi, Japan, Sept. 7–11, 2008.
11. G. Mazzanti. "Analysis of the Combined Effects of Load Cycling, Thermal Transients and Electro-thermal Stress on Life Expectancy of High Voltage ac Cables." *IEEE Trans. Power Del.*, Vol. 22, No. 4, pp. 2000–2009, October 2007.
12. G. Mazzanti. "The Combination of Electrothermal Stress, Load Cycling and Thermal Transients and Its Effects on the Life of High Voltage ac Cables." *IEEE Trans. Dielectr. Electr. Insul.*, Vol. 16, No. 3, pp. 1168–1179, August 2009.
13. G. Mazzanti, G.C. Montanari, and L. Dissado. "Electrical Aging and Life Models: The Role of Space Charge." *IEEE Trans. Dielectr. Electr. Insul.*, Vol. 12, pp. 876–890, October 2005.
14. T.W. Dakin. "Electrical Insulation Deterioration Treated as a Chemical Rate Phenomenon." *AIEE Trans.*, Vol. 67, pp. 113–122, 1948.
15. S. Glasstone, K.J. Laidler, and H.E. Eyring. *The Theory of Rate Processes*. McGraw-Hill, New York, 1941.
16. S.N. Zhurkov. "Kinetic Concept of the Strength of Solids." *Int. J. Fract. Mech.*, Vol. 1, pp. 311–323, 1965.
17. J.O. Bostrom, N. Hampton, H. Lennartsson, E. Marsden, and U. Nilsson. "Electrical Stress Enhancement of Contaminants In XLPE Insulation Used for Power Cables." *IEEE Electr. Insul. Mag.*, Vol. 19, No. 4, pp. 6–11, July–August 2003.
18. W.G. Weissenberg, U. Rengel, and R. Scherer. "EHV XLPE Cable Systems up to 400 kV—More Than 10 Years Field Experience." CIGRE Paper No. B1-102, 2004.
19. J. Kjellqvist, G. Goethel, S.J. Han, A. Mendelsohn, and M. Mangnus. "New Semiconductive Products Providing Higher Performance Cables." Proceedings of 19th International Conference on Electricity Distribution, CIRED, Vienna, May 21–24, 2007.
20. L.A. Dissado, G. Mazzanti, and G.C. Montanar. "Elemental Strain and Trapped Space Charge in Thermoelectrical Aging of Insulating Materials. Part 1: Elemental Strain under Thermo-electrical-Mechanical Stress." *IEEE Trans. Dielectr. Electr. Insul.*, Vol. 8, No. 6, pp. 959–965, December 2001.
21. L.A. Dissado, G. Mazzanti, and G.C. Montanari. "Elemental Strain and Trapped Space Charge in Thermoelectrical Aging of Insulating Materials. Life Modeling." *IEEE Trans. Dielectr. Electr. Insul.*, Vol. 8, No. 6, pp. 966–971, December 2001.
22. G. Mazzanti, G.C. Montanari, and F. Civenni. "Model of Inception and Growth of Damage from Microvoids in Polyethylene-based Materials for HVDC Cables—Part 1: Theoretical Approach." *IEEE Trans. Dielectr. Electr. Insul.*, Vol. 14, No. 5, pp. 1242–1254, October 2007.
23. G. Mazzanti, G.C. Montanari, and F. Civenni. "Model of Inception and Growth of Damage from Microvoids in Polyethylene-based Materials for HVDC Cables—Part 2: Parametric Investigation and Data Fitting." *IEEE Trans. Dielectr. Electr. Insul.*, Vol. 14, No. 5, pp. 1255–1263, October 2007.

24. G.C. Montanari, G. Mazzanti, and L. Simoni. "Progress in Electrothermal Life Modeling of Electrical Insulation over the Last Decades." *IEEE Trans. Dielectr. Electr. Insul.*, Vol. 9, No. 5, pp. 730–745, 2002.

25. EPRI Report No. 1008720. DC Cable Systems with Extruded Dielectrics. Palo Alto, California, December 2004.

26. M. Jeroense, M. Bergkvist, A. Gustafsson, H. Rapp, T. Sörqvist, and J. Svahn."Increased Voltage for the HVDC Light® Product Range—A Complete Solution." Paper B.4.7, pp. 425–430, JICABLE 2007.

27. T. Kubota, Y. Takahashi, S. Sakuma, M. Watanabe, M. Kanaoka, and H. Yamanouchi. "Development of 500-kV XLPE Cables and Accessories for Long Distance Underground Transmission Line. Part I. Insulation Design of Cables." *IEEE Trans. Power Del.*, Vol. 9, No. 4, pp. 1738–1749, October 1994.

28. N. Takeda, S. Izumi, K. Asari, A. Nakatani, H. Noda, M. Yamaguchi, and M. Tan. "Development of 500-kV XLPE Cables and Accessories for Long-Distance Underground Transmission Lines. Part IV: Electrical Properties of 580-kV Extrusion Molded Joints." *IEEE Trans. Power Del.*, Vol. 11, No. 2, pp. 635–643, April 1996.

29. L. Simoni. *Fundamentals of Endurance of Electrical Insulating Materials.* CLUEB, Bologna, Italy, 1983.

30. G. Mazzanti and G.C. Montanari. "A Comparison between XLPE and EPR as Insulating Materials for HV Cables." *IEEE Trans. Power Del.*, Vol. 12, No. 1, pp. 15–28, January 1997.

31. T. Worzyk. *Submarine Power Cables.* Spinger, Berlin, 2009.

32. R.N. Hampton. "Some of the Considerations for Materials Operating under High-Voltage, Direct-Current Stresses." *IEEE Electr. Insul. Mag.*, Vol. 24, No. 1, pp. 5–13, January–February 2008.

33. E. Chiodo and G. Mazzanti. "Theoretical and Practical Aids for the Proper Selection of Reliability Models for Power System Components." *Int. J. Reliability Safety*, Vol. 2, Nos. 1/2, pp. 99–128, 2008.

34. E. Chiodo and G. Mazzanti. "Mathematical and Physical Properties of Reliability Models in View of Their Application to Modern Power System Components." In *Innovations in Power Systems Reliability*, G.J. Anders and A. Vaccaro, Eds. p. 59–140, Springer, London, 2011.

35. W. Nelson. *Applied Life Data Analysis*, Wiley, New York, 1982.

36. W. Mosch and W. Hauschild. *Statistical Techniques for HV Engineering*. Peter Peregrinus, London, 1992.

37. G.C. Montanari and M. Cacciari. "A Probabilistic Life Model for Insulating Materials Showing Electrical Threshold." *IEEE Trans. Dielectr. Electr. Insul.*, Vol. 24, No. 1, pp. 127–134, 1989.

38. E. Occhini. "A Statistical Approach to the Discussion of the Dielectric Strength in Electric Cables." *IEEE Trans. Power App. Syst.*, Vol. 90, No. 6, pp. 2671–2682, December 1971.

39. M. Marzinotto, G. Mazzanti, and C. Mazzetti. "Comparison of Breakdown Performances of Extruded Cables via the Enlargement Law." Proc. 2006 Conference on Electrical Insulation and Dielectric Phenomena, pp. 760–763, Kansas City, Missouri, 2006.

40. M. Marzinotto. "On the Application of the Enlargement Law to Cable Lines." Proceedings of IEEE Power Tech 2005, St. Petersburg, Russia, June 27–30, 2005.

41. M. Marzinotto, C. Mazzetti, M. Pompili, and P. Schiaffino. "EPR Lifetime under Impulsive Voltage Stress." Proceedings of IEEE Conference on Electrical Insulation and Dielectric Phenomena IEEE–CEIDP 2005, Nashville, Tennessee, October 16–19, 2005.

42. M. Marzinotto and G. Mazzanti. "The Application of the Enlargement Law to HVDC Cable Lines." Proceedings of 2010 IEEE Conference on Electrical Insulation and Dielectric Phenomena (IEEE CEIDP 2010), pp. 506–509, West Lafayette, Indiana, October 17–20, 2010.

43. M. Marzinotto and G. Mazzanti. "A Deeper Insight into the Application of the Enlargement Law to HVDC Cables." Proceedings of 2011 IEEE Conference on Electrical Insulation and Dielectric Phenomena (IEEE CEIDP 2011), pp. 505–509, Cancun (Mexico), October 16–19, 2011.

44. E.S. Cooper, L.A. Dissado, and J.C. Fothergill. "Application of Thermoelectric Aging Models to Polymeric Insulation in Cable Geometry." *IEEE Trans. Dielectr. Electr. Insul.*, Vol. 12, No. 1, pp. 1–10, February 2005.

45. G. Mazzanti. "Effects of the Combination of Electro–thermal Stress, Load Cycling and Thermal Transients on Polymer-Insulated High Voltage ac Cable Life." Proc. 2005 IEEE Power Engineering Society (PES) General Meeting, San Francisco, California, June 12–16, 2005.

46. IEC 60216-1. Electrical Insulating Materials—Properties of Thermal Endurance—Part 1: Ageing Procedures and Evaluation of Test Results, edition 5.0, July 2001.

47. IEC 60505. Evaluation and Qualification of Electrical Insulation Systems, Edition 3.0, October 2004.

48. H.S. Endicott, B.D. Hatch, and R.G. Sohmer. "Application of the Eyring Model to Capacitor Aging Data." *IEEE Trans. Comp. Parts*, Vol. 12, pp. 34–41, March 1965.

49. G. Mazzanti, G.C. Montanari, and L. Simoni. "Study of the Synergistic Effect of Electrical and Thermal Stresses on Insulation Life." Proc. 1996 Conference on Electrical Insulation and Dielectric Phenomena (CEIDP), pp. 684–687, San Francisco, California, October 1996.

50. G. Mazzanti, G.C. Montanari, L. Simoni, and M.B. Srinivas. "Combined Electro-Thermo-Mechanical Model for Life Prediction of Electrical Insulating Materials." Proceedings of 1995 Conference on Electrical Insulation and Dielectric Phenomena (CEIDP), pp. 274–277, Virginia Beach Virginia, October 1995.

51. E. Ildstad and S.T. Hagen. "Electrical Treeing and Breakdown of Mechanically Strained XLPE Cable Insulation." Conf. Rec. of 1992 IEEE Int. Symp. on El. Insul., Baltimore, MD, pp. 135–139, June 1992.

52. G.C. Montanari and G. Mazzanti. "From Thermodynamic to Phenomenological Multi-stress Models for Insulating Materials without or with Evidence of Threshold." *J. Phys. D: Appl. Phys.*, Vol. 27, pp. 1691–1702, 1994.

53. L.A. Dissado. "Predicting Electrical Breakdown in Polymeric Insulators: From Deterministic Mechanisms to Failure Statistics." *IEEE Trans DEI*, Vol. 9, No. 5, pp. 860–875, October 2002.

54. H.R. Zeller and W. Schneider. "Electrofracture Mechanics of Dielectric Aging." *J. Appl. Phys.*, Vol. 56, pp. 455–459, 1984.

55. H.R. Zeller. "Breakdown and Prebreakdown Phenomena in Solid Dielectrics." *IEEE Trans. Electr. Insul.*, Vol. 22, No. 2, pp. 115–122, April 1987.

56. T. Baumann, B. Fruth, F. Stucki, and H.R. Zeller. "Field–Enhancing Defects in Polymeric Insulators Causing Dielectric Ageing." *IEEE Trans. Electr. Insul.*, Vol. 24, No. 6, pp. 1071–1076, December 1989.

57. S.A. Boggs. "Theory of Defect-Tolerant Dielectric Systems." *IEEE Trans. Electr. Insul.*, Vol. 28, No. 3, pp. 365–371, June 1993.

58. G. Jiang, J. Kuang, and S. Boggs. "Critical Parameters for Electrical Tree Formation in XLPE." *IEEE Trans. Power Del.*, Vol. 13, No. 2, pp. 292–296, April 1998.

59. L.A. Dissado and J.C. Fothergill. *Electrical Degradation and Breakdown in Polymers*. Peregrinus, London, 1992.

60. K. Kaneko, T. Mizutani and Y. Suzuoki. "Computer Simulation on Formation of Space Charge Packets in XLPE Films." *IEEE Trans. Dielectr. Electr. Insul.*, Vol. 6, No. 2, pp. 152–158, April 1999.

61. A. See, L.A. Dissado and J.C. Fothergill. "Electric Field Criteria for Charge Packet Formation and Movement in XLPE." *IEEE Trans. Dielectr. Electr. Insul.*, Vol. 8, No. 6, pp. 859–866, December 2001.

62. L.A. Dissado, G. Mazzanti, and G.C Montanari. "The Incorporation of Space Charge Degradation in the Life Model for Electrical Insulating Materials." *IEEE Trans. Dielectr. Electr. Insul.*, Vol. 2, No. 6, pp. 1147–1158, December 1995.

63. G. Mazzanti, G.C. Montanari, and L.A. Dissado. "A Space-Charge Life Model for AC Electrical Aging of Polymers." *IEEE Trans. Dielectr. Electr. Insul.*, Vol. 6, No. 6, pp. 864–875, December 1999.

64. D. Kahaner, C. Moler, and S. Nash. *Numerical Methods and Software*. Prentice-Hall, Englewood Cliffs, N.J., 1989.

65. J.P. Crine. "A Molecular Model to Evaluate the Impact of Aging on Space Charges in Polymer Dielectrics." *IEEE Trans. Dielectr. Electr. Insul.*, Vol. 4, No. 5, pp. 487–495, October 1997.

66. G.C. Montanari. "The Electrical Degradation Threshold of Polyethylene Investigated by Space Charge and Conduction Current Measurements." *IEEE Trans. Dielectr. Electr. Insul.*, Vol. 7, pp. 309–315, 2000.

67. J.L. Augè, C. Laurent, D. Fabiani, and G.C. Montanari. "Investigating dc Polyethylene Threshold by Space Charge, Current and Electroluminescence Measurements." *IEEE Trans. Dielectr. Electr. Insul.*, Vol. 7, pp. 797–803, 2000.

68. G.C. Montanari. "Bringing an Insulation to Failure: The Role of Space Charge." *IEEE Trans. Dielectr. Electr. Insul.*, Vol. 18, No. 2, pp. 339–364, 2011.

69. AEIC CS9–06 "Specification for Extruded Insulation Power Cables and Their Accessories Rated above 46 kV through 345 kV ac." Association of Edison Illuminated Companies, December 2006.

70. G. Mazzanti, G.C. Montanari, and S. Serra. "Theory of Inception Mechanism and Growth of Defect-Induced Damage in Polyethylene Cable Insulation." *J. Appl. Phys.*, Vol. 98, No. 3, pp. 034102.1–034102.15, August 2005.

71. P.H.F. Morshuis and J.J. Smit. "Partial Discharges at DC Voltage: Their Mechanism, Detection and Analysis." *IEEE Trans. Dielectr. Electr. Insul.*, Vol. 12, No. 2, pp. 328–340, 2005.

72. J. Lewis. "Charge Transport, Charge Injection and Breakdown in Polymeric Insulators." *J. Phys. D: Appl. Phys.*, Vol. 23, pp. 1469–1478, 1990.

73. T.J. Lewis, J.P. Llewellyn, and M.J. van der Sluijs. "Electrically Induced Mechanical Strain in Insulating Dielectrics." IEEE CEIDP, pp. 328–333, Arlington, Texas, October 1994.

74. C.L. Griffiths, J. Freestone, and R.N. Hampton. "Thermoelectric Aging of Cable Grade XLPE." IEEE CEIDP, pp. 578–582, Atlanta, Georgia, June 1998.

75. T.J. Lewis. "Ageing–A Perspective." *IEEE EI Mag.*, Vol. 17, No. 4, pp. 6–16, July–August 2001.

76. T.J. Lewis. "Polyethylene under Electric Stress." *IEEE Trans. Dielectr. Electr. Insul.*, Vol. 9, No. 5, pp. 717–729, October 2002.

77. N.F. Mott. "Electrons in Non-Cristalline Materials." In *Electronic and Structural Properties of Amorphous Semiconductors*, P.G. LeComber and J. Mort, Eds. Academic, London, 1973.

78. J.P. Crine. "On the Interpretation of Some Electrical Aging and Relaxation Phenomena in Solid Dielectrics." *IEEE Trans. Dielectr. Electr. Insul.*, Vol. 12, No. 6, pp. 1089–1107, December 2005.

79. D. Muto, S. Kondo, K. Kaneko, Y. Suzuoki, T. Mizutani, and T. Itaya. "Space Charge Characteristics of Modified High-Density Polyethylene for DC Cables." IEEE ISEIM, pp. 545–548, Toyohashi, Japan, September 1988.

80. A. Cavallini, D. Fabiani, G. Mazzanti, and G.C. Montanari. "A Life Model Based on Space Charge Quantities for HVDC Polymeric Cables Subjected to Voltage Polarity Inversions." *IEEE Trans. Dielectr. Electr. Insul.*, Vol. 9, No. 4, pp. 514–523, August 2002.

81. G.C. Montanari. "Relation between Space Charge and Polymeric Insulation Ageing: Cause and Effect." *IEE Proc. Sci. Meas. Technol.*, Vol. 50, No. 2, pp. 53–57, March 2003.

82. G.C. Montanari and D. Fabiani. "Evaluation of dc Insulation Performance Based on Space-Charge Measurements and Accelerated Life Tests." *IEEE Trans. Dielectr. Electr. Insul.*, Vol. 7, No. 3, pp. 322–328, June 2000.

83. F.N. Lim, R.J. Fleming, and R.D. Naybour. "Space Charge Accumulation in Power Cable XLPE Insulation." *IEEE Trans. Dielectr. Electr. Insul.*, Vol. 6, No. 2, pp. 273–281, April 1999.

84. G. Mazzanti, G.C. Montanari, and F. Palmieri, "Quantities Extracted from Space-Charge Measurements as Markers for Insulation Ageing." *IEEE Trans. Dielectr. Electr. Insul.*, Vol. 10, No. 2, pp. 198–203, April 2003.

85. B. Garros, C. Audry, H. Schadlich, G.C. Montanari, and I. Ghinello. "Evaluation of Insulation Degradation of Stressed XLPE Cables." *REE, Rev. l'Electr. l'Electron.* pp. 44–48, 1999.

86. Abou-Dakka, A. Bulinski, and S. Bamji. "Space Charge Evolution in XLPE and EPR under Uniform Electric Fields." IEEE CEIDP, pp. 136–139, Atlanta, Georgia, October 1998.

87. G. Platbrood, S. Agnel, A. Tourreille, and P. Bulens."Comparison of New Cables and Aged Cables out of the Grid by the Thermal Step Method." IEEE CEIDP, pp. 396–399, Cancun, Mexico, October 2002.

88. G.C. Montanari, G. Mazzanti, F. Palmieri, A. Motori, G. Perego, and S. Serra. "Space Charge Trapping and Conduction in LDPE, HDPE and XLPE." *J. Phys. D: Appl. Phys.*, Vol. 34, pp. 2902–2911, October 2001.

89. G. Mazzanti, G.C. Montanari, and J.M. Alison. "A Space-Charge Based Method for the Estimation of Apparent Mobility and Trap Depth as Markers for Insulation Degradation. Theoretical Basis and Experimental Validation." *IEEE Trans. Dielectr. Electr. Insul.*, Vol. 10, No. 2, pp. 187–197, April 2003.

90. A. See, J.C. Fothergill, L.A. Dissado, G. Stevens, G.C. Montanari, C. Laurent and G. Teyssedre, "Evaluation of the Onset of Space Charge and Electroluminescence as a Marker for Crosslinked Polyethylene Aging", *IEEE CSC, Tours*, pp. 185–188, 2001.

91. P.H.F. Morshuis, J. Beyer and G.C. Montanari. "Space Charge and Conduction Current Measurements as a Means to Determine Thresholds for Electrical Aging of HVDC Insulating Materials," CIGRÉ, Paris, 2000.

92. J.L. Auge, C. Laurent, D. Fabiani, G.C. Montanari. "Investigating dc Polyethylene Threshold by Space Charge and Electroluminescence Measurements." *IEEE Trans. Dielectr. Electr. Insul.*, Vol. 7, pp. 797–803, 2000.

93. L.A. Dissado, C. Laurent, G.C. Montanari, P. H. F. Morshuis. "Demonstrating a Threshold for Trapped Space Charge Accumulation in Solid Dielectrics under dc Field." *IEEE Trans. Dielec. Electr. Insul.*, Vol. 12, No. 3, pp. 612–620, June 2005.

94. C. Laurent. "Charge Dynamics in Polymeric Materials and its Relation to Electrical Aging," 2012 IEEE Conference on Electrical Insulation and Dielectric Phenomena (IEEE CEIDP 2012), pp. 1–20, Montreal (Canada), October 13th–17th 2012 (Whitehead Memorial Lecture).

95. G.C. Montanari and M. Cacciari, "Electrical life threshold models for insulating materials subjected to electrical and multiple stresses. Probabilistic approach to generalized life models." *IEEE Trans. Dielectr. Electr. Insul.*, Vol. 27, No. 5, pp. 987–999, October 1992.

CHAPTER 7

MAIN REALIZATIONS OF HVDC EXTRUDED CABLE SYSTEMS IN THE WORLD

This chapter describes in some detail the practical realizations of HVDC extruded cable systems by presenting the main HVDC extruded systems that have been installed and are currently in service worldwide. The list of these operational HVDC extruded cable systems is not long at present, but it has grown quickly in the last few years and is expected to grow even faster in the next decade.

The contents of this chapter—enriched with several contributions received from a few cable system manufacturers—is as follows:

- An overview of some quantitative data relevant to the main HVDC extruded cable systems that are currently in operation worldwide (Section 7.1)
- A detailed illustration of each one of the projects listed in Section 7.2

7.1 OVERVIEW

As pointed out in Chapters 1 and 2, HVDC cable systems at present are either paper insulated or extruded cable systems. Among paper insulated cable types, mass-impregnated non draining (MIND) cables have been those most extensively used around the world since their early beginning in the 1950s. The main advantage of the MIND cable systems, and more in general of DC systems using lapped insulation, is their well-established reliability for bulk power transmission [1–6]. Nevertheless, much of recent research and development efforts are being devoted to extruded solutions [3, 7–9], mainly for two reasons:

1. The concerns about the leakage of impregnants in paper insulated cables;
2. The advantages seen in extruded insulation (see Chapters 1 and 2). An important advantage of extruded cable systems—highlighted in particular in [1]—concerns the accessories and is the rather short jointing time needed compared to MIND cable systems, especially if prefabricated joints are used; this feature is very appealing for land cable installations where many joints are required.

Extruded Cables for High-Voltage Direct-Current Transmission: Advances in Research and Development, First Edition. Giovanni Mazzanti and Massimo Marzinotto.

Extruded cables were widely tested by the joint Japanese industry and by the European industry. Then, in 1999 a European manufacturer realized the first commercial HVDC extruded cable system, in the world, the Gotland link, as pointed out in previous chapters. Since then, this manufacturer has installed more than 2000 km of both underground and undersea[1] HVDC extruded cables [10, 11], together with 1000 prefabricated joints and many terminations [12]. The voltage rating of such systems ranges from 80 to 200 kV, while the rated power has increased from the 50 MW of the Gotland project to the 500 MW of the East-West Interconnector [1, 7–15].

Actually, for having a more complete picture of installed HVDC extruded cable systems worldwide, two more links must be added:

1. the Trans Bay Cable in California, realized by another European manufacturer [16–21], with voltage and power ratings of ±200 kV and 400 MW, respectively;
2. the Hokkaido–Honshu intertie in Japan, realized by a Japanese manufacturer, with voltage and power ratings of ±250 kV and 600 MW, respectively.

Overall, more than 2600 km of HVDC extruded cable systems have been installed in the world to date (see Table 7.1), and many more projects are ongoing (see Table 7.2) [10, 11, 16–22].

A further gradual increase in both power and voltage is being implemented in practice nowadays. Transmission powers in excess of 400 MW are more suitable to higher transmission voltage, and, in fact, the third generation of HVDC extruded cable systems, with an operating voltage level above 300 kV, has been qualified by successfully completing type tests and prequalification tests [1, 12]. An example of the progression of voltage and power rating, conductor cross section, and insulation thickness with time is shown in Figure 7.1 (after [23]).

This new voltage level paves the way for power transmission up to 1000 MW, as shown in Figure 7.2 (after [9]). According to [1] today's highest voltage laboratory-qualified HVDC extruded cable systems are 500-kV rated cables owned by Japanese companies [24–27], but no commercial projects were developed from those cable systems yet.

In this enduring run toward higher voltages and powers, some new HVDC extruded cable systems to be commissioned in the future are outstanding for their significance. They are listed in Table 7.2(a) (after [22]) and 7.2(b) (after [18]). As far as the projects listed in Table 7.2(a) are concerned, it can be emphasized that the DolWin1 project, expected to be in service in 2013, will be the first HVDC extruded cable system at the 320-kV level in the world, with a remarkable further increase of transmitted power up to the 800-MW level. As to the projects in Table 7.2(b), they all feature top levels of rated voltage and power, especially the last one, with the highest transmitted power among all projects listed in Table 7.2.

[1] The first installations were land cables, but the latest projects—the Cross Sound, the Troll A, the Estlink and East-West Interconnector—are undersea projects with 340 m as the present record laying depth for extruded DC cables [15].

Table 7.1 Main HVDC Extruded Cable Systems in Service All Over the World

Project	Customer	Country	Sea/Land	Year	Length (km)	Voltage (kV)	Power (MW)	Cond. (mm^2)
Hokkaido-Honshu	EPDC	Japan	Land	2012	0.5	±250	600	600
Hokkaido-Honshu	EPDC	Japan	Sea	2012	44	±250	600	600
Hokkaido-Honshu	EPDC	Japan	Land	2012	0.8	±250	600	600
East-West Interconnector	EirGrid	Ireland-Wales	Land	2012	75	±200	500	—
East-West Interconnector	EirGrid	Ireland-Wales	Sea	2012	186	±200	500	—
Valhall	BP	Norway	Sea	2011	292.0	150	78	—
BorWin1	Transpower	Germany	Sea	2011	243.3	±150	400	1200
BorWin1	Transpower	Germany	Sea	2011	22.6	±150	400	1600
BorWin1	Transpower	Germany	Land	2011	155.3	±150	400	2300
Trans Bay	California TSO	United States	Sea	2010	160.0	±200	400	1100
Trans Bay	California TSO	United States	Land	2010	10.0	±200	400	1100
Estlink	Nordic Energy Link	Estonia–Finland	Sea	2006	150.0	±150	—	1000
Estlink	Nordic Energy Link	Estonia–Finland	Land	2006	62.0	±150	350	2000
Troll A	Statoil	Norway	Sea	2004	284.0	±80	2 × 41	300
Troll A	Statoil	Norway	Sea	2004	0.5	±80	2 × 41	630
Cross Sound	Transenergie–US	United States	Sea	2002	83.2	±150	—	1300
Cross Sound	Transenergie–US	United States	Sea	2002	1.5	±150	330	630
Murraylink	Transenergie–US	Australia	Land	2002	223.2	±150	220	1200
Murraylink	Transenergie–US	Australia	Land	2002	136.8	±150	220	1400
Murraylink	Transenergie–US	Australia	Land	2002	1.5	±150	220	630
DirectLink	Transenergie–US	Australia	Land	1999	390.0	±84	3 × 60	630
GotLight	Gotland Energy	Sweden	Land	1998	140.0	±80	50	340
Total (Sea)					**1467.1**			
Total(Land)					**1195.5**			
Grand total					**2662.6**			

Sources: Compiled from [10,11,16–22].

Table 7.2 **HVDC Extruded Cable Systems to be Commissioned in the Future**

(a)

Project	In service	Power (MW)	DC Voltage (kV)	Station Location	Transmission	Application	Customer
DolWin1, Germany	2013	800	±320	Dörpen West, 380 kV DolWin alpha, 155 kV	165 km	Offshore wind	TenneT Offshore GmbH, Germany
NordBalt, Sweden–Lithuania	2015	700	±300	Klaipeda, 330 kV Nybro, 400 kV	450 km	Grid interconnection	SvK. Sweden; LITGRID turtas, Lithuania
DolWin2, Germany	2015	900	±320	Dörpen-West 380 kV DolWin beta 155 kV	135 km	Grid interconnection	TenneT Offshore GmbH, Germany

(b)

Name	From	To	Voltage (kV)	Power (MW)	Marine Length (km)	Land Length (km)	Total Length (km)
BorWin2	Offshore BorWin Beta	Onshore Diele (GER)	±300	800	123.5	76.5	200
HelWin1	Offshore HelWin Alpha	On-shore Büttel (GER)	±250	276	85	45	130
SyllWin1	Offshore SylWin Alpha	Onshore Büttel (GER)	±320	900	159	45.5	205
INELFE	Baixa (FRA)	Sta. Llogaia (SPA)	2 × ±320	2000	—	64	64

Sources: (a) After [22] and (b) after [18].

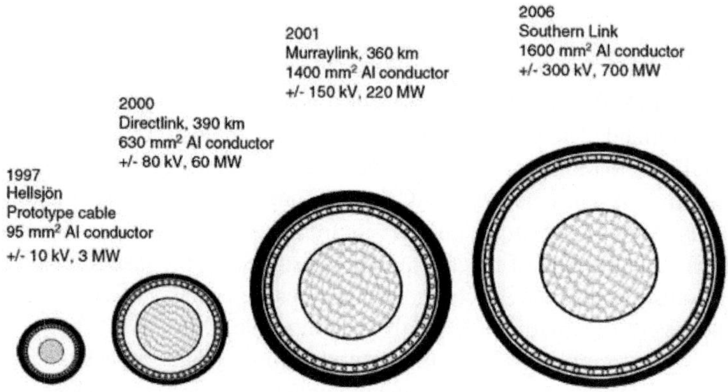

Figure 7.1 Progression of conductor cross section and insulation thickness of HVDC extruded cables as cable voltage and power increase with time (courtesy of ABB).

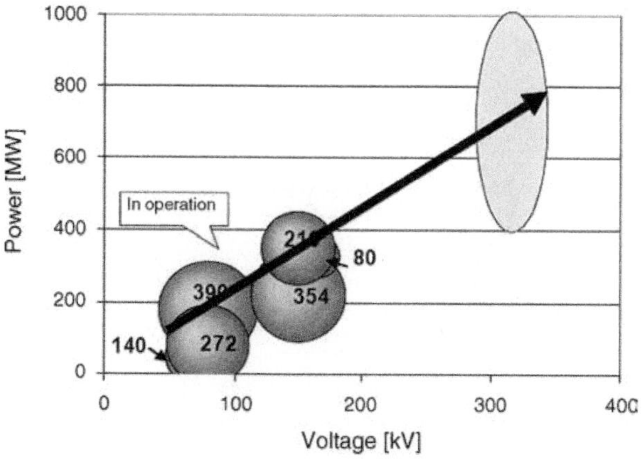

Figure 7.2 Foreseen increase in power and voltage for HVDC extruded cable systems. Every bubble represents an existing commercial project. The size of the bubble refers to the amount of cable delivered. The numbers denote the number of cable kilometers of the projects (courtesy of ABB).

7.2 EXTRUDED SYSTEMS IN SERVICE

In [28], dating back to 2004, it was observed that the service experience of HVDC extruded systems was in its infancy. The situation at present is changing, since several HVDC extruded cable systems—mentioned in Section 7.1—are already in service, while others are either being commissioned or at the planning stage. For this reason it is worthwhile examining HVDC extruded cable systems in operation, thereby also describing some practical problems tackled at the design, manufacturing, and installation stages.

7.2.1 Gotland Link

The need for the Gotland link came from the fact that a large capacity of wind-generated power (~40 MW) was installed at Näs on the southwest coast of the Swedish island of Gotland. The peak demand in this region is much lower, and this pushed the local utilities to transport efficiently the wind power energy to the town of Visby, ~70 km northward. The existing HVAC line was not sufficient to carry the full production capacity. Moreover, voltage quality requirements called for the compensation of the remarkable amount of reactive power consumed by the installed wind power; static var compensators and synchronous machines were analyzed as possible solutions, but they were regarded as too costly or unfeasible [29].

These issues were compared with the several technical and environmental advantages of extruded HVDC cable systems in bipolar configuration (see Chapters 1 and 2), in particular: low energy losses; very low static magnetic field, being the two underground cables laid close to each other; and the ability of the HVDC system to compensate for the variation of wind power generation while keeping a stable voltage level and minimizing the losses. For these reasons in 1997 the Gotland Link project (see Fig. 7.3*a* after [30])—the world's first HVDC extruded cable link—was contracted, consisting of a 72-km-long bipolar cable system rated 50 MW at ±80 kV

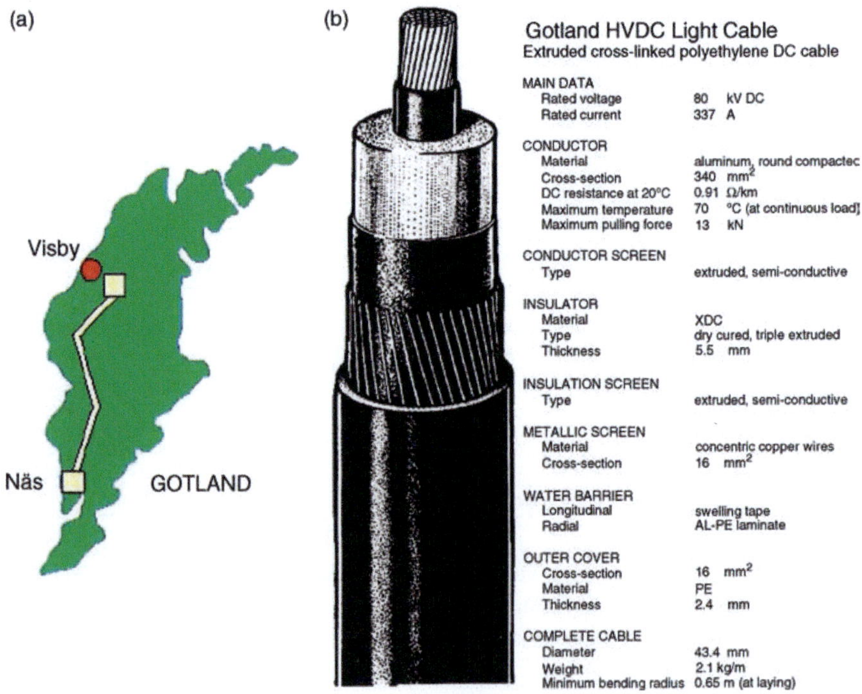

Figure 7.3 The Gotland link: (*a*) route of the link and (*b*) the 340-mm^2 Gotland extruded HVDC cable (courtesy of ABB).

DC voltage level. Voltage source converters (VSC) were used for this link, both because of the vulnerability of HVDC extruded insulation to voltage polarity reversals (see Chapters 2 and 3) and because of the several operational benefits of HVDC VSC systems, for example, independent control of active and reactive power, rapid change of power direction, and superior power quality control [29].

An exhaustive development effort singled out space-charge accumulation, electrical resistivity, and breakdown strength as the key characteristics for choosing the peroxide cross-linkable insulation system (see Chapter 5). When the contract for the Gotland Link project was signed in late 1997, the development of the extruded cable system was at the stage of testing the 95-mm^2 cable prototypes and accessories by means of heating cycle and impulse voltage type tests, as described in [30] and discussed in Chapter 5 as well (see Sections 5.6.2 and 5.7.1). Along with these types of tests the project engineering was carried out and the final design of the cable was selected (see Fig. 7.3*b* after [30]), thereby attaining small—less than 44 mm in diameter—but robust cables with low weight (just above 2 kg/m), high flexibility, and mechanical strength [30]. The outer PE cover is made of medium-density polyethylene, giving a robust cable easily handled in the factory and onsite. Thus, the extruded HVDC cables could be laid by ploughing, yielding only minor environmental impact and reducing the installation time and cost. The first length of cable started being manufactured in late May 1998, and the full cable system with all accessories was type-tested before being accepted and delivered [30].

Since the extrusion process of the selected cable design allows the delivery of any length of practical interest, the cable line route was planned so as to place a joint pit next to all major road crossings. On the whole, 24 route sections from 1200 to 3600 m long were singled out, thus 48 cable drums weighing between 4.5 and 9.4 tons were delivered to Gotland. In total, 142.8 km of cable were manufactured in six batches between June and September 1998. The customer greatly appreciated the flexibility in delivery lengths and the short production time [30].

As already mentioned, thanks to the combined robustness and flexibility of the design, the cables were installed by ploughing (see Fig. 7.4*a* after [30]), an

(a) (b)

Figure 7.4 Installation of the Gotland link: (*a*) cable ploughing and (*b*) installation of joints on site (courtesy of ABB).

economical and environmentally friendly technique that eases and speeds up the granting of permits. The cables, installed in parallel in bipolar configuration, were first laid on the soil along the route, then ploughed down 65 cm with the use of standard tools. Wherever rocks or boulders were close to the ground surface, the required embedding depth was ensured by means of rock cutting prior to ploughing. At the same time, a fiber-optic duct was installed, thereafter blowing the fiber-optic cable through the duct and connecting it properly. As soon as the cables were installed, cable joints began to be assembled (see Fig. 7.4*b* after [30]). The complete cable system passed the after-installation test in the early 1999 [30].

7.2.2 Murraylink

The Australian Murraylink (see Fig. 7.5*a*) connects the transmission grids of the Australian states of Victoria (Vic) and South Australia (SA), linking two newly built AC/DC converter stations and covering 177 km between the towns of Berri (SA) and Red Cliffs (Vic). This makes the Murraylink the longest underground HVDC extruded cable line in the world, and perhaps the longest underground cable system worldwide. This cable system has voltage and power rating equal to ± 150 kV and 200 MW, respectively, and utilizes extruded power cables with two aluminum conductor sizes, that is, 1200 and 1400 mm^2 [7, 31]. The cable dielectric is a modified XLPE insulation compound [28]. The converter stations are based on VSC converters that employ IGBT transistor valves and pulse width modulation for optimal control of the transmission of active and reactive power. Design details were reported in Sections 5.6.3 and 5.7.2. Other details of the project are as follows [7].

As can be argued, the Murraylink is a bipolar cable system. Each of the two HVDC extruded cables (see Fig. 7.5*b*) that connect the converter stations has one single polarity throughout the whole service life. The current flow direction sets the power flow direction in such a way that full power reversal can be obtained within tens of seconds. The cable line follows mainly a straight interstate road through a rural area with scattered population and crosses the Murray River near Berri. The area is semiarid and deemed sensitive to human impact. Hence, the customer applied

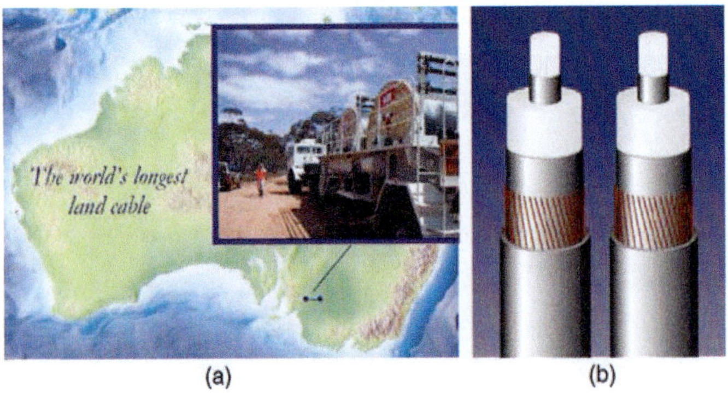

Figure 7.5 Australian Murraylink: (*a*) the route of the link and (*b*) the Murraylink bipolar cables (courtesy of ABB).

a rigorous and complex Environmental Management Plan for reducing the environmental impact during installation and operation of the cable link as much as possible. The plan was in turn split into subplans that accounted for different environmental aspects (e.g., revegetation, biodiversity preservation, and noise control) with such a care that, as an example, the number of trees and bushes to be cut was specified in the development approval and agreed on before installation. This effort led to an environmentally friendly cable installation that was awarded the 2002 National Case Earth Award for environmental excellence [7].

Permitting and environment preservation were made easier by maximizing the use of natural material for backfill via the development of an ad hoc installation gear (see Fig. 7.6*a* after [7]), consisting in a leading truck that pulled a low-bed trailer with two cable drums for the positive and negative cable, respectively; after that, a large chain-saw excavator dug up an ~1400-mm-deep and ~500-mm-wide cable trench. The excavated material was sieved into a fine and a coarse fraction. The cables were taken off the lead truck at the same time and transported on overhead roller guides to the rear of the excavator vehicle. Cables and fine refill went down into the cable trench through a square metallic box to guide them into the specified position. The coarse soil fraction was stockpiled along the cable trench. A backhoe followed the excavator vehicle and pushed the coarse material back into the cable trench. At the end, very little impact was seen in the natural vegetation. In a subsequent cleanup operation the soil reinstatement was completed (Fig. 7.6*b* after [7]).

The equipment proved to be very efficient. The average laying speed was over 1000 m per day with a peak 3000 m per day, excluding cleanup. This is a substantial improvement compared to the figures reported earlier. The joints, erected in compliance with strict cleanliness requirements, were assembled after the cables had been laid and buried except for the joint pits. The complete DC transmission line consisting of 2×177-km cables and almost 400 premolded joints was tested successfully at the 210-kV DC voltage level for 15 min in July 2002 [7]. Murraylink has been in operation since 2002 and has featured an availability of 98.5%, including planned maintenance [32].

(a) (b)

Figure 7.6 Installation of the Murraylink: (*a*) configuration of installation equipment and (*b*) after reinstatement (courtesy of ABB).

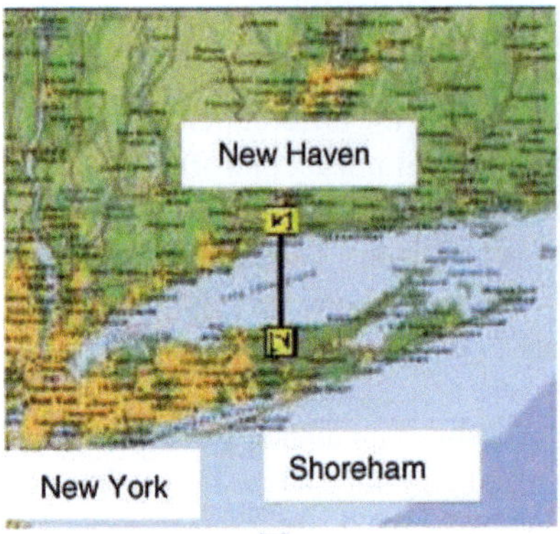

(a)

Cable data

Voltage	+/- 150 kV DC
Power	330 MW
Length	2 x 42 = 84 km
Conductor	1300 mm^2 Cu
Insulation	Polymeric
Weight	30 kg/m
Customer	TransEnergieUS
Year	2002

(b)

Figure 7.7 Cross Sound HVDC link: (*a*) the route of the link and (*b*) Cross Sound cable data (courtesy of ABB).

7.2.3 Cross Sound Cable (CSC)

The Cross Sound Cable Project is a ±150-kV, 330-MW undersea intertie linking the grids of Connecticut (at New Haven) and Long Island (at Shoreham) in the United States. The link is ~42 km long, of which 40 km consist of undersea cables (see Fig. 7.7*a*). Protection against shipping activities was achieved by burying the two HVDC extruded undersea cables at a depth of 1.80 m into the seabed, thereby also mitigating the environmental impact of the installation on marine life [28]. Figure 7.7*b* shows the main cable data relevant to the Cross Sound Cable Project.

The Cross Sound cable has a threefold purpose: (1) interconnecting two electricity markets; (2) answering the urgent electricity demand of the Long Island community; and (3) improving voltage and reactive power control of the Connecticut and New England power systems, thereby increasing supply stability and reliability.

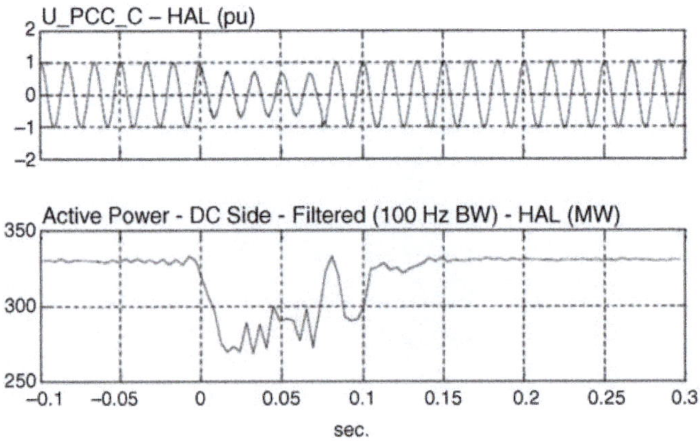

Figure 7.8 Dynamic response to a line fault in the 345-kV grid on March 17, 2005 (courtesy of ABB).

(a)

Cable data

Voltage	+/- 80 kV DC
Power	2 x 40 MW
Route length	68 km
Conductor	300 mm^2 Cu
Water depth	350 m
Cable length	4 x 68 km = 272 km
Weight	15 kg/m
Customer	Statoil SF, Norway
Year	2004

(b)

Figure 7.9 Troll A platform: (*a*) view of the platform with HVDC VSC converter module and (*b*) Troll A HVDC cable data (courtesy of ABB).

This is illustrated in Figure 7.8, where a transient is shown such that a classic HVDC system would most likely have failed and needed to be restarted:

> The upper graph shows a voltage drop after a line fault in the 345-kV AC grid.
>
> The lower graph shows the corresponding active power transmitted through the DC circuit, showing how after only 0.15 s the originally transmitted power of 330 MW is reached again, thus stabilizing supply on both sides of the system.

As a proof of this stabilizing action, the supply to Long Island was not affected by the vast 2003 blackout, thanks to the stabilizing action of the Cross Sound Cable [32].

7.2.4 Troll A Platform

A completely different feature of the VSC HVDC technology is emphasized by a project commissioned at the end of 2004 with the aim of supplying—for the first time ever—power from land to the Troll A gas platform (see Fig. 7.9*a*), located within the largest gas field in the North Sea.[2] VSC HVDC, different from CSC HVDC (see Section 2.4.2), does not require any short-circuit power from the neighboring grid for commutation. This allows connecting weak grids (e.g., on islands) to a stronger system, as well as supplying passive loads—a feature exploited for the Troll A platform [32].

The Troll A link consists of two bipolar 40-MW HVDC cable lines of ~70-km length (the main data relevant to the Troll A transmission system are listed in Fig. 7.9*b* [33]). The HVDC VSC converter on the platform also drives the motors of the compressors supplied by the HVDC cable lines. This integration led to the

[2] The traditional way to supply these offshore platforms is by gas turbines on the platform, which is disadvantageous both for safety reasons and because of the need of operation and maintenance staff on the platform [32].

"longest speed-variable drive in the world," thereby achieving a very compact design on the platform (for details, see [32]).

7.2.5 Estlink

Estlink is a VSC HVDC system with a capacity of 350 MW, a DC voltage of ±150 kV, 2 × 74-km-long undersea cables and 2 × 31-km-long land cables that connects the Estonian power system to the Finnish grid and thus the Nordel system. The Estlink route is shown in Fig. 7.10*a*, while Figure 7.10*b* shows the main cable data [34]).

The Estlink was ordered in 2004 and completed after less than 20 months, by the end of 2006. The delivery was accelerated, thanks to (1) the high degree of standardization, modularity, and prefabrication reached since the beginning of the VSC HVDC technology in the 1990s, and (2) the light weight of cables, in combination with cable joints that can be set much faster than with mass-impregnated cables. The 71 km of sea cables were laid in 10 days only as one piece each [32].

The cable route starts with a land section of 2 × 9-km-long cables from the Harku converter station in Estonia to the shore. Thereafter, 2 × 74-km-long undersea cables stretch beneath the Gulf of Finland to the Finnish shore, where finally 2 × 22-km-long land cables reach the other converter station at Espoo. The Estonian land cable delivery consisted of 24 cable drums and weighted ~200 tons, while the Finnish land cable delivery consisted of 60 cable drums and weighed ~500 tons. The land cables were sent in 22 shipments across the Baltic Sea to the installations sites [34].

The Estlink undersea cable was produced in long lengths (thereby minimizing jointing operations) and loaded directly from the factory onto the laying vessel. The cable, weighing ~3850 metric tons, was laid in one section and buried 1 m into the sea floor to achieve protection against fishing gear and ship anchors. In order to minimize the magnetic field and enable cable burying by water jetting in one single run, the two HVDC extruded cables were bundled together.

As far as cable system accessories are concerned, the following were employed for the Estlink [34]:

(a)

Cable data

Voltage	+/- 150 kV DC
Power	350 MW
Length	2 x 74 km submarine cable 2 x 31 km land cable
Conductor	1000 mm² Cu submarine cable 2000 mm² Al land cable
Insulation	Polymeric
Weight	26 kg/m submarine cable 11 kg/m underground cable
Customer	AS Nordic Energy Link
Year	2006

(b)

Figure 7.10 Estlink: (*a*) the route of the link and (*b*) Estlink cable data (courtesy of ABB).

Cable terminations matching the HVDC VSC converters

Flexible factory joints for the sea cable, with the same properties as the cable itself

Prefabricated joints for the land cable

Undersea cable repair joints (as contingency)

Transition joints between undersea and land cable.

The system was commissioned at the beginning of 2007 [34].

7.2.6 BorWin 1

As pointed out above, another advantage of VSC HVDC compared to CSC HVDC is its black-start capability, that is, the possibility to start the system without power supply from the neighboring grid (see Section 2.4.2). This feature, combined with the static var converter (SVC) functionality provided by the converters, makes VSC HVDC the ideal solution for connecting offshore wind farms located far away from the coast to the onshore grid. For this reason, a 400-MW VSC HVDC extruded system was selected when the world's largest offshore wind farm cluster called Borkum 2 started to be connected to the grid in northern Germany in 2007. According to the connection layout, AC power from the 80 wind turbines goes to an AC platform from which a 1-km-long 170-kV XLPE-insulated undersea cable supplies an offshore HVDC VSC converter station platform. From the converter station platform, two 1200-mm^2 HVDC extruded undersea cables run parallel to the island of Norderney, where they are connected to two 1600-mm^2 HVDC extruded undersea cables. Those cables run to a transition point onshore where they are jointed to two 2300-mm^2 HVDC extruded underground cables, which cover a distance of 75 km to the HVDC VSC converter station at the Diele substation. At Diele the power is converted to AC and fed into the 380-kV transmission grid. Overall, two 75-km-long land cables and two 128-km-long undersea cables connect the existing 400-kV substation at Diele in Lower Saxonia with the offshore platform for the Borkum 2 wind farm cluster [32, 35].

The BorWin 1 link route is shown in Figure 7.11*a*, whereas Figure 7.11*b* shows the main relevant cable data. It must be pointed out that this link employs both land and undersea cables, thus special "transition" joints are required [12].

Cable data		
Voltage	+/- 150 kV DC, 170 kV AC	
Power	400 MW	
Length	2x125 km submarine HVDC	cable (Cu)
	2x75 km underground HVDC	cable (Al)
	1 km 3-core submarine XLPE cable (Cu)	
Customer	transpower, Germany	
Year	2011	

Figure 7.11 BorWin 1 link (*a*) the route of the link and (*b*) BorWin 1 cable data (courtesy of ABB).

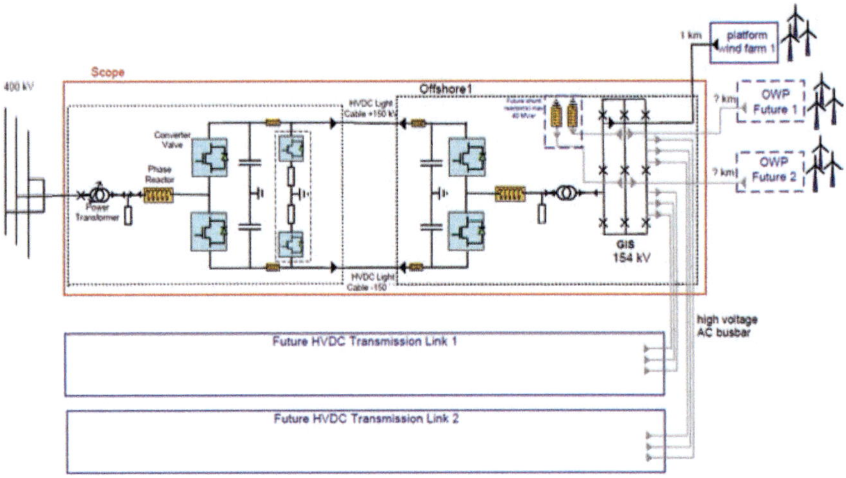

Figure 7.12 Modular extension concept for BorWin 1 offshore wind farm connection (courtesy of ABB).

Figure 7.12 shows how the system can be broadened following a modular concept, should more wind farms be built in the cluster in the future [32].

7.2.7 The Trans Bay Project

The Trans Bay Project is a ±200-kV, 400-MW, ~2 × 85-km-long undersea extruded HVDC cable system installed in the San Francisco Bay, California; see Figure 7.13. More precisely, the cable line consists of 80 km of undersea cable and 5 km of land

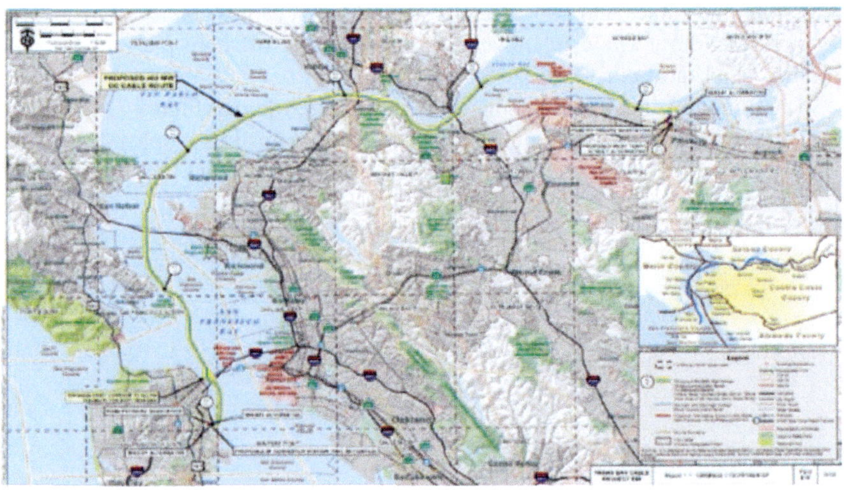

Figure 7.13 Trans Bay Project route (courtesy of Prysmian).

cable. The Trans Bay Cable connects Pittsburg with San Francisco across the San Francisco Bay and has been installed by a partnership between the cable manufacturer (that has manufactured and laid-out the cable) and the converter manufacturer (engaged since September 2007 in providing two HVDC VSC converter stations based on the MMC—modular multilevel—technology) [21]. Cable insulation is an XLPE compound for HVDC. At the time of commissioning it was the first extruded HVDC cable system project at the 200-kV level implemented in the world for commercial power transmission. Operation has started in March 2010 [16–21].

Details about the Trans Bay Cable design were reported at Section 5.6.4. Prequalification tests, carried out according to CIGRE TB 219 procedure [36], were successfully concluded for a ±200-kV DC level. Type tests using the Trans Bay Cable HVDC cable and accessories were also performed.

The cables were transported from Italy by means of the deep-water cable installer ship *Giulio Verne*, which also laid the cable route in sufficiently deep waters (i.e., >10 m). In shallow water the installation was completed with a properly equipped "shallow water cable installer barge." Cable laying and burying was performed using a hydroplough (see Figs 7.14 and 7.15).

The two HVDC cables—plus a fiber-optic cable—were simultaneously laid and buried in a bundle configuration to minimize environmental impact. The bundle cables were fastened together with ropes and straps applied before approaching the laying sheave (see Fig. 7.16). The bundle is approximately 25 cm in diameter and has been laid at a depth of ~1.8 m beneath the seabed over the whole route.

A cross section of HVDC XLPE insulated bipolar cable bundle for the Trans Bay Cable Project is illustrated in Figure 7.16; the cable features a 1100-mm^2 copper conductor, extruded XLPE insulation, lead sheath, and steel wires armor.

Significant preparation work was needed to locate the cable route. A detailed marine survey of the bay was carried out, including:

- Bathymetry
- Seafloor mapping (with side scan sonar)

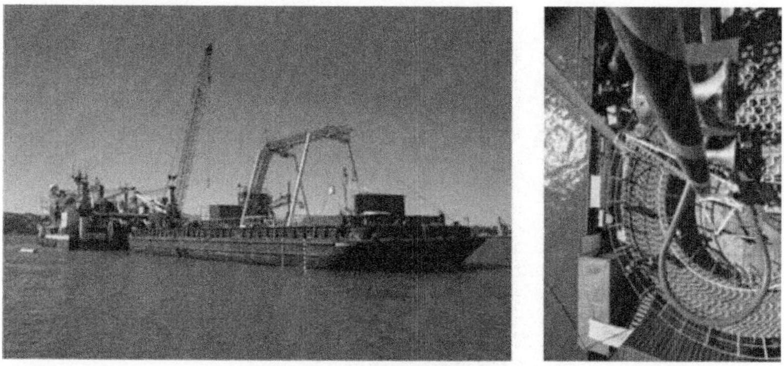

Figure 7.14 Laying activities of the Trans Bay Cable in San Francisco Bay (courtesy of Prysmian).

Figure 7.15 Facilities used for the laying activities of the Trans Bay Cable in San Francisco Bay: (*a*) view of the ship *Giulio Verne* (Deep Water Cable Installer); (*b*) turntable aboard the ship *Giulio Verne*; (*c*) Barge (Shallow Water Cable Installer); and (*d*) hydroplow (courtesy of Prysmian).

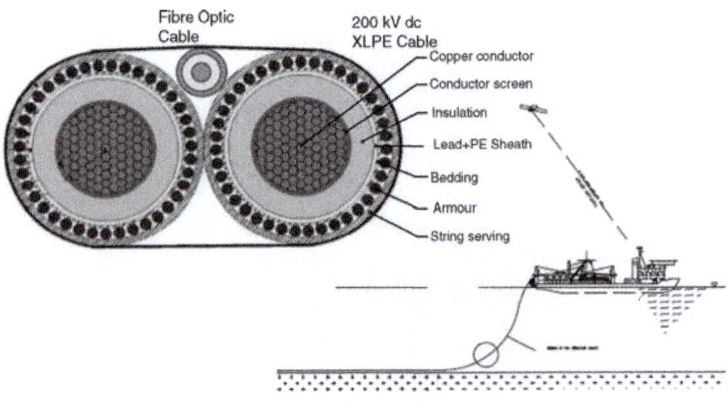

Figure 7.16 Cross section of HVDC cable for the Trans Bay Cable Project (courtesy of Prysmian).

- Subbottom profiling
- Magnetometer analysis
- Sediment sampling (with relevant geotechnical analysis)

7.2.8 The Hokkaido–Honshu Intertie

This intertie represents the first HVDC project in the world with a DC-XLPE insulated cable operated in a system that requires the polarity reversal of the cable in order to invert the direction of the power between the two terminal converter stations. In fact, such cable is part of a mixed overhead/cable line link that interconnects two current source converter stations located in Hakodate and Kamikita (see Figure 7.17a).

The Hokkaido–Honshu intertie went in operation during the late '70s with ±250 kV SCFF cables in between two overhead line sections. Recently, with the aim to retrofit the intertie, new ±250 kV DC-XLPE insulated cables have replaced the old ones to be put in operation with the existing converter stations. The DC-XLPE compound used for these new cables is improved with nanofillers with the aim to take under control the negative effects of space charges accumulation, keep the electrical resistivity high even at high temperature and improve the DC breakdown strength. J-Power Systems has been believing to nanofilled compounds for DC application

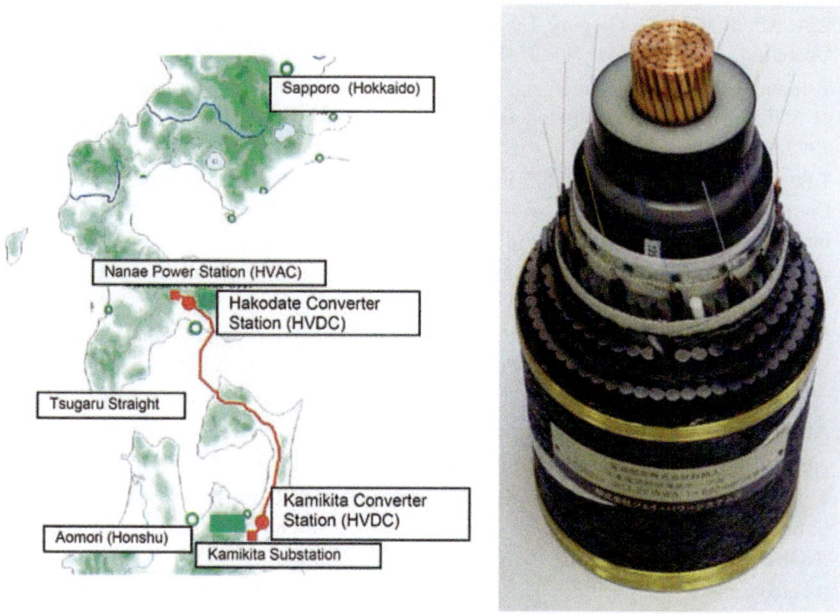

(a) (b)

Figure 7.17 The ±250 kV –600 MW Hokkaido–Honshu intertie: (a) intertie route; (b) a picture of the power cable (optical fibers are also visible) (published under courtesy of Electric Power Development Co., Ltd. and J-Power Systems Corporation").

since the '80s when they started the developments of such materials, and tested DC 500-kV cable with this material.

This cable successfully passed the qualification tests in accordance with [36] with a maximum conductor temperature of 90°C and a lightning impulse test both without (+750 kV and −750 kV) and with a superimposed DC voltage (+530 kV with −250 kV and −530 kV with +250 kV). A picture of the cable is reported in Fig. 7.17b. The cable went in operation at the end of 2012.

REFERENCES

1. M. Jeroense. "HVDC, the Next Generation of Transmission. Highlights with Focus on Extruded Cable Systems." Proc. 2008 International Symposium on Electrical Insulating Materials (ISEIM 2008), pp. 10–15, Yokkaichi, Japan, Sept. 7–11, 2008.
2. R.N. Hampton. "Some of the Considerations for Materials Operating under High-Voltage, Direct-Current Stresses." *IEEE Electr. Insul. Mag.*, Vol. 24, No. 1, pp. 5–13, January–February 2008.
3. F. Ruter, J.L. Parpal, and S.G. Swingler. "A review of HVDC Extruded Cable Systems." CIGRÉ 2000 Joint Panel Session—Effect of Electrical Stresses on Insulating Materials, Paper P2-03, 2000.
4. EPRI Report EL-3973. Design Limits of Oil/Paper High-Voltage Direct-Current Cable, 1985.
5. C.A. Arkell and B.G. Gregory. Design of Self-Contained Oilfilled Cables for UHV D.C. Operation. *CIGRÉ*, Paper No. 21-301, 1984.
6. T. Inoue, M. Mitani, I. Shigetoshi, Y. Maekawa, O. Fujii, and M. Shimada. "Installation of 500 kV DC PPLP-Insulated Oil-Filled Submarine Cable." *Jicable*, pp. 515–520, 1999.
7. T.B. Worzyk, M. Jeroense, T. Sörqvist, and M. Farr. "The Murraylink Project—The First Commercial 150 kV Extruded HVDC Cable System." International Conference on Insulated Power Cables, Paper A.7.4., pp. 217–222, JICABLE '03.
8. Brochure CIGRE 496. Recommendations for Testing DC Extruded Cable Systems for Power Transmission at a Rated Voltage up to 500kV. CIGRE WG B1-32, April 2012.
9. M. Jeroense, M. Bergkvist, A. Gustafsson, H. Rapp, T. Sörqvist, and J. Svahn. "Increased Voltage for the HVDC Light® Product Range—A Complete Solution." Paper B.4.7, pp. 425–430, JICABLE 2007.
10. ABB. Reference List High Voltage Cables, 2GM7001, 2009.
11. ABB. Reference List High Voltage Submarine Cables, 2GM7005, 2009.
12. T. Sörqvist, T. Christen, M. Jeroense, V. Mondiet, and R. Papazyan. "HVDC-Light® Cable Systems—Highlighting the Accessories." 21st Nordic Insulation Symposium (NORDIS 2009), Gothenburg (Sweden), June 15–17, 2009.
13. B. Jacobson, Y. Jiang-Häfner, P. Rey, G. Asplund, M. Jeroense, A. Gustafsson, and M. Bergkvist. "HVDC with Voltage Source Converters and Extruded Cables for up to ±300 kV and 1000 MW." Paper B4-105 CIGRE Session 2006, Canada, Montréal, October 1–4, 2006.
14. B. Normark."Cross Sound Goes Underground." Power Eng. Int., December, 2002.
15. M. Bergkvist, O. Hansson, M. Jeroense, and C. Sonesson. "Development of Extruded Single Core Cables for Deep Waters." Paper B.7.1, JICABLE 2007.

16. M. Marelli, A. Orini, and E. Zaccone. "Cavi sottomarini in corrente continua: soluzioni per il Mediterraneo." Proceedings of the 2009 AEIT Conference, Catania (Italy), September 27–29, 2009 (in Italian).
17. R. Bazzocchi, and G. Cordioli. "Recenti sviluppi della tecnica HVDC." Proceedings of the 2009 AEIT Conference, Catania (Italy), September 27–29, 2009 (in Italian).
18. F. Pittelli, M. Marelli, and E. Zaccone. "Cavi in corrente continua per uso terrestre." Atti del Convegno Nazionale AEIT 2011, Milano, June 27–29, 2011 (in Italian).
19. N. Kuljaca, M. Marelli, A. Micheletti, and E. Zaccone. "Cavi sottomarini per impianti di generazione off-shore—stato dell'arte e sviluppi futuri." Atti del Convegno Nazionale AEIT 2011, Milano, June 27–29, 2011 (in Italian).
20. M. Bacchini R. Grampa, M. Marelli, T. Westerweller, and N. Kelley. "200 kV DC Extruded Cables Crossing the San Francisco Bay." CIGRÉ Session Paper B1-105, Paris 2010.
21. T. Westerweller, K. Friedrich, U. Armonies, A. Orini, D. Parquet, and S. Wehn. "Trans Bay Cable—World's First HVDC System Using Multilevel Voltage-Sourced Converter." CIGRÉ Session Paper B4-101, Paris 2010.
22. DolWin2. 900 MW HVDC Light Transmission: The World's Largest Wind Power Grid Connection. ABB AB Grid Systems—HVDC, POW-0075 rev.0.
23. HVDC Light®: Technical Background. Prepared by Sea Breeze Olympic Converter, LP, 2007.
24. Y. Maekawa, K. Watanabe, S. Maruyama, and H. Hirota."Research and Development of DC ±500 kV Extruded Cables." CIGRÉ 2002, 21-203.
25. T. Yamanaka, S. Maruyama, and T. Tanaka. "The Development of DC +/−500 kV XLPE Cable in Consideration of the Space Charge Accumulation." Proceedings of the 7th International Conference on Properties and Applications of Dielectric Materials P3-44, Nagoya, pp. 689–694, June 1–5, 2003.
26. S. Maruyama, N. Ishii, T. Yamanaka, T. Kimura and H. Tanaka,"Development of 500 kV DC XLPE Cable System and Its Pre-qualification Test." Paper A.7.2., Jicable 2003.
27. S. Maruyama, N. Ishii, M. Shimada, S. Kojima, H. Tanaka, M. Asano, T. Yamanaka, and S. Kawakami."Development of a 500-kV DC XLPE Cable System." Furukawa Rev., No. 25, pp. 47–52, 2004.
28. EPRI Report No. 1008720. DC Cable Systems with Extruded Dielectrics. Palo Alto, California, December 2004.
29. G. Asplund. "HVDC Using Voltage Source Converters. A New Way to Build Highly Controllable and Compact HVDC Substations." Presented at the International Council on Large Electric Systems (CIGRÉ), 2000 Session, Panel 2, Paper 2-04, Paris, France, 2000.
30. M. Byggeth, K. Johannesson, L. Palmqvist, J. Jonsson, C. Törnkvist, C. Liljegren, and U. Axelsson. "The Development of an Extruded HVDC Cable System and Its First Application in the Gotland HVDC Light Project." Paper B7-5, pp. 538–542, 5th International Conference on Insulated Power Cables (JiCable'99), Versailles, France, June 20–24, 1999.
31. HVDC Light®. "Murraylink Connecting Victoria and South Australia". ABB Power Technology Products AB, SEHVC–M043E, 2003.
32. J. Kreusel. "Survey Paper—Part 2. Integrated AC/DC Transmission Systems—Benefits of Power Electronics for Security and Sustainability of Power Supply." Proc. PSCC 2008, Glasgow, July 14–17, 2008.

33. HVDC Light®. Offshore Wind Farm Link. "Troll A Platform." ABB Power Technology Products AB, SEHVC–M062E, 2004.

34. HVDC Light® Submarine Cable Link. "Estlink—Connecting Estonia to the Nordic Network." ABB AB High Voltage Cables, SEHVC –M061E, 2007.

35. HVDC Light® Offshore Wind Farm Link. "The World's Most Remote Offshore Wind Farm BorWin1." 2011-02, 2GM5067 GB.

36. Brochure CIGRE 219. "Recommendations for Testing DC Extruded Cable Systems for Power Transmission at a Rated Voltage up to 250kV." CIGRE, February 2003; also published as CIGRE Working Group 21.01. "Testing DC Extruded Cable Systems for Power Transmission up to 250 kV." *Electra*, No. 206, pp. 46–53, February 2003.

INDEX

Extruded Cables for High-Voltage Direct-Current Transmission: Advances in Research and Development, First Edition. Giovanni Mazzanti and Massimo Marzinotto.

IEEE Press Series on Power Engineering

Books in the IEEE Press Series on Power Engineering

1. *Principles of Electric Machines with Power Electronic Applications*, Second Edition
   M.E. El-Hawary
2. *Pulse Width Modulation for Power Converters: Principles and Practice*
   D. Grahame Holmes and Thomas Lipo
3. *Analysis of Electric Machinery and Drive Systems*, Second Edition
   Paul C. Krause, Oleg Wasynczuk, and Scott D. Sudhoff
4. *Risk Assessment for Power Systems: Models, Methods, and Applications*
   Wenyuan Li
5. *Optimization Principles: Practical Applications to the Operations of Markets of the Electric Power Industry*
   Narayan S. Rau
6. *Electric Economics: Regulation and Deregulation*
   Geoffrey Rothwell and Tomas Gomez
7. *Electric Power Systems: Analysis and Control*
   Fabio Saccomanno
8. *Electrical Insulation for Rotating Machines: Design, Evaluation, Aging, Testing, and Repair*
   Greg Stone, Edward A. Boulter, Ian Culbert, and Hussein Dhirani
9. *Signal Processing of Power Quality Disturbances*
   Math H.J. Bollen and Irene Y.H. Gu
10. *Instantaneous Power Theory and Applications to Power Conditioning*
    Hirofumi Akagi, Edson H. Watanabe and Mauricio Aredes
11. *Maintaining Mission Critical Systems in a 24/7 Environment*
    Peter M. Curtis
12. *Elements of Tidal-Electric Engineering*
    Robert H. Clark
13. *Handbook of Large Turbo-Generator Operation and Maintenance*, Second Edition
    Geoff Klempner and Isidor Kerszenbaum
14. *Introduction to Electrical Power Systems*
    Mohamed E. El-Hawary
15. *Modeling and Control of Fuel Cells: Distributed Generation Applications*
    M. Hashem Nehrir and Caisheng Wang
16. *Power Distribution System Reliability: Practical Methods and Applications*
    Ali A. Chowdhury and Don O. Koval
17. *Economic Market Design and Planning for Electric Power Systems*
    James Momoh and Lamine Mili

18. *Operation and Control of Electric Energy Processing Systems*
James Momoh and Lamine Mili
19. *Restructured Electric Power Systems: Analysis of Electricity Markets with Equilibrium Models*
Xiao-Ping Zhang
20. *An Introduction to Wavelet Modulated Inverters*
M.A. Rahman and S.A. Saleh
21. *Probabilistic Transmission System Planning*
Wenyuan Li
22. *High Voltage and Electrical Insulation Engineering*
Ravindra Arora and Wolfgang Mosch
23. *Maintaining Mission Critical Systems in a 24/7 Environment*, Second Edition
Peter Curtis
24. *Doubly Fed Induction Machine: Modeling and Control for Wind Energy Generation Applications*
G. Abad, J. Lopez, M.A. Rodriguez, L. Marroyo, and G. Iwanski
25. *High Voltage Protection for Telecommunications*
S. Blume
26. *Smart Grid: Fundamentals of Design and Analysis*
James Momoh
27. *Electromechanical Motion Devices*, Second Edition
Paul Krause, Oleg Wasynczuk, and Steven Pekarek
28. *Electrical Energy Conversion and Transport: An Interactive Computer-Based Approach*, Second Edition
George G. Karady and Keith E. Holbert
29. *ARC Flash Hazard and Analysis and Mitigation*
Jay C. Das
30. *Handbook of Electrical Power System Dynamics: Modeling, Stability, and Control*
Mircea Eremia and Mohammad Shahidehpour
31. *Analysis of Electric Machinery and Drive Systems,* Third Edition
Paul Krause, Oleg Wasynczuk, Scott Sudhoff, and Steven Pekarek
32. *Extruded Cables for High Voltage Direct Current Transmission: Advances in Research and Development*
Giovanni Mazzanti and Massimo Marzinotto

Printed and bound by CPI Group (UK) Ltd, Croydon, CR0 4YY
17/09/2026
14964284-0002